Y0-CCH-782

Optics of the Atmosphere

WILEY SERIES IN PURE AND APPLIED OPTICS

Advisory Editor
Stanley S. Ballard, University of Florida

ALLEN AND EBERLY • *Optical Resonance and Two-Level Atoms*
BOND • *Crystal Technology*
CATHEY • *Optical Information Processing and Holography*
CAULFIELD AND LU • *The Applications of Holography*
FRANCON AND MALLICK • *Polarization Interferometers*
GERRARD AND BURCH • *Introduction to Matrix Methods in Optics*
HUDSON • *Infrared System Engineering*
JUDD AND WYSZECKI • *Color in Business, Science, and Industry*, Third Edition
KNITTL • *Optics of Thin Films*
LENGYEL • *Introduction to Laser Physics*
LENGYEL • *Lasers, Second Edition*
LEVI • *Applied Optics, A Guide to Optical System Design*, Volume I
LOUISELL • *Quantum Statistical Properties of Radiation*
McCARTNEY • *Optics of the Atmosphere*; *Scattering by Molecules and Particles*
MOLLER AND ROTHSCHILD • *Far-Infrared Spectroscopy*
PRATT • *Laser Communication Systems*
SHULMAN • *Optical Data Processing*
WILLIAMS AND BECKLUND • *Optics*
ZERNIKE AND MIDWINTER • *Applied Nonlinear Optics*

Optics of the Atmosphere

Scattering by Molecules and Particles

EARL J. McCARTNEY

John Wiley & Sons, New York / London / Sydney / Toronto

Library of Congress Cataloging in Publication Data

McCartney, Earl J., 1908–
Optics of the atmosphere.

(Wiley series in pure and applied optics)
Bibliography: p.
Includes index.
1. Meteorological optics. 2. Light—Scattering.
I. Title.
QC976.S3M3 551.5′271 76-10941
ISBN 0-471-01526-1

Printed in the United States of America

10 9 8 7 6 5 4 3 2 1

To Marie

Preface

This book is addressed to those who seek a better understanding of the scattering processes that occur in the atmosphere and the resulting effects on the atmospheric propagation of light. Persons having such needs or interests are found in the fields of atmospheric physics, meteorology, air pollution control, and military surveillance. The intended readers may be scientists, designers, development engineers, or technicians—all of whom may be variously concerned with electrooptical or visual instrumentation and the measurement of atmospheric properties. Or the reader might be, apart from his vocation, that fortunate person who is an inquiring student of our physical environment. As an engineer concerned for many years with problems related to the atmospheric propagation of light I have had these prospective readers in mind during the preparation of the book.

A current treatment of atmospheric scattering has available the enormous amount of information generated in recent years. During the past 3 decades investigations of the terrestrial environment have ranged from the top of the atmosphere to the bottom of the ocean. These activities continue, and the resulting growth of knowledge is comparable to a continuing explosion. Well to the front of this growth move atmospheric physics and meteorology, which have become both interdisciplinary and global in their breadth of interest and scope of subject matter. Sharing in the growth of these parent subjects is atmospheric optics, which treats the relationships between the physical properties of the atmosphere and the propagation of light. These relationships are exhibited in the observable processes known as *scattering*, *absorption*, *emission*, and *refraction*.

The objectives of this book are to bring together the relevant information on the physical properties of the atmosphere and the theories and principles of scattering, and to interpret and present this information in forms suitable for direct use. We thus mediate between specialized knowledge in the fields of optics and atmospheric physics and the needs of persons who are not specialists throughout these broad fields. Following this middle-road philosophy for readers having a diversity of backgrounds naturally entails a problem of presentation. Something new and worthwhile on one aspect or another of the subject must be communicated to each reader, who may be a specialist on some other aspect.

An intermediate technical level of presentation is employed in the book. I believe that this level is appropriate to the needs of a technically oriented beginner at any point, without being so elementary that it annoys another reader who may be an expert at that point. Efforts have been made to provide treatments that are sufficiently theoretical to serve as bases for further studies and sufficiently practical to aid in solving specific problems. This volume, as its title implies, deals only with scattering. Considerations of book size and desired unity of subject matter have imposed this sensible restriction. It is planned to cover atmospheric absorption and emission in a second volume, employing a style of treatment similar to that of the present volume.

I am conscious of my debts to many persons. First of course are the many workers in atmospheric optics who have been publishing the results of their studies and experiments since the time of Lord Rayleigh. Genuine inspiration was obtained from Humphreys' *Physics of the Air* (1940), Middleton's *Vision through the Atmosphere* (1952), Johnson's *Physical Meteorology* (1954), and van de Hulst's *Scattering by Small Particles* (1957). Additional insights and viewpoints were gained from the technical papers published by Hulburt and his associates at the Naval Research Laboratory, by Elterman and numerous other workers at the Air Force Cambridge Research Laboratories, by Deirmendjian and Penndorf, and by many others whose publications are referenced in the text.

Specific help closer to home must be acknowledged. The overall work was planned and preliminary versions written while I was a research engineer at the Sperry Gyroscope Division of the Sperry Rand Corporation. At that time, Dr. L. W. Holmboe, E. W. Cheatham, T. C. Hutchison, and R. W. Jagoe provided much encouragement and assistance. I express my strong feeling of gratitude for the help graciously given by Mrs. Jimmie Aushman of the Sperry Engineering Library for several years. Many of the technical papers and reports cited in the bibliography were obtained with her assistance. Too often we take for granted the indispensable services performed by a good technical library.

I am pleasantly obligated to my good friends W. I. Thompson, III, of the U.S. Department of Transportation and H. M. Rathjen of the Systems Group of Thompson Ramo Woolridge for their enthusiasm and encouragement. My appreciation goes to Mrs. Mildred Anderson for her careful typing of the manuscript through several revisions. I wish to thank Professor Stanley S. Ballard, advisory editor of the Wiley Series in Pure and Applied Optics, for his thoughtful reading of the text and for many helpful suggestions. Above all, I am grateful to Beatrice Shube of John Wiley & Sons for expert advice over a lengthy time span and for what must have been sheer faith that the task would be completed. I count myself fortunate in having had the opportunity to write this book.

EARL J. MCCARTNEY

Rockville Centre, New York
March 1976

Contents

Optics of the Atmosphere

1

An Overview of Atmospheric Scattering

From earliest times our principal knowledge of the physical world has been derived from the basic act of seeing.

> Light brings us news of the Universe. Coming to us from the sun and the stars, it tells us of their existence, their positions, their movements, their constitutions and many other matters of interest. Coming to us from objects that surround us more nearly it enables us to see our way about the world: we enjoy the forms and colors that it reveals to us, we use it in the exchange of information and thought. If the meaning of the word is extended, as may be done with every right and reason, to cover the wide range of radiations which are akin to it and yet are not visible to the eye, then light is also the great conveyor of energy from place to place in the world and in the universe .... [Sir William Bragg (c1940)]

As earth dwellers we are immersed in an ocean of atmosphere several hundred miles deep, and we spend our lives at these depths. The light bringing us information from a distance, illuminating our surroundings, and bearing energy to make our world habitable, performs these functions only by traveling in this atmospheric ocean. The characteristics of the light, notably its color, intensity, polarization, and spatial distribution, are altered by this encounter with the atmosphere.

Atmospheric optics is the province of all alterations of the characteristics of light as it propagates in the atmosphere. The alterations are the consequences, or perhaps the facts, of interactions between the energies carried by the light waves and the materials of the atmosphere. The interactions fall into three classes: scattering, absorption, and emission. This book deals only with scattering over the broad spectral region from about 0.3 μm (near ultraviolet) to about 100 μm (extreme infrared).

The objective of this chapter, as its title implies, is to present a broad view of atmospheric scattering without detailed regard to actual scattering mechanisms. This view will provide readers with a perspective of this large subject—a context in which later discussions of specifics should be meaningful. In terms of analogy, this chapter may be likened to a tour prospectus, giving the traveler a map of the route, a description of the country, and notices of the stopovers, all as preparation for his trip.

1.1 THE ROLE OF SCATTERING

Most of the light that reaches our eyes comes not directly from its sources but indirectly by means of scattering. Unless we are looking at a source such as the sun, a flame, or an incandescent filament within a clear bulb, the light by which we see an object has been scattered. Our terrestrial environment abounds in scattering. The ocean surface, the land surface, and the multitude of objects around us—most material things are visible by reason of the light they reflect. The property of *reflectance*, exhibited by all liquid and solid materials at their surfaces, is a particular manifestation of *scattering* by the closely packed atoms and molecules constituting these materials. Scattering is an *observable* of the interaction of light and matter, and it occurs at all wavelengths in the electromagnetic spectrum.

As we shift our gaze from the ground to the hemisphere of the sky, we see many striking examples of scattering. In this domain it assumes many forms and takes on the scope and variability of the atmosphere itself. The interactions involve all atmospheric materials. Molecules of the various gases, small particles having a wide range of sizes and an astonishing diversity of sources and compositions, water droplets whose size range is greater than three orders of magnitude—these are the optical materials of the atmosphere. These materials and their changing distributions in time and space, over the earth's surface and from the surface to the upper reaches of the atmospheric envelope, are central to our subject.

The visual phenomena of scattering are numerous and impressive. Although they have been familiar to us since early years, this familiarity should not breed indifference. These phenomena, however, are only a small part of the total, just as the visual spectrum itself is only a small part of the total spectrum. Scattering by haze and clouds markedly influences the amount of solar energy reaching the lower levels of the atmosphere and the earth's surface. In turn, scattering also influences the amount of radiant energy that leaves the earth for space. Thus scattering and atmospheric absorption-emission jointly control the heat budget of the earth and, as a result, the average temperature of our environment.

Finally, we hope that many readers will look with renewed interest at the many optical effects displayed on a grand scale across the sky. The awakening colors at sunrise, the unfathomable blue of a clear sky, the moving shapes and tints of clouds, the thickening of haze into fog, the rainbow, and particularly the red bow at sunset—these are to be studied and enjoyed. As Hulburt (1956) suggested after admiring the twilight sky, Nature dips freely into her great bag of optical tricks to produce her best effects. Readers wishing to take rewarding excursions in these directions can hardly do better than to follow Minnaert (1954). Atmospheric optics is unique among physical sciences in providing beautiful examples of its subject matter, and opportunities of observation are open to all.

1.2 ELEMENTS OF RADIOMETRY AND PHOTOMETRY

Radiometry and photometry are the twin disciplines concerned with measuring the attributes of light, within the context of carefully specified spectral and geometric constraints. The basic attribute is the rate at which electromagnetic energy, per stated spectral band, flows through a defined solid angle or area in particular directions. These geometric constraints may be associated with a material volume, a material surface, a conceptual surface, or free space itself. All other attributes devolve on this basic one.

We believe that the clearest meaning of an optical theory is achieved when it is expressed in radiometric or photometric terms. In turn, this allows the formulation of verifiable concepts. In a similar manner, the practical meaning of any optical phenomenon is found in its radiometric or photometric measurement. No branch of optics needs to be tied more closely to these twin disciplines than does scattering. For all these reasons, before entering on a study of scattering, we review the standard nomenclature of radiometry and photometry used in this book.

1.2.1 Standard Nomenclature

This book employs the practice of *American National Standard: Nomenclature and Definitions for Illuminating Engineering, ANSI Z7.1-1967*, listed in the references as ANSI (1967). This standard has evolved over a period of half a century and has undergone numerous revisions. It incorporates the firm recommendations of the Nomenclature Committee of the Optical Society of America, as summarized by MacAdam (1967). Further, it reflects general agreement with the nomenclature proposals of the International Commission on Illumination (CIE), the International Electrotechnical Commission (IEC), the International Organization for Standardization (ISO), and the SUN Commission of the International Union for Pure and Applied Physics. These agreements indicate levels of standardization, authority, and acceptance which transcend those of the traditional, differing nomenclatures employed in the past. As a convenience for the reader, the principal radiometric quantities of

ANSI Z7.1 are reviewed in this section, and several emendations useful in atmospheric scattering are presented.

Readers who wish to go beyond the material of this section are encouraged to study the following works. The annotated bibliography by Nicodemus (1969) is a valuable guide to some very extensive literature. His paper has been reprinted in AIP (undated) along with treatments by EK (1965), Meyer-Arendt (1968), Muray et al. (1971), Nicodemus (1967, 1968, 1970), and Spiro et al. (1965). Other papers have been published by Bell (1959) and Levin (1968). Chapter-length treatments in books devoted to more inclusive subjects can be found in Hardy and Perrin (1932), Sears (1949), Middleton (1952), Wolfe and Nicodemus (1965), Nicodemus (1967), Levi (1968), Stanley (1968), Hudson (1969), Williams and Becklund (1972), and Engstrom (1974). Book-length treatments are given by Moon (1936), Forsythe (1937), Walsh (1958), and Drummond (1970).

1.2.2 Geometric Considerations

The propagation path of an electromagnetic wave in free space, ignoring any bending due to a gravitational field, is a straight line. Wave fronts, or surfaces of uniform phase, remain parallel to each other and perpendicular to the direction of travel. Indeed, such a path between a light source and a receiver provides the only operational or direct method of establishing a straight line. According to Fermat's principle, the time required for light to travel this path is minimum, and in free space this is the only possible path. Thus the terms *free-space optical path* and *straight line* are equivalent. Straight-line propagation is difficult to demonstrate rigorously by experiment; in this respect it is reminiscent of the Fitzgerald–Lorentz contraction hypothesis. Straight-line propagation also holds for a material medium, provided that no gradient of refractive index exists normal to the path. Any such gradient causes a bending of the path toward the direction of the gradient. In the atmosphere, vertical gradients practically always exist, while horizontal gradients usually occur only infrequently and weakly.

Geometrically speaking, sources are classified as either *point* or *extended sources.* The first type radiates a spherical wave, while *each* elemental area of the second type radiates a spherical wave. The criterion for distinguishing between the two types is the ratio of the greatest projected dimension of the source to the path distance of interest, that is the plane angle subtended by the source. Astronomical objects are familiar examples. Stars, subtending angles less than 1 arc-sec, are point sources while the sun and moon, subtending angles of about 31 arc-min, are extended sources. In practice, however, it is often convenient to observe the following rule. When the path distance is at least 20 times greater than the source projected dimension, the source can be regarded as a point, with an error of 1% or less. In this case the *irradiance*, that is, the quantity of radiant flux incident per unit area of a surface normal to the path, at any distance from the source can be computed directly from the inverse-square law described below. When the prescribed dimension/distance criterion is not met, or when greater accuracy is wanted, the irradiance must be found by integrating over the solid angle subtended by the source. The method of integration is explained in Walsh (1953) and in many textbooks on geometric optics.

Extended sources usually do not emit radiant flux uniformly with direction. Instead, most types of surfaces exhibit, to a varying extent, the cosine characteristic illustrated in Figure 1.1*a*. Here the radiant flux element $d\Phi$ emitted within the elemental solid angle $d\omega$ varies with the angle θ from the surface normal according to

$$d\Phi_\theta = d\Phi_n \cos\theta \tag{1-1}$$

This is Lambert's cosine law, and surfaces whose elements obey this law, either for emission or for reflection, are called *lambertian* or perfectly diffuse surfaces. A lambertian surface is approximated by an optically

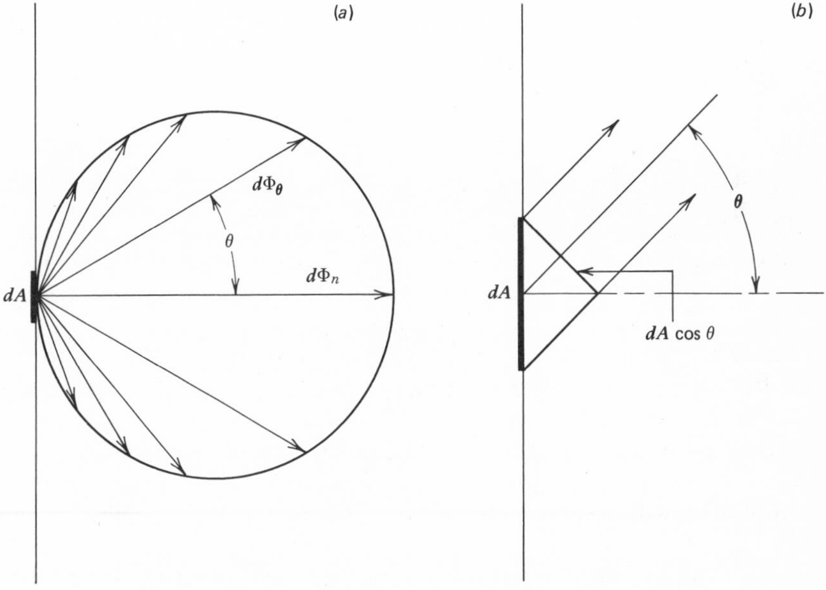

Figure 1.1 Variation in (*a*) flux reflected or emitted by a diffuse surface and (*b*) source projected area, with angle from the normal. From Sears (1949), *Optics*, 3d. ed. (Courtesy of Addison-Wesley Publishing Co.)

rough or matte surface such as that of a blotter. Important also is the dependence of the effective emitting or reflecting area on the angle θ the direction of interest makes with the surface normal, as shown in Figure 1.1*b*. In all such cases the effective area is the source area projected onto a plane normal to this direction.

As a consequence of the spherical wave emitted by a point source, the area of the expanding wave front increases as the square of the distance traveled. The irradiance of the surface then must vary inversely as the square of the distance from the source. The surface in question may be either a real one or a conceptual one representing the cross section of a light beam. In the general case the irradiance E at distance x is given by

$$E = \frac{I}{x^2} \cos \theta \tag{1-2}$$

where I is the source intensity, or radiant flux per unit solid angle, and θ is the angle between the surface normal and the direction to the source. Equation (1-2) expresses two basic laws of radiometry: the inverse-square law and the cosine law. Any beam of light, no matter how small its divergence angle, is subject to the inverse-square law. When the source is very distant or when beam-forming elements are used with a local source, the bounding rays of a selected beam may exhibit a very small divergence angle. The spatial scale of a particular experiment may then permit the divergence to be ignored, and the beam is said to be *collimated.* A beam of starlight is a nearly perfect example of collimated light from a distant source.

1.2.3 Radiometric Quantities

The radiometric quantities are listed in Table 1.1, along with their symbols, dimensions, and units. The basic symbols are the same as those for the corresponding photometric quantities listed in Table 1.2 and discussed in Section 1.2.5. When it is necessary to distinguish between the symbols, the subscript e or v is added; for example, radiant energy is denoted by Q_e and luminous energy by Q_v. Any of the radiometric quantities may be restricted to a narrow spectral band by adding the word *spectral* to the name, and then adding to the symbol a subscript λ for a spectral concentration. In general, the subscript λ means that the quantity is differential with respect to wavelength. A λ in parentheses indicates that the quantity is being considered as a function of wavelength.

A spectral radiometric quantity can be stated in terms of photons instead of the customary energy and power units, that is, the joule and the watt. Since ANSI Z7.1 makes no provision for this, but is not restrictive, we follow Muray et al. (1971) and denote photon flux by adding a

TABLE 1.1 STANDARD SYMBOLS, DIMENSIONS, AND UNITS OF THE RADIOMETRIC QUANTITIES

Symbol	Quantity	Dimension	Common unit
Q_e	Radiant energy	ML^2T^{-2}	Joule (J)
W_e	Radiant density	$ML^{-1}T^{-2}$	Joule per cubic meter ($J\ m^{-3}$)
Φ_e	Radiant flux	ML^2T^{-3}	Watt (W) or joule per second ($J\ sec^{-1}$)
	Radiant flux density at a surface		
M_e	Radiant exitance (radiant emittance)	MT^{-3}	Watt per square centimeter ($W\ cm^{-2}$)
E_e	Irradiance	MT^{-3}	Watt per square meter ($W\ m^{-2}$)
I_e	Radiant intensity	ML^2T^{-3}	Watt per steradian ($W\ sr^{-1}$)
L_e	Radiance	MT^{-3}	Watt per square centimeter per steradian ($W\ cm^{-2}\ sr^{-1}$) Watt per square meter per steradian ($W\ m^{-2}\ sr^{-1}$)

Source: ANSI (1967). (Courtesy of Illuminating Engineering Society.)

subscript q to the symbol. The radiometric quantity so designated is spectral of course, because the energy of a photon depends on the frequency. The dependence is expressed by

$$Q_q = h\nu = \frac{hc}{\lambda} \tag{1-3}$$

where h is Planck's constant, ν is the frequency, λ is the wavelength, and c is the speed of light. Radiant energy is then stated as a number of photons at a specified wavelength per unit spectral interval. Radiant power is stated the same way, but with the added phrase "per unit time."

The quantities in Table 1.1 are now discussed briefly. Radiant energy Q is perhaps the most basic of all. Because the radiant energy emitted by most types of sources depends strongly on the wavelength and the spectral interval considered, it is often necessary to deal with the spectral radiant energy Q_λ. This is defined by

$$Q_\lambda = \frac{dQ}{d\lambda} \tag{1-4}$$

which means that, as the spectral interval is made indefinitely smaller, the

TABLE 1.2 STANDARD SYMBOLS, DIMENSIONS, AND UNITS OF THE PHOTOMETRIC QUANTITIES

Symbol	Quantity	Dimension	Common unit
K	Luminous efficacy	—	Lumen per watt (lm W^{-3})
V	Luminous efficiency	—	Numeric (0 to 1)
Q_v	Luminous energy (quantity of light)	ML^2T^{-2}	Lumen-hour (lm-hr) Lumen-second (lm-sec), talbot (T)
W_v	Luminous density	$ML^{-1}T^{-2}$	Lumen-second per cubic meter (lm-sec m^{-3})
Φ_v	Luminous flux	ML^2T^{-3}	Lumen (lm)
	Luminous flux density at a surface		
M_v	Luminous exitance (luminous emittance)	MT^{-3}	Lumen per square foot (lm ft^{-2})
E_v	Illuminance (illumination)	MT^{-3}	Footcandle (lm ft^{-2}) Phot (lm/cm^{-2}) Lux (lm m^{-2})
I_v	Luminous intensity (candlepower)	ML^2T^{-3}	Candela (lm sr^{-1})
L_v	Luminance (photometric brightness)	MT^{-3}	Candela per unit area Stilb (cd cm^{-2}) Nit (cd m^{-2}) Lambert (cd π cm^{-2}) Footlambert (cd π ft^{-2}) Apostilb (cd π m^{-2})

Source: ANSI (1967). (Courtesy of Illuminating Engineering Society.)

amount of radiant energy becomes ever smaller until the ratio takes on the limiting value Q_λ. Thus Q_λ strictly means the radiant energy per unit wavelength interval at the wavelength λ. Identical reasoning applies to each spectral quantity formed from the entries in Table 1.1.

A finite amount of energy, however, requires a finite spectral width; an infinitely narrow interval contains no energy. The units for any spectral quantity must specify the interval, which is usually made the same as the wavelength unit. The radiant energy over a spectral interval is found from the integration

$$Q = \int_{\lambda_1}^{\lambda_2} Q(\lambda)\, d\lambda \tag{1-5}$$

which yields the total energy when the limits are extended to 0 and ∞. When the spectral distribution of the energy or power is that of a blackbody, a need for integration usually can be avoided by the use of tabulated functions available in the literature.

Radiant density w is the electromagnetic energy content of a unit volume of space at an instant and is defined by

$$w = \frac{dQ}{dV} \tag{1-6}$$

The proviso "at an instant" takes account of the fact that the energy is transported through the unit volume at the velocity of light. The product of energy density and velocity is the radiant power or radiant flux per unit area, which can be taken as either end face of the unit volume. Although this radiant power per unit area has the same units as radiant exitance and irradiance in the table, physically it is somewhat different. Actually, it is flux in transit from a source. This quantity is considered further in a later discussion of irradiance.

Radiant flux Φ is the quantity usually referred to by the term *radiation*, which we restrict to the *process* of initiating and sending forth electromagnetic energy by means of either waves or photons. Radiant flux is defined by

$$\Phi = \frac{dQ}{dt} \tag{1-7}$$

which expresses the time rate of energy flow past a given point. The term *radiant power* is a synonym for radiant flux, within the spirit of ANSI Z7.1. Radiant power is the quantity detected and measured by most types of sensors, including the eye. Integrating types of sensors such as photographic film, however, respond to energy, which is the product of exposure time and power.

Radiant exitance M, formerly called radiant emittance, and irradiance E are the two aspects of radiant flux at a surface. As the names suggest, the first refers to the radiant flux leaving a surface either by emission or reflection, while the second refers to the radiant flux incident on a surface. The quantities are defined by

$$M \text{ or } E = \frac{d\Phi}{dA} \tag{1-8}$$

which means that, as the area is made indefinitely smaller, the amount of radiant flux also becomes smaller until the ratio takes on a limiting value.

The solid angle through which the flux leaves or reaches the surface is not specified for either quantity. In the case of emission or of diffuse reflection, the total solid angle is usually 2π sr, or a hemisphere. In the case of irradiance, the solid angle may vary from this value (e.g., when a horizontal surface is exposed to the entire sky) down to a very small value (e.g., when a surface intercepts a collimated beam).

In treating atmospheric optics a convenient quantity is the flux per unit cross-sectional area of a light beam. The term *flux density* often has been applied to this quantity. The term *density*, however, strongly implies a volume and therefore is not consistent with the dimensions of flux per unit area. Also, confusion with the quantity radiant density defined above is possible. For these reasons the term *flux density* is not used in this book. To avoid introducing a new term, we call attention to the fact that the very concept of cross-sectional area may refer to a *virtual* surface in space normal to the axis of the beam, as well as to a material surface. We then may visualize that the virtual or conceptual surface is irradiated by the flux of the beam and so may speak of *beam irradiance.* This term as used herein will always mean the flux per unit cross-sectional area of the beam and refers to the flux in transit.

Radiant intensity I is the radiant flux proceeding from a point source per unit solid angle in the direction considered. It is defined by

$$I = \frac{d\Phi}{d\omega} \tag{1-9}$$

where ω is the solid angle subtended by an area of interest on a plane normal to the direction of the source. As this angle is made indefinitely smaller, the ratio takes on a limiting value. Many types of point sources do not emit isotropically. For example, the intensity of light scattered by small particles varies greatly with direction. In measuring such light, it is important to keep the receiver acceptance angle small to avoid averaging. Actually, because I is differential with respect to ω, it should be interpreted as the radiant flux within an infinitesimal cone centered about the direction of interest.

The concept of intensity is also applicable to an elemental area of an extended source, hence to a finite area at a sufficient distance. The intensity varies in some manner with the angle between the direction considered and the normal to the surface. If the surface is lambertian, the variation in intensity obeys the cosine law illustrated in Figure 1.1*a*. Substituting for $d\Phi$ in (1-9) the value of $d\Phi_\theta$ from (1-1), we have

$$I_\theta = \frac{d\Phi_n \cos\theta}{d\omega} \tag{1-10}$$

This result is important for the concept of radiance discussed below.

Radiance L applies to an extended surface which may be either an emitter or a reflector. The radiance of such a surface in any direction θ is defined as the ratio of the radiant flux leaving an element of the surface to the product of the projected area of the element and the solid angle as each of these two quantities is made indefinitely small. In symbols we have

$$L_\theta = \frac{d^2\Phi}{(dA\cos\theta)\,d\omega} = \frac{dI_\theta}{dA\cos\theta} \tag{1-11}$$

where the second part, which is obtained by the use of (1-9), shows that the radiance in any direction is equal to the intensity in that direction per unit projected area. For a lambertian surface and any direction from the normal, (1-11) can be written from (1-10) as

$$L_\theta = \frac{dI_n\cos\theta}{dA\cos\theta} = \frac{dI_n}{dA} \tag{1-12}$$

Thus the radiance of a lambertian surface is independent of the observing direction. A familiar photometric example is the sun which, although spherical, appears to be a flat disk of uniform (nearly so) luminance.

1.2.4 Radiance Considered Further

The above definition of radiance is satisfactory for many situations but is confusing when applied to a volume source such as the sky. Actually, the sky has no area and certainly is at no definite distance. This difficulty is avoided by defining radiance in terms of the effect produced at the receiver on a reference plane normal to the direction of observation. This alternative viewpoint, highly important to atmospheric optics, is developed by Middleton (1952), Levin (1968), and Baker (1974). The end result, assuming no attenuation by the medium, is

$$L = \frac{d^2\Phi}{dA'\,d\omega'} = \frac{dE}{d\omega'} \tag{1-13}$$

Here L is the radiance of the source, Φ is the radiant flux at the normal receiving surface, dA' is an area element at this surface, $d\omega'$ is the solid angle subtended at this surface by an area element of the source, and E is the irradiance of the surface. Paraphrasing Middleton (1952): The radiance of an extended source in a given direction is equal to the irradiance produced at a surface normal to this direction divided by the solid angle subtended at this surface by the source. Evidently the smaller the portion of the source considered, that is, the more nearly the infinitesimal form (1-13) is approached, the greater the validity of this viewpoint. In comparing (1-11) and (1-13), the first equation defines radiance in terms of the sending end, while the second defines it in terms of the receiving end. The two definitions are equivalent. Thus when sky radiance is stated in terms of W $\text{cm}^{-2}\ \text{sr}^{-1}$, the reader may think of the amount of radiant power incident per unit area of a normal receiving plane at his eye per steradian of sky.

An important consequence of Lambert's law is the relationship between radiant exitance M and radiance L. By writing (1-11) as

$$d^2\Phi = L\, d\omega\, dA \cos\theta \tag{1-14}$$

and referring to Figure 1.2, it is clear that the flux emitted by a surface element dA in the direction θ goes into the solid angle $d\omega$ lying between θ and $\theta + d\theta$. All this flux is incident at the corresponding zone on the hemisphere. In terms of θ the area ds of this zone is

$$ds = 2\pi r^2 \sin\theta\, d\theta \tag{1-15}$$

while in terms of ω it is

$$ds = r^2\, d\omega \tag{1-16}$$

so that

$$d\omega = 2\pi \sin\theta\, d\theta \tag{1-17}$$

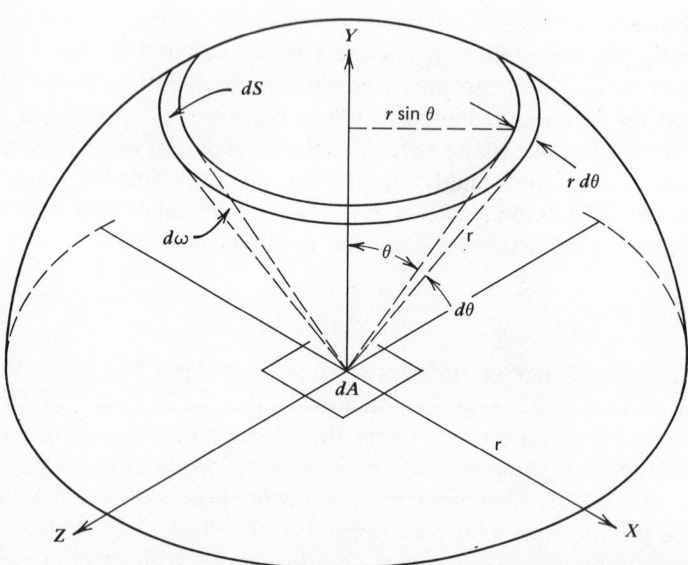

Figure 1.2 Geometry for integrating the radiance of a lambertian surface to find the radiant exitance.

Substitution of (1-17) into (1-14) and integration over the hemisphere give

$$d\Phi = 2\pi L\, dA \int_0^{\pi/2} \sin\theta \cos\theta \, d\theta$$
$$= \pi L\, dA \qquad (1\text{-}18)$$

Combining this with (1-8), we have

$$M = \pi L \qquad (1\text{-}19)$$

Thus the total radiant flux emitted into a hemisphere by a unit area of a lambertian surface, that is, the radiant exitance of the surface, is equal to π times its radiance, not 2π times its radiance.

1.2.5 Response of the Eye

Photometry is the division of radiometry in which radiant flux is evaluated, not in absolute terms such as joules or watts but in terms of its ability to evoke a visual sensation. Flux so evaluated is called luminous flux, but the geometric parameters of distance, area, and angle are the same in photometry as in radiometry. To each radiometric quantity there corresponds a photometric quantity whose name is qualified by the adjective luminous. Because the eye was the original sensor, visual considerations were paramount in the early developments of optics. Photometry thus preceded the larger field of radiometry. In the ultraviolet and infrared regions photoelectric and thermal sensors are essential of course. Although instrumentation employing such sensors, as well as photographic film, has largely replaced the eye in photometric work, the objective of the measurements usually is to obtain results that agree with visual results. The eye therefore remains the ultimate judge of light in the narrow spectral region from about 0.38 to 0.76 μm. Its responses to various types and levels of stimuli are the reference standards.

The human eye and associated neural system constitute the most remarkable optical sensor of all time. The focusing properties and millions of separate retinal receptors arrange the incoming stimuli into a replica of the original spatial scene, whether it be 10 in. or 10 m away. The ability to image a wide field with virtually no distortion probably is unmatched by any other optical sensor. The capabilities of adaptation enable the eye to function at starlight levels of illuminance as well as at sunlight levels—an astounding dynamic range greater than 1 to 1 million. A thought-provoking investigation of the eye as an optical sensor is presented by Rose (1948, 1973). The relatively long time constant of the eye (approximately 0.1 sec) produces a persistence of vision which makes possible the cinema and without which all motion viewed by fluorescent lighting would exhibit intolerable stroboscopic effects. Finally, its selective, psychophysical response to wavelength wraps the bare visual sensation in a coat of many colors, providing us with additional information and immeasurably enriching our aesthetic appreciation.

The response of the eye to radiant flux at different wavelengths is shown in Figure 1.3. The curves thus reveal the relative efficiency of monochromatic flux, having constant power at each wavelength, in producing the sensation of luminance (brightness) in the eye of a "standard observer." Curve a, corresponding to the values in Appendix D, refers to *photopic vision*, which exists when the field luminance is greater than about 10^{-2} cd m^{-2}. This is approximately the luminance of a clear moonlit sky near the horizon. Curve b applies when the luminance is

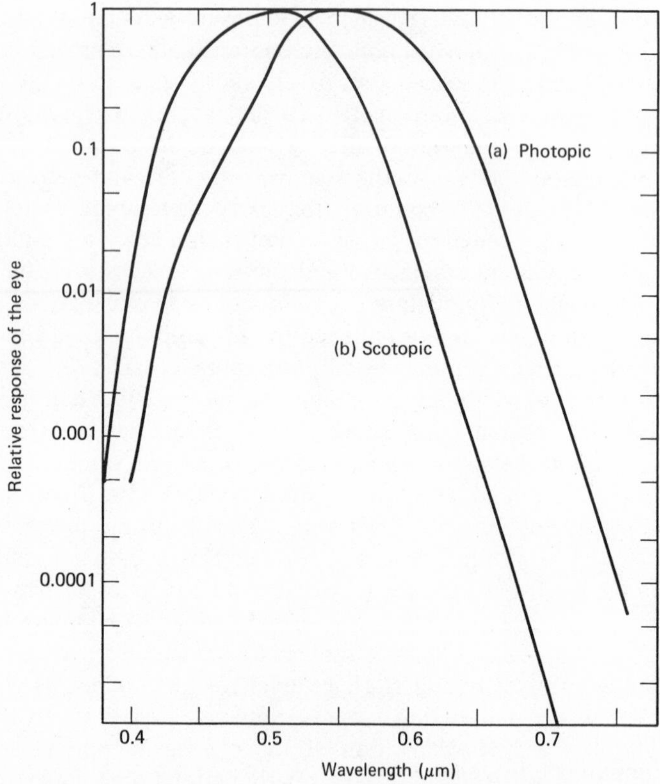

Figure 1.3 Relative spectral response of the eye at photopic and scotopic levels of luminance. These responses define the luminous efficiency of radiant flux. From Walsh (1953).

below this value. At such low levels the eye becomes dark-adapted, which requires several minutes at luminances near 10^{-2} cd m^{-2}, and the vision is *scotopic*. The eye is far more sensitive in scotopic than in photopic vision. Each curve has been normalized to unity at its maximum, so that it shows the relative luminous efficiency of radiant flux. The shift of the maximum to shorter wavelengths as the field luminance is decreased produces changes in the relative luminance of different colors. This is known as the *Purkinje effect*, and the luminance range over which it occurs is called the *Purkinje range*.

1.2.6 Photometric Quantities

The photometric quantities are listed in Table 1.2; each one corresponds to a particular radiometric quantity in Table 1.1. In photometry, the unit of luminous flux (power) is the lumen, just as the watt is the unit of radiant flux (power). The lumen is defined, however, not in terms of power as such, but in terms of a standard source. Originally this source was a "standard candle," whose very name suggests that photometry is a traditional art. As time passed, an "international candle" became the reference source, and this was variously reproduced in the form of calibrated gas flames and carbon filament lamps. The lumen is now defined as the luminous flux emitted into 1 sr by a source whose luminous intensity is 1 cd. The candela is defined as one-sixtieth of the luminous intensity of a blackbody radiator, having an area of 1 cm^2, at the temperature of solidifying platinum (2042 K). So defined, the candela is nearly the equivalent of the old international candle. A definitive account of these standards is given by Jones and Preston (1969).

Referring now to the first entry in Table 1.2, we note the relationship between the lumen and the watt. The luminous efficacy K of radiant flux is defined by

$$K = \frac{d\Phi_v}{d\Phi_e} \tag{1-20}$$

and is expressed in lumens per watt. In photopic vision, K has its maximum value at 0.555 μm. The candela as defined above has a value such that, at this wavelength,

$$\begin{aligned} 1\ \text{W} &= 680\ \text{lm} \\ 1\ \text{lm} &= 1.47 \times 10^{-3}\ \text{W} \end{aligned} \tag{1-21}$$

A calculation of these values is given by Sears (1949).

The luminous efficiency V of radiant flux is defined by

$$V_\lambda = \frac{K_\lambda}{K_{max}} \tag{1-22}$$

This has a maximum value of unity at 0.555 μm, as in Figure 1.3 and Appendix D. At other wavelengths the number of lumens in a sample of radiant flux must be found from

$$\Phi_{v,\lambda} = 680\, V_\lambda \Phi_{e,\lambda} \tag{1-23}$$

where $\Phi_{e,\lambda}$ is the spectral radiant flux in watts. When a range of wavelengths is involved, the luminous flux is found from

$$\Phi_{v,\lambda} = 680 \int_{\lambda_1}^{\lambda_2} V(\lambda)\Phi_e(\lambda)\, d\lambda \tag{1-24}$$

The integral must be solved by graphical or numerical methods, because $V(\lambda)$ is not an analytic function of wavelength.

The remaining quantities in Table 1.2 are now described only briefly, because the previous discussions of the matching radiometric quantities are largely applicable. Luminous energy Q_v, corresponding to radiant energy Q_e, is defined by

$$Q_v = \int_{380\,\mu\mathrm{m}}^{760\,\mu\mathrm{m}} K(\lambda)Q_e(\lambda)\, d\lambda \tag{1-25}$$

and the unit of measurement is the product of luminous flux (in lumens) and a selected unit of time. Luminous density is defined by

$$w_v = \frac{dQ_v}{dV} \tag{1-26}$$

Luminous flux Φ_v expresses the time rate at which luminous energy flows past a given point, according to

$$\Phi_v = \frac{dQ_v}{dt} \tag{1-27}$$

The unit of luminous flux is the lumen, whose relationship to the unit of radiant flux (the watt) is described above. The remaining quantities are defined in terms of the lumen.

Luminous exitance M_v refers to the luminous flux leaving a surface either by emission or reflection, while illuminance E_v refers to the luminous flux incident on a surface. This surface may be either real or conceptual, such as the cross section of a beam. Exitance and illuminance are defined by

$$M_v \text{ or } E_v = \frac{d\Phi_v}{dA} \tag{1-28}$$

Although illuminance is the counterpart of irradiance, ANSI Z7.1 prefers but does not insist on the term *illumination* for illuminance. This book, however, employs the term *illuminance* for the quantity defined by (1-28) and restricts the term *illumination* to the process itself. The basic measures of illuminance are illustrated in Figure 1.4, where a point source is emitting 1 lm into 1 sr, so that its luminous intensity is 1 cd. The units in Table 1.2 may be correlated with this diagram, where we see the basis for the widely used but misleading term *footcandle.* Conversion factors for the units of illuminance are given in Table 1.3. Analogously to the use of the term *beam irradiance,* as defined in Section 1.2.3, the term *beam illuminance* as used in this book refers to the luminous flux per unit cross-sectional area of a light beam.

Luminous intensity is defined by

$$I_v = \frac{d\Phi_v}{d\omega} \tag{1-29}$$

The unit is the candela, defined in terms of the standard source and equivalent to 1 lm sr^{-1}. Thus an isotropic point source (although such a source scarcely exists) having a luminous intensity of 1 cd would emit a total luminous flux of 4π lm. Very often the luminous intensity of a beam-projecting source such as a searchlight is expressed in terms of candlepower (a traditional unit). One candlepower is equal to 1 cd. The concept of luminous intensity follows directly from the concept of a point source, as may be appreciated from Figure 1.4. Luminous intensity is equally applicable to an elemental area of a surface, as in Figure 1.1*a*;

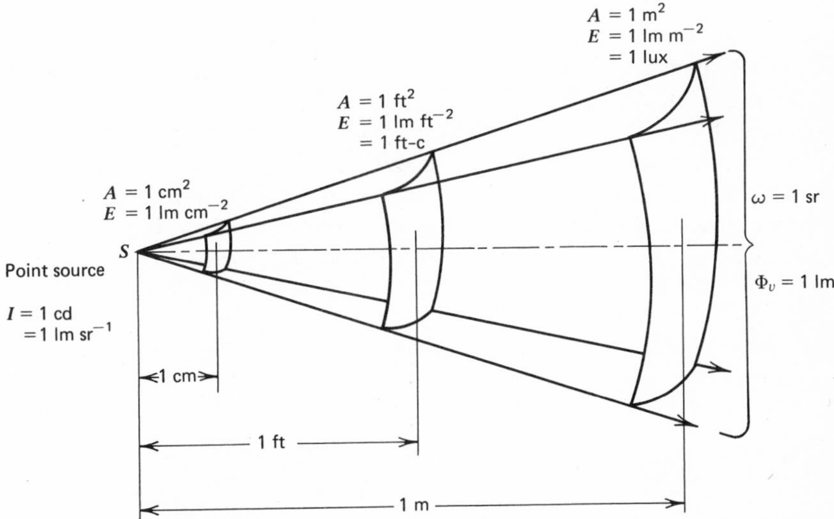

Figure 1.4 The basic measures of illuminance. From Sears (1949), *Optics*, 3d. ed. (Courtesy of Addison-Wesley Publishing Co.)

TABLE 1.3 ILLUMINANCE (ILLUMINATION) CONVERSION FACTORS

1 lm = 1/680 lightwatt	1 W-sec = 1 J = 10^7 ergs
1 lm-hr = 60 lm-min	1 phot = 1 lm cm^{-2}
1 ftcandle = 1 lm ft^{-2}	1 lux = 1 lm m^{-2}

Number of → Multiplied by ↘ Equals number of ↓	Footcandles	Lux*	Phots	Milliphots
Footcandles	1	0.0929	929	0.929
Lux*	10.76	1	10,000	10
Phots	0.00108	0.0001	1	0.001
Milliphots	1.076	0.1	1,000	1

*The International Standard (SI) unit.
Source: ANSI (1967). (Courtesy of Illuminating Engineering Society.)

hence by extension it is applicable to a finite area at a sufficient distance. In this manner the candela is defined as noted above.

Luminance L_v, the counterpart of radiance L_e, refers to an extended source and is defined by

$$L_v = \frac{d^2\Phi_v}{(dA \cos\theta)\, d\omega} = \frac{dI_v}{dA \cos\theta} \tag{1-30}$$

When the surface is lambertian, L_v is independent of direction, as may be seen from (1-12). Luminance was formerly called *brightness*, a name now reserved for the basic visual sensation produced by a source. The term *photometric brightness*, however, is synonymous with luminance. When the extended source is the sky, its luminance may be interpreted from the viewpoint used for sky radiance. That is, sky luminance is equal to the lumens incident per unit area on a plane surface at the observer's location normal to the direction of interest per steradian of sky. Thus, analogously to (1-13),

$$L_V = \frac{d^2\Phi v}{dA'\, d\omega'} = \frac{dE_v}{d\omega'} \tag{1-31}$$

Luminance is related to luminous exitance by

$$\pi L_v = M_v \tag{1-32}$$

which is the counterpart of (1-19). For example, a surface that has a luminance of 1 cd m^{-2} has a luminous exitance of π lm m^{-2}. Several units

TABLE 1.4 LUMINANCE (PHOTOMETRIC BRIGHTNESS) CONVERSION FACTORS

1 nit = 1 cd m^{-2}
1 sb = 1 cd cm^{-2}
1 apostilb (international) = 0.1 mL = 1 blondel
1 apostilb (German Hefner) = 0.09 mL
1 L = 1000 mL

Number of → Multiplied by ↘ Equals number of ↓	Foot-lamberts	Candela* per square meter	Milli-lamberts	Candelas per square inch	Candelas per square foot	Stilbs
Footlamberts	1	0.2919	0.929	452	3.142	2,919
Candelas per square meter (nit)*	3.426	1	3.183	1,550	10.76	10,000
Millilamberts	1.076	0.3142	1	487	3.382	3,142
Candelas per square inch	0.00221	0.000645	0.00205	1	0.00694	6.45
Candelas per square foot	0.3183	0.0929	0.2957	144	1	929
Stilbs	0.00034	0.0001	0.00032	0.155	0.00108	1

*International System (SI) unit.

Source: ANSI (1967). (Courtesy of Illuminating Engineering Society.)

are commonly used for expressing luminance; the conversion factors are listed in Table 1.4. The area conversion units in Appendix C also may be found convenient.

1.3 PRINCIPAL CHARACTERISTICS OF SCATTERING

With the radiometric principles of the preceding section in mind, we now look at the basic scattering process and consider the several types of scattering that occur when a light wave is incident on a particle. This is done first in terms of a single isolated particle, and the effects of particle size relative to wavelength are described. The ideas are then extended to the case of a group of particles scattering in proximity to each other, thus creating composite effects. Requirements are stated for single, independent, incoherent scattering, which is the type treated in this book. Finally, several distinctions are drawn between scattering and absorption.

1.3.1 Nature of the Scattering Process

Scattering is the process by which a particle—any bit of matter—in the path of an electromagnetic wave continuously (1) abstracts energy from the incident wave, and (2) reradiates that energy into the total solid angle centered at the particle. The particle is a point source of the scattered (reradiated) energy. For scattering to occur, it is necessary that the refractive index of the particle be different from that of the surrounding medium. The particle is then an optical discontinuity, or inhomogeneity, to the incident wave. When the atomic nature of matter is considered, it is clear that no material is truly homogeneous in a fine-grained sense. As a result, scattering occurs whenever an electromagnetic wave propagates in a material medium. In the atmosphere the particles responsible for scattering run the size gamut from gas molecules to raindrops, as listed in Table 1.5. The wide ranges of size and concentration are noteworthy.

TABLE 1.5 PARTICLES RESPONSIBLE FOR ATMOSPHERIC SCATTERING

Type	Radius (μm)	Concentration (cm^{-3})
Air molecule	10^{-4}	10^{19}
Aitken nucleus	10^{-3}–10^{-2}	10^{4}–10^{2}
Haze particle	10^{-2}–1	10^{3}–10
Fog droplet	1–10	100–10
Cloud droplet	1–10	300–10
Raindrop	10^{2}–10^{4}	10^{-2}–10^{-5}

About each particle the intensity of the scattered radiant energy, hereafter called the *scattered intensity*, forms a characteristic three-dimensional pattern in space. If the particle is isotropic, the pattern is symmetric about the direction of the incident wave. The form of the pattern depends strongly on the ratio of particle size to wavelength of the incident wave, as illustrated by the three examples in Figure 1.5. In Figure 1.5*a* the relatively small particle tends to scatter equally into the forward and rear hemispheres. When the particle is larger, as in Figure 1.5*b*, the overall scattering is greater and is more concentrated in the forward direction. For a still larger particle, as in Figure 1.5*c*, the overall scattering is even greater. Most of it is now concentrated in a forward lobe, and secondary maxima and minima appear at various angles. Further increases in particle size produce patterns of even greater complexity. In all cases the form of the pattern is influenced by the relative refractive index, that is, the ratio of the refractive index of the particle to that of the medium surrounding the particle.

Scattering is explained in terms of the electromagnetic wave theory and the electron theory of matter. Briefly stated here but elaborated in later chapters, the electric field of the incident or primary wave sets into

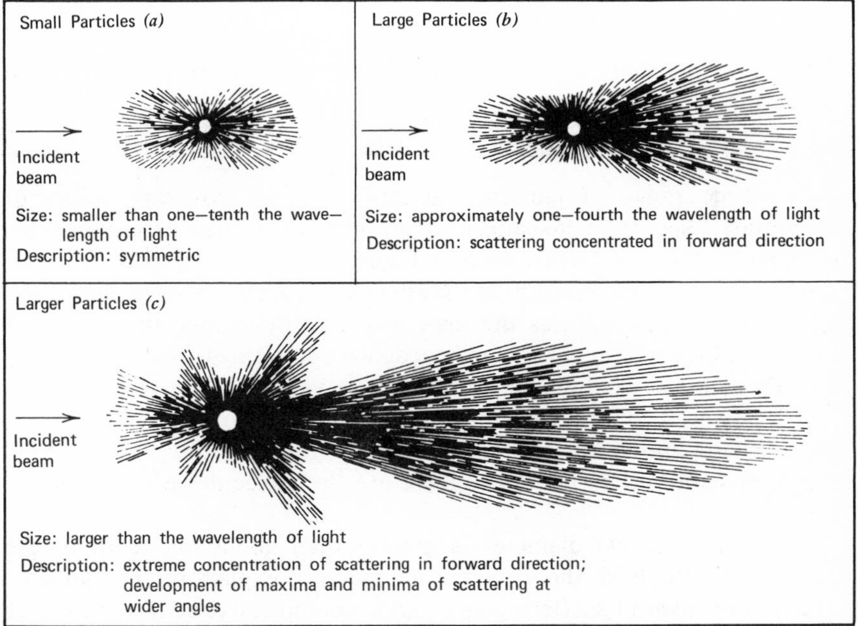

Figure 1.5 Angular patterns of scattered intensity from particles of three sizes. (*a*) Small particles, (*b*) large particles, (*c*) larger particles. From Brumberger et al. (1968).

oscillation the electric charges of the particle, be it a molecule or a cloud droplet. The oscillating charges constitute one or more electric dipoles which radiate secondary, spherical waves. Since the charges oscillate synchronously with the primary wave, the secondary waves have the same frequency, hence wavelength, as the primary wave, and they bear fixed phase relations to it. Timewise, the scattering process is a continuous one and, when averaged over complete cycles, produces no net change in the internal energy states of the particle. Spectrally, the process is also continuous, although it is strongly dependent on wavelength for a particle of given size. This may be inferred from Figure 1.5.

Additional characteristics are noted. Even if the incident light is unpolarized, the scattered light is polarized to some extent. The type and degree of polarization depend on the optical properties of the particle, the polarization of the incident light, and the direction in which the scattered light is observed. When the particle is isotropic, the scattered intensity referred to a particular polarization is a function of particle size, particle relative refractive index, and wavelength of the incident light. These are the three parameters of scattering. To the extent that they are known, the scattering pattern can be accurately predicted from theory. Speaking broadly, the scattering process has two observable aspects. When the scattered intensity is of concern, as in studies of skylight, we deal with *angular scattering*. When the total flux removed from a light beam is of concern, as in studying the attenuation of sunlight, we deal with *total scattering*.

1.3.2 Types of Scattering

The wide range of particle size in Table 1.5, covering orders of magnitude, suggests that scattering itself may show large variations. This is indeed a fact, as can be seen in Figure 1.5. When the particle is far smaller than the wavelength, the scattering is called *Rayleigh scattering.* This name commemorates the man who first developed the theory of scattering by very small isotropic particles. Scattering of this type varies directly as the second power of the particle volume and inversely as the fourth power of the wavelength. Equal amounts of flux are scattered into the forward and back hemispheres, as in Figure 1.5*a*. The principal Rayleigh scatterers in the atmosphere are the molecules of atmospheric gases.

When the particle diameter is greater than about one-tenth of the wavelength, Rayleigh theory is not adequate to explain the phenomena. The greater overall scattering and pattern complexity, as in Figure 1.5*b* and *c*, require for their explanation the theory developed by Mie. Although his theory is strictly applicable only to isotropic spheres, it is customary to employ the term *Mie scattering* even though the particles may be somewhat irregular in shape. The full Mie theory is expressed as a mathematical series embracing all particle sizes; the first term of the series is equivalent to the Rayleigh expression. For spheres of great relative size, such as raindrops illuminated by visual light, the Mie theory can be closely approximated by the principles of reflection, refraction, and diffraction. Every particle in the atmosphere is actually a Mie scatterer, but we apply the term only to particles larger than Rayleigh scatterers.

Attention is called to a third type of scattering which, under certain conditions, accompanies Rayleigh scattering. As noted previously, this type of scattering occurs without change in frequency. However, when the incident light is nearly monochromatic, or alternatively consists of line spectra, a careful analysis of the scattered light reveals weak spectral lines not present in the incident light. Such changes in frequency are the result of changes in the energy levels of the molecules. The changes or transitions take place concurrently with Rayleigh scattering and produce frequencies greater and less than the frequency of the principally scattered light. The frequency shifts are related to the differences between the permitted energy levels, and they provide data for identifying the molecular species. This phenomenon is *Raman scattering*, named for the Indian physicist who first investigated it. This type of scattering requires quantum theory for its explanation and is not treated in this book.

1.3.3 Scattering by Many Particles

In most practical cases we are concerned with the scattering by all the particles within a given volume of space. When the average separation distance is several times the particle radius, each particle is considered to scatter independently of all the others. This means that each scattering pattern, such as those shown in Figure 1.5, is unaffected by the neighboring scattering. This is *independent scattering.* The separation criterion is easily satisfied in all the meteorological conditions typified by the particles listed in Table 1.1, as simple calculations will show. Consequently, independent scattering prevails in the atmosphere. The criterion is not met by the closely packed atoms and molecules of high-pressure gases, liquids, and solids. Therefore independent scattering does not obtain in such media.

We consider now independent scattering. When the particles are randomly arranged and randomly moving, no coherent phase relationships exist between the separately scattered waves. Hence no interferences among these waves can be discerned, and their intensities rather than their amplitudes are additive. This is *incoherent scattering.* If the particles are identical, the composite or resultant intensity pattern is the same as that from a single particle. The randomness requirements are met by all atmospheric molecules and particles. The requirements are not met by the more regularly arranged atoms and molecules of liquids and solids, so their scattering has coherent aspects. In a liquid, for example, mutual interferences suppress most of the lateral scattering. The scattering of x rays by the atoms of a crystal lattice provides a more striking example. Here the interferences are sufficiently strong to produce major maxima and minima of intensity at certain angles, giving rise to the term *x-ray diffraction.*

In the discussion to this point it has been tacitly assumed that the particle is exposed only to the light of an incident or direct beam. That is, *single scattering* has been assumed. No account has been taken of the fact that each particle in a scattering volume is exposed to and also scatters a small amount of the light already scattered by the other particles. This light, very weak by comparison with that of the direct beam, reaches a given particle from many directions, as suggested by Figure 1.6. Hence some of the light that has been first-scattered may be rescattered one or more times before emerging from the scattering volume. This is called *secondary* or *multiple scattering.* Although multiple scattering has little effect on the total amount of light removed from the direct beam, it may significantly alter the composite pattern of scattered intensity due to all the particles.

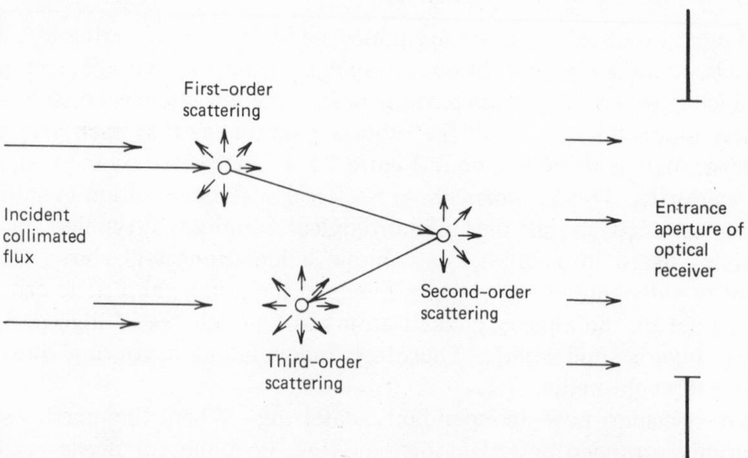

Figure 1.6 Multiple scattering, or rescattering of once-scattered light. The fluxes from all orders of scattering are incident upon all particles from all directions.

This characteristic of multiple scattering can be appreciated from Figure 1.5. For example, visualize that the pattern in Figure 1.5*c* is overlaid with a multitude of similar but far weaker patterns having all orientations in the plane of the figure. It becomes clear that the composite pattern, while still retaining the principal features of the original, exhibits fewer and smaller variations in intensity as a function of angle. In the extreme, as with a very turbid medium, all sense of the direct beam is lost, and the scattered light tends to reach an observer rather uniformly from all directions. This tendency is manifested in a dense fog.

1.3.4. Distinctions Between Scattering and Absorption

Scattering must be distinguished from absorption. Both processes remove flux from a given beam of light, but the similarity ends there. As already noted, scattering is explained in terms of the wave theory of light, and it produces no net change in the internal energy states of the molecules. In contrast, absorption requires quantum theory for its explanation and does produce changes in the energy states. Three forms of such internal energy exist: rotational, vibrational, and electronic arrangement. These forms are additional to the kinetic energy of molecular translation, which plays an indirect but essential role in absorption and also in the associated process called *emission.*

Within the small domains of molecular space and action, each form of internal energy is quantized to discrete, permitted values or levels. The incident radiant energy also must be regarded as quantized, and only whole quanta can be accepted by the molecule. In absorbing a quantum of energy the molecule thereby undergoes a transition from a lower to a higher state of one of the three internal energy forms. Timewise, absorption is a discontinuous process because of the quantizations. Spectrally the process is selective, not continuous, because only those quanta can be absorbed whose energies are equal to the differences between the permitted levels.

Absorption is only the first part of a cycle which is completed by *emission.* As a consequence of molecular motions and collisions, molecules endlessly exchange internal energy for translational energy, and vice versa. Molecules already excited to upper levels are deexcited or *relaxed* to lower levels, on a statistical basis. Conversely, molecules at lower levels are excited to upper levels by collisions, again on a statistical basis. Most of the upper levels are inherently unstable, however, and molecules occupying these levels undergo transitions to lower levels by emitting quanta of radiant energy. These downward transitions may be either spontaneous or stimulated. The first type predominates in the atmosphere, while the second type is the operation basis of lasers. The emitted quanta have energies equal to the differences between the initial and final levels, so emission is as spectrally selective as absorption. Being quantized, emission by an individual molecule is a discontinuous process.

From the foregoing summary, several basic distinctions between scattering and absorption-emission become apparent. These distinctions should be kept in mind. Absorption and emission by gases are not discussed further in this volume. Absorption by particles, however, is dealt with to the extent that several quantities developed in scattering theory are employed as measures of absorption. This allows both scattering and absorption by particles to be considered jointly as a process called *extinction.*

1.4 SCATTERING IN THE ATMOSPHERE

Scattering in the atmosphere has three principal manifestations or observables. These are (1) the diffuse light from the hemisphere of sky over an observer, (2) the attenuation or weakening of the flux in a beam of light, and (3) the restriction on the visibility of a distant object. The first two observables are the direct consequences of angular and total scattering, respectively. These two aspects of the scattering process are treated in this section as measurable phenomena, without detailed examination of the process itself. Accordingly, they are discussed in terms of measurement concepts and associated geometry. The third observable is an interesting combination of the first two and is treated in Section 1.5.

1.4.1 Angular Scattering and Polarization

In order to describe the observable features of angular scattering in a quantitative way, we first specify a geometry for measuring the intensity and polarization of the scattered light. Referring to Figure 1.7, consider that a unit volume of space at point O contains a suspension of scatterers. These may be either gas molecules or small bits of matter in macroscopic form. Their exact properties are ignored here, because detailed mechanisms are not being considered at this stage. The suspension itself is assumed to meet the requirements for independent, incoherent, single scattering summarized in Section 1.3.3.

Now let a collimated beam of nearly monochromatic light be projected along the X axis so that it is incident on the unit volume, and let a detector of radiant flux be located at point D. The distance OD is sufficiently great, compared to the unit volume, that the latter can be treated as a point source. The X axis and the line OD form the *observation plane,* often called the *scattering plane.* The direction of OD in this plane is specified by angle θ, measured from the *forward direction* of the incident light.

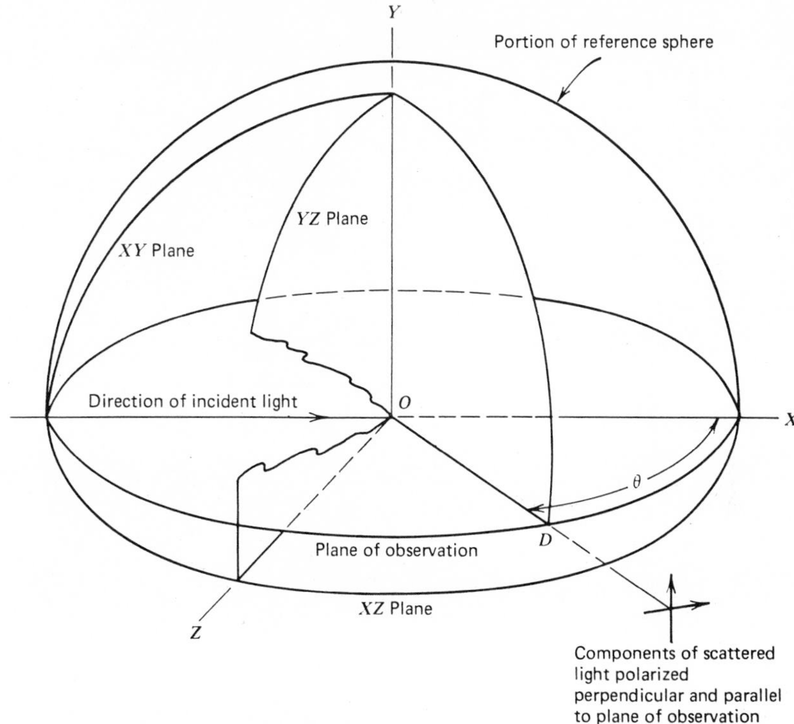

Figure 1.7 Geometry of angular scattering. Scatterer is at point O, observation direction is OD, and observation (scattering) angle is θ.

Angle θ is the *observation angle*, often called the *scattering angle*, and may lie between 0 and 360 deg. Because of symmetry in the scattering pattern, it is usually sufficient to deal only with values of θ between 0 and 180 deg. In some treatments in the literature, the scattering angle is measured from the reverse direction of the incident light.

Let the detector D be rotated, in the plane DOX, about O in order to measure the scattered intensity as a function of θ. Theory predicts and experiment confirms that the spectral scattered intensity $I_\lambda(\theta)$ from the unit volume is given by

$$I_\lambda(\theta) = \beta_\lambda(\theta) E_\lambda \tag{1-33}$$

where E_λ is the spectral irradiance at the unit volume due to the incident light. Because scattering depends strongly on wavelength, expressions such as (1-33) apply strictly only to a narrow spectral interval, as indicated by the subscript λ. This should be kept in mind, since the subscript is often omitted for simplicity.

The proportionality constant $\beta(\theta)$ in (1-33), with the angular dependence always indicated, is the *volume angular scattering coefficient*; hereafter it is called the *angular coefficient.* It defines the angular scattering property of a medium per unit path length of the irradiating beam and per steradian centered about the direction of observation. If, as usual, a solid angle is deemed dimensionless, the dimension of $\beta(\theta)$ is L^{-1} and its value is stated as a number per unit of path length. When the incident irradiance has the value unity, the numerical value of $\beta(\theta)$ represents the intensity of scattered flux in watts per steradian passing through an elemental solid angle in the direction θ per unit path length of scattering medium. The unit path length actually represents a volume, because a dimensional factor L^{-2} is introduced into (1-33) by the units of E. Hence $I(\theta)$ refers to this volume, which should be kept as a unit volume to avoid confusion. This can be done by making the area unit for which E is stated consistent with the length unit employed for $\beta(\theta)$.

When it is necessary to distinguish between scattering by molecules and particles, the subscript m or p may be used with $\beta(\theta)$. Values of $\beta(\theta)$ for common atmospheric constituents cover a wide range. Consider first the molecular scattering by standard air, that is, air at 0 °C and a pressure of 1 atm. At $\lambda = 0.55\ \mu$m and $\theta = 180$ deg, the value of β_m (180 deg), usually designated $\beta_m(\pi)$, is $1.36 \times 10^{-6}\ m^{-1}\ sr^{-1}$. If the incident irradiance is $10\ W\ m^{-2}$, the value of $I(\pi)$ is only $1.36 \times 10^{-5}\ W\ sr^{-1}$. Since there are about 2.7×10^{25} molecules per cubic meter, the contribution of each molecule is notably small. In contrast, for a cumulus cloud having 2×10^{8} droplets per cubic meter, the value of $\beta_p(\pi)$ is about $10^{-3}\ m^{-1}\ sr^{-1}$.

Scattered light always exhibits some type and degree of polarization which can be resolved into two linear, orthogonal components. A natural reference plane for specifying these components is the observation plane *DOX* in Figure 1.7. The component whose electric vector is perpendicular to this plane is the perpendicular component, designated by the subscript $\perp$. Similarly, the component whose electric vector is parallel to this plane is the parallel component, designated by the subscript $\|$. Thus the scattered light consists of, and can be experimentally resolved into, the components $I_\perp(\theta)$ and $I_\|(\theta)$. Likewise, the incident light can be regarded as the sum of two noncoherent components $E_\perp$ and $E_\|$ with respect to the observation plane.

The intensity and polarization characteristics of scattered light are frequently expressed in terms of the four Stokes parameters of polarization. Although most books on physical optics treat the subject of polarization, very few deal with these parameters, which are quite basic. Originated by the English physicist Stokes (1852), they completely describe any type of light. Detailed treatments of these parameters can be found in the books by Shurcliff (1962) and Shurcliff and Ballard (1964). Additional information is given by Mueller (1948), Walker (1954), and Priebe (1969).

1.4.2 Total Scattering and Attenuation

The relationship between total scattering and attenuation of the flux in a light beam is straightforward. Referring to Figure 1.8, let a collimated beam of near-monochromatic light be projected into a suspension of particles. Each particle, by angularly scattering flux in all directions, removes that amount of flux from the beam. We are not here concerned with the particulars of that flux, hence we ignore the very small amount sent in the exact direction of the beam. In a path differential distance dx the fractional amount of flux totally scattered from the beam is represented by

$$\frac{dE_\lambda}{E_\lambda} = \beta \, dx \tag{1-34}$$

Here E_λ is the spectral irradiance at lamina dx, and β is a quantity proportional to the totally scattered flux per unit length of path for a given value of E_λ.

The beam irradiance at a distance along the path is found by integrating (1-34) between the limits 0 and x, which gives

$$E_x = E_0 \exp(-\beta x) \tag{1-35}$$

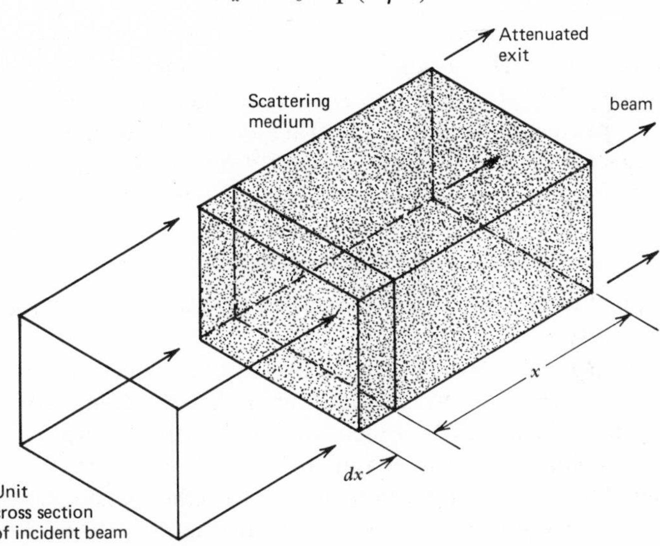

Figure 1.8 Attenuation of light by scattering. The particles of the medium scatter a fraction of the incident light in all directions.

where E_0 is the beam irradiance at $x = 0$. Although the subscript λ has been deleted, it is understood that both E and β are spectral quantities. Equation (1-35) is Bouguer's exponential law of attenuation, discovered by him about 1729 and rediscovered later by Lambert. It is important to note that Bouguer's law does not take into account any forward-scattered flux, whose effect on transmission is described in Section 1.4.4. Also, the law strictly applies only to collimated beams. Practically, however, it is usually applied to beams that are approximately collimated. The inverse-square law for diverging beams is taken into account by writing

$$E_x = \frac{I_0 \exp(-\beta x)}{x^2} \tag{1-36}$$

where I_0 is the intensity of the source, assumed to be a point. This is Allard's law, discovered by him about 1876. Its application to the visibility of point sources of light is explained in Section 1.5.5.

The proportionality constant β in the above expressions is the *volume total scattering coefficient*, hereafter referred to as the *total coefficient*. It defines the ability of a unit volume of a suspension to scatter totally light of a specified wavelength. Recalling the meaning of the angular coefficient $\beta(\theta)$, we see that

$$\beta = \int_0^{4\pi} \beta(\theta)\, d\omega \tag{1-37}$$

where $d\omega$ is an element of the total solid angle ω about the unit volume of scatterers. The coefficient β has the dimension L^{-1} and its value is expressed as a number per unit path length. Various such units have been used in stating the value of β; the conversion factors in Appendix C may be found convenient. The total scattering by molecules and particles can be distinguished by the subscripts m and p. The two effects are directly additive, as far as attenuation is concerned, and where necessary we can write

$$\beta_{sc} = \beta_m + \beta_p \tag{1-38}$$

In subsequent treatments where subscripts are not employed, the context identifies the type of scatterer.

Values of β cover a wide range for ordinary atmospheric constituents. For molecular scattering by standard air, β has the value $0.012\ \text{km}^{-1}$ at $\lambda = 0.55\ \mu\text{m}$. For clear air, in which small haze particles are the principal scatterers, β has a value of about $0.17\ \text{km}^{-1}$ at $0.55\ \mu\text{m}$. At the same wavelength but in dense fog, where droplets are the significant scatterers, β has a value near $40\ \text{km}^{-1}$. The resulting attenuation over a 1-km path is almost astronomical; substitution in (1-35) shows that the ratio E_x/E_0 is only about 10^{-17}.

When the molecules or particles of the medium absorb light in addition to scattering it, each process separately removes flux from the beam. The two effects are additive where attenuation is concerned and are expressed by the extinction coefficient

$$\beta_{ex} = \beta_{sc} + \beta_{ab} \tag{1-39}$$

where β_{ab} is an absorption coefficient which is analogous to β_{sc} including the dimension L^{-1}. When extinction is being considered, β_{ex} is used in expressions such as (1-35). The energy associated with β_{ab} does not in a direct sense propagate from the beam as does the scattered energy. Instead, it raises the temperature of the medium in the first instance, with emission ultimately occurring.

1.4.3 Additional Measures of Attenuation

Occasionally the attenuation characteristic of a scattering or extinction path is expressed by the quantity *mean free path*. Referring to (1-35), we see that when $x = 1/\beta$ the equation becomes

$$E_x = E_0 \exp(-1) \tag{1-40}$$

and the beam irradiance is reduced to 36.8% of the initial value. The value of x that brings this to pass is called the mean free path. This is reminiscent of the concept of the molecular mean free path in the kinetic theory of gases, where on the average only exp (−1) of the molecules travel that distance without undergoing collisions. Here it obviously means that only exp (−1) of the flux travels that distance without being scattered.

The attenuation of microwaves is usually expressed in decibels per kilometer, and the following relation permits this to be done for optical attenuation. The decibel is defined:

$$\text{Decibel} = 10 \log \frac{\text{emergent power}}{\text{incident power}} \tag{1-41}$$

From (1-35) the ratio of emergent to incident radiant flux for a 1-km path is just

$$\frac{E_x}{E_0} = \exp(-\beta) \tag{1-42}$$

Combining the two expressions gives

$$\text{db km}^{-1} = -4.343\beta \qquad \text{km}^{-1} \tag{1-43}$$

The attenuation of a path is sometimes expressed in terms of the *optical density*. The unit of optical density is defined by

$$D = \log \frac{\text{incident power}}{\text{emergent power}} = \log \frac{1}{\tau} \tag{1-44}$$

for a path of specified length. Setting (1-44) equal to (1-42), taking natural logarithms of both sides, and assuming a 1-km path, we have

$$D \text{ km}^{-1} = 0.4343\beta \qquad \text{km}^{-1} \tag{1-45}$$

Attenuation also can be expressed in *nepers*, a name intended to commemorate the inventor of natural logarithms. This unit is defined:

$$\text{Neper} = \ln \left(\frac{\text{emergent power}}{\text{incident power}}\right)^{1/2} \tag{1-46}$$

Rewriting (1-35) for a 1-km path as

$$\ln \frac{E_x}{E_0} = -\beta \tag{1-47}$$

and comparing this with (1-46), we see that

$$\text{Neper km}^{-1} = -\beta/2 \qquad \text{km}^{-1} \tag{1-48}$$

which shows that the two measures are dimensionally identical and differ by a factor of two.

1.4.4 Transmittance and Optical Thickness

The amount of attenuation is often stated in terms of the quantity known as *transmittance* τ and defined:

$$\tau = \frac{\text{emergent power}}{\text{incident power}} = \frac{E_x}{E_0} = \exp(-\beta x) \tag{1-49}$$

Frequently in the literature transmittance is denoted by the symbol T, and sometimes by T. For a 1-km path, two useful relationships are obtained by eliminating the exponential from (1-49):

$$\begin{aligned} \tau \quad & \text{km}^{-1} = \text{antilog}(-0.4343\beta) \qquad \text{km}^{-1} \\ \beta \quad & \text{km}^{-1} = -2.303 \log \tau \qquad \text{km}^{-1} \end{aligned} \tag{1-50}$$

where the condition $0 < \tau < 1$ always holds. List (1966) provides tabulated values of τ km^{-1}, known as the *transmissivity*, as a function of β km^{-1} and values of β km^{-1} as a function of τ km^{-1}. The tabulations cover wide ranges of values.

The discussion thus far has dealt only with the direct transmission of original, unscattered flux from the source. No account has been taken of any forward-scattered flux such as that in the main lobe of Figure 1.5*c*, which may enter a receiver field of view. Such flux adds to the direct unscattered flux, thereby producing a greater receiver response than would otherwise obtain. This type of transmission, which involves both the direct flux and a small portion of the forward-scattered flux, is *diffuse transmission.* The additional flux entering the receiver comes from a spatial volume, usually of double-conical shape, which surrounds the source-receiver line. This volume is defined by the angular divergence of the source beam, the angular field of view of the receiver, and the source-receiver distance. For a given beam irradiance, the additional flux depends on these factors, as well as on the receiver aperture area, the concentration of scatterers in the volume being considered, and the angular pattern of scattered intensity. The reader will find it instructive to make a simple sketch of the situation.

Evidently the diffuse transmittance of a path is greater than the direct transmittance as calculated from Bouguer's law (1-35). For a given source-receiver geometry the difference between the transmittances becomes progressively greater for scattering by molecules, haze, and fog or cloud. This is because the forward narrow-angle scattering increases rapidly with particle size, as shown in Figure 1.5. The influence exerted by the angular field of view for values greater than 25 deg has been investigated by Stewart and Curcio (1952), Gibbons (1958, 1959), and Eldridge and Johnson (1958, 1962). Much smaller fields, typically only a few degrees in width, usually are employed in measuring the nominally direct transmittance. For such fields, Middleton (1949) both analyzed and measured the effects of source-beam angle and receiver angular field of view. A general but informative discussion of this subject is given by Hodkinson (1966b). The crux of the matter is of course the applicability limit of Bouguer's law; actual laboratory data on the limit are presented by Zuev (1974). In this book the term *transmittance* means direct transmittance unless stated otherwise.

A widely used concept in atmospheric optics is *optical thickness* (or optical depth), which is applicable to any path characterized by an exponential law. From (1-35) it is clear that attenuation is determined by the product βx, but not by either factor singly. The optical thickness T for a path length x thus is defined by

$$\mathrm{T}_m,\ \mathrm{T}_p, \text{ or } \mathrm{T}_{ex} = (\beta_m,\ \beta_p, \text{ or } \beta_{ex})x \tag{1-51}$$

which implies that the coefficient β, whether for molecular or particle scattering or for particle extinction, has a constant value over the distance x. Such a path is a *homogeneous path,* and horizontal paths are usually considered homogeneous. When β is a function of x, which is nearly always true for a vertical or a slant path, we write

$$\mathrm{T}_x = \int_0^x \beta(x)\, dx$$

The optical thickness is used directly in the attenuation formula (1-35), thus

$$E_x = E_0 \exp(-T) \tag{1-52}$$

1.5 VISIBILITY ALONG AN ATMOSPHERIC PATH

The term *visibility* is commonly used in stating how well we can see generally under given meteorological conditions. According to Huschke (1959), the term as employed in weather-observing practice has two dissimilar meanings: the greatest distance at which it is possible to see and identify, with the unaided eye, (1) in the daytime, a prominent dark object against the horizon sky, and (2) at night, an unfocused, moderately intense light source. Both meanings involve observer or subjective factors; this is particularly true for the recognition requirement. Daytime estimates of visibility are actually subjective evaluations of the attenuation of contrast, while nighttime estimates are subjective evaluations of the attenuation of beam illuminance. In all cases, however, scattering is the primary determinant of visibility. The first meaning of the term was originally investigated by Kochsmeider (1924), and the second by Allard (1876). Duntley (1948a, b) and Middleton (1952) provide standard treatments of the subject, while additional insights can be obtained from Hulburt (1941), Johnson (1954), Fleagle and Businger (1963), Tverskoi (1965), and Zuev (1974). Recent investigations are described by Gordon et al. (1975) and Porch et al. (1975). Practically all aspects of visibility and its measurement are covered in the annotated bibliography by Ingrao (1973).

1.5.1 The Airlight

In observing an extensive landscape, particularly from an elevated station, we are quickly aware that the features appear progressively lighter as our attention shifts from the foreground toward the horizon. This phenomenon, known as the *airlight*, is created by angular scattering of environmental light toward the observer by the atmospheric molecules and particles within his cone of vision. In general environmental light consists of direct sunlight, diffuse skylight, and ground-reflected light. Referring to Figure 1.9, consider the element of solid angle $d\omega'$, whose axis OH is horizontal. In this direction the atmospheric scattering properties can be assumed to be invariant with distance. It is evident from the figure that at any point along OH all values of the scattering angle θ are involved in the totality of light scattered toward the observer. At any value of θ, however, the angular coefficient $\beta(\theta)$ is proportional to the total coefficient β, as can be appreciated from (1-37).

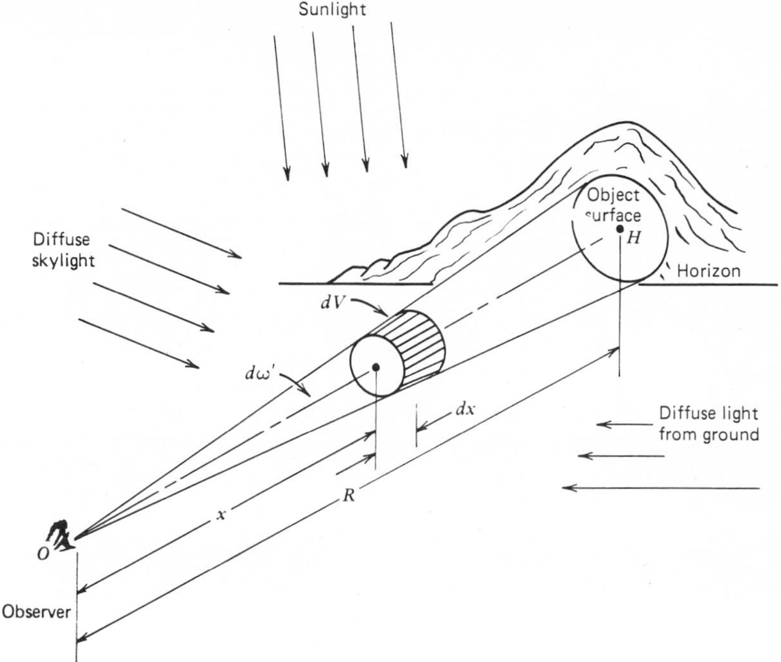

Figure 1.9 Source of the airlight between the observer and an object, and apparent luminance of an object due to the airlight. From Middleton (1952).

Consider next the volume element dV, shown as a lamina. Clearly,

$$dV = d\omega' x^2\, dx \tag{1-53}$$

Whatever the distribution of the environmental illuminance over dV, the luminous intensity of this element is

$$dI = dV\, \sigma\beta \tag{1-54}$$

where σ is a proportionality factor which is a function of the illuminance and the angular scattering properties of the materials within dV. Happily this factor cancels from the argument later. The illuminance produced at the observer by the light scattered from dV is, from (1-36),

$$dE = dI x^{-2} \exp(-\beta x) \tag{1-55}$$

which takes into account both the inverse-square law and the attenuation due to total scattering.

Invoking now the viewpoint of luminance (radiance) expressed by (1-31), we divide both sides of (1-55) by $d\omega'$ to obtain the luminance of

dV. The result is

$$dL = \left(\frac{dI}{d\omega'}\right) x^{-2} \exp(-\beta x) \tag{1-56}$$

In terms of x and $d\omega'$, (1-54) is

$$dI = d\omega' \, x^2 \sigma\beta \, dx \tag{1-57}$$

Substituting this expression for dI in (1-56) gives

$$dL = \sigma\beta \exp(-\beta x) \, dx \tag{1-58}$$

for the apparent luminance of dV. The line of sight (or cone of rays within the element $d\omega'$) extends from the observer outward through the entire atmosphere. The luminance of the entire path, which the observer interprets as the luminance L_h of the horizon, is found by integrating (1-58) between the limits $x = 0$ and $x = \infty$. The integration yields simply

$$L_h = \sigma \tag{1-59}$$

for the horizon luminance.

Consider next that the base of the cone $d\omega'$ terminates at the surface of a black object rising above the horizon, as in Figure 1.9. The object is at distance R from the observer. The luminance of the conical volume so formed, which the observer interprets as the luminance of the object, is found by integrating (1-58) between the limits $x = 0$ and $x = R$. This gives

$$L_o = \sigma[1 - \exp(-\beta R)] \tag{1-60}$$

Substituting for the factor σ its value from (1-59), we have

$$L_o = L_h[1 - \exp(-\beta R)] \tag{1-61}$$

Interpreting this, when the object is close to the observer, with R effectively equal to zero, the luminance L_o of the object is also zero. As R increases, L_o also increases until, as R becomes very large, L_o becomes equal to the horizon luminance L_h, in the limit. This is the maximum apparent luminance the object can attain.

1.5.2 Object-Background Contrast

We distinguish an object from its surroundings because of differences in luminance and chromaticity between various parts of the viewed field. Such differences are usefully expressed as *contrasts*. For our purposes, luminance contrasts are more important than chromaticity factors in determining visibility. Assume that an isolated object is viewed against a uniform, extended background. The object-background contrast C is defined by

$$C = \frac{L_o - L_b}{L_b} \tag{1-62}$$

where L_o and L_b are, respectively, the luminances of the object and the background. If $L_o < L_b$, the contrast is negative and becomes -1 for an ideal black object if the background has any luminance at all. If $L_o \gg L_b$, as in the case of a light source seen against the night sky, the contrast may have a large positive value.

The reduction in contrast by the airlight is a familiar circumstance in the outdoor world. At any distance R, the contrast of an ideally black object against the horizon sky is found from (1-62) by substituting L_h for L_b and (1-61) for L_o. This gives

$$C = -\exp(-\beta R) \tag{1-63}$$

The argument of the exponential is identical to the optical thickness T defined by (1-51). Equation (1-63) can also be written in the convenient form

$$\ln C = -\beta R \tag{1-64}$$

Evidently the greatest distance at which the observer can discern the black object depends on the minimum value of contrast he can perceive, that is, on the luminance-contrast threshold of his eye. This matter is examined in the following section.

In deriving (1-63), for simplicity an ideally black object, that is, one of zero inherent luminance, was assumed. The full theory, as developed by Duntley (1948a) and Middleton (1952), is not so restrictive. They write (1-62) in a form defining an inherent contrast C_0 between two adjacent objects, or between an object and its background:

$$C_0 = \frac{L_0 - L_0'}{L_0'} \tag{1-65}$$

where L_0 and L_0' are the luminances of the objects seen close at hand. Similarly, they define the apparent contrast C_R at a distance R by

$$C_R = \frac{L_R - L_R'}{L_R'} \tag{1-66}$$

where L_R and L_R' are the apparent luminances of the objects seen at a distance R. These investigators then show that

$$C_R = C_0 \left(\frac{L_0'}{L_R'}\right) \exp(-\beta R) \tag{1-67}$$

Considering a horizontal line of sight and letting L_0' and L_R' represent the luminances of the horizon at distances 0 and R, respectively, we see that

$$\frac{L_0'}{L_R'} = 1$$

because the horizon luminance does not change as we go toward or away from the objects. The horizon luminance is an *equilibrium value.* Equation (1-67) then becomes

$$C_R = C_0 \exp(-\beta R) \tag{1-68}$$

This is the general expression for the attenuation of contrast for objects having *any* inherent contrast, and not just the value -1. Recent investigations in which scattering coefficients were determined by observing the contrasts of distant mountains against the sky are described by Porch et al. (1975).

1.5.3 Visual Thresholds of Perception

The eye has a very nonlinear response to changes in stimulus. This characteristic is undoubtedly associated with its ability to function well at luminance values differing by factors as large as 10^6. A general law of sensation, formulated by Weber in 1834 and known by his name, states: "The increase of stimulus necessary to produce a just perceptible increment of sensation bears a constant ratio to the whole stimulus." Application of this law to vision was investigated in 1859 by Fechner, who found that, when the field luminance changed from L_1 to L_2, the change in sensation was proportional to $\log L_2 - \log L_1$. In other words, as the stimulus increased in geometric progression, the sensation increased in arithmetic progression.

For purposes here, the "just perceptible increment of sensation" in the statement of Weber's law is the *visual threshold of perception.* Actually we are interested in two kinds of thresholds associated with two types of light sources in surrounding luminous fields. These types are (1) a luminous area and (2) a point source of illuminance. The two kinds of thresholds are discussed in the following paragraphs.

The first kind is a luminance-contrast threshold. In the present context this refers to the perception of a small target area within a large surrounding field. Let the luminances of target and field be made nearly equal, until the observer can just perceive a difference between the two. Denoting the luminances by L and $L + \Delta L$, the threshold ϵ is defined:

$$\epsilon = \frac{(L + \Delta L) - L}{L} = \frac{\Delta L}{L} \tag{1-69}$$

where $\Delta L / L$ is the ratio of the least perceptible increment of field luminance to the field luminance itself. This ratio is called Fechner's fraction. Since the time of Fechner many investigations of this threshold have been made. In general they have shown that ϵ remains fairly constant for all values of L greater than about 1 cd m^{-2}, and for targets having angular subtenses greater than 1 deg. The value of ϵ is also affected by the presence of other stimuli in the field, by the psychophysical condition of the observer, and by the desired probability or assurance that the target has indeed been perceived, or *detected.*

Blackwell (1946) presents extensive test data (the Tiffany Foundation data) on luminance-contrast thresholds. The tests employed targets that were brighter or darker than the surrounding field which formed a luminous background. The data represent thousands of observations made by many persons under controlled conditions. The angular diameters of the targets ranged from 0.6 to 360 arc-min; the field luminances varied from less than 10^{-5} to about 400 cd m^{-2}; and the resulting thresholds extended from less than 0.01 to more than 100. Middleton (1952) summarizes and discusses these experiments and data.

A small portion of the Blackwell data most relevant to the subject of visual range is plotted in Figure 1.10. The original data, which correspond to a detection probability of 50%, have been multiplied by the factor 1.62 to bring the probability up to about 90%. At field luminances greater than about 1 cd m^{-2} the thresholds for the two largest targets remain essentially constant. Noteworthy are the rapid increases in threshold for lesser luminances and smaller targets. The discontinuities in the curves at luminances near 2×10^{-3} cd m^{-2} mark the transition from foveal vision

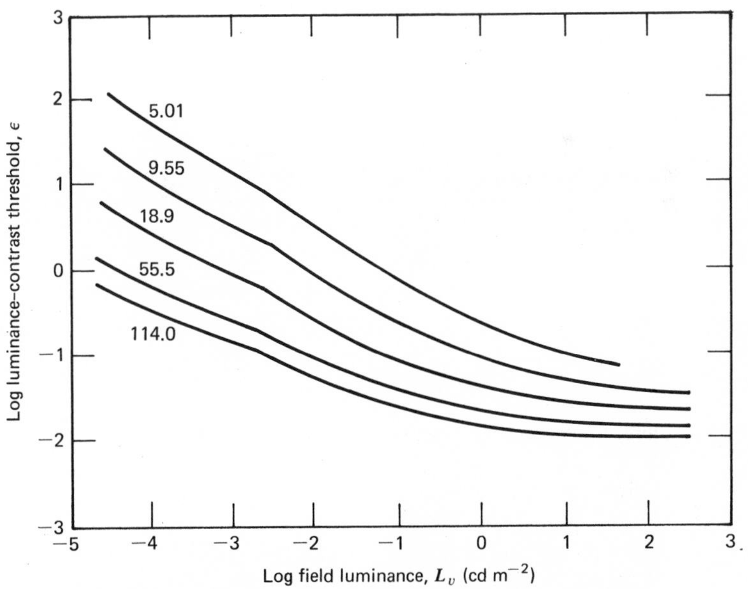

Figure 1.10 Luminance-contrast threshold as a function of field luminance. The number alongside each curve gives the angular subtense of the target in arc-minutes. Data from Blackwell (1946).

(photopic or light-adapted) to parafoveal (scotopic or dark-adapted) vision. Several sky conditions can be correlated with the values of luminance along the abscissa scale from the examples given in Table 1.6.

The second kind of visual threshold important in atmospheric optics is the threshold for point sources of light. A point source is defined as a stimulus that affects the eye only in proportion to its intensity. The maximum diameter of a point source observed with the unaided eye can vary from 1 arc-min or even less, at field luminances greater than about 1 cd m^{-2}, to 10 arc-min for dark-adapted parafoveal vision. The threshold is basic to the visual detection of signal lights and beacons and to the visibility of stars. It is stated not in terms of luminance contrast as for extended sources, but in terms of the illuminance produced at the eye by the point source. Such thresholds have been investigated by Green (1932) and Knoll et al. (1946). The Blackwell (1946) data for very small targets are also applicable when treated by the method of Middleton (1952), who discusses and compares the results of all three investigations.

Data from these investigations are plotted in Figure 1.11. Curves *A* and *B* correspond to a practical certainty of detection, while curve *C* corresponds to a 50% probability. If the latter curve is raised by 0.3 log units to represent a practical certainty, the three curves agree well at the lower values of luminance. In the experiments that were the basis for curve *A*, it was found that the threshold for the point source is governed only by the field luminance in that vicinity, provided there is no greater luminance in the field. The star magnitudes corresponding to the values of illuminance

TABLE 1.6 APPROXIMATE LUMINANCE OF THE SKY NEAR THE HORIZON FOR VARIOUS CONDITIONS*

Condition	Candelas per square meter
Clear day	10^4
Overcast day	10^3
Heavily overcast day	10^2
Sunset, overcast day	10
One-quarter hour after sunset, clear	1
One-half hour after sunset, clear	10^{-1}
Fairly bright moonlight	10^{-2}
Moonless, clear night sky	10^{-3}
Moonless, overcast night sky	10^{-4}

*Extreme meteorological conditions may cause the luminance to vary by a factor of 10 or more.
Source: Middleton (1952).

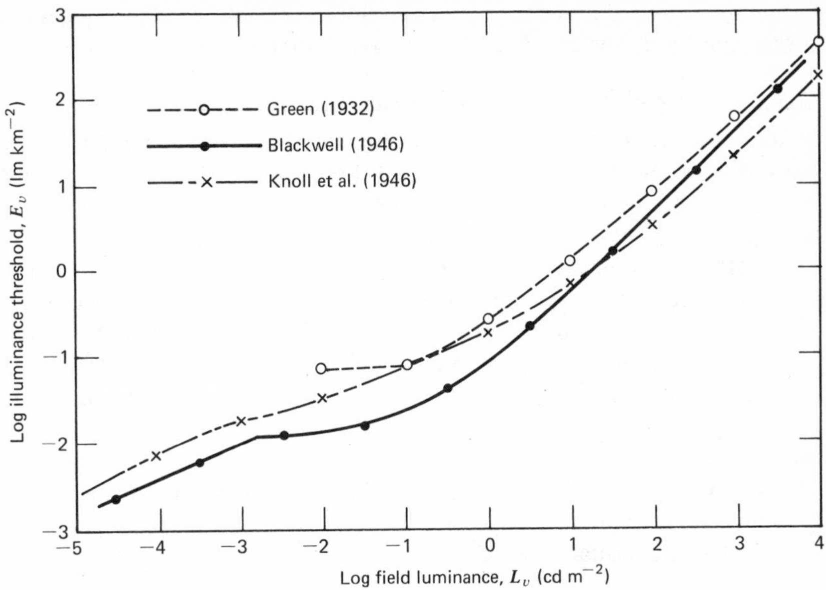

Figure 1.11 Illuminance threshold as a function of field luminance.

along the ordinate scale can be found from the data listed in Table 1.7. At a field luminance of about 10^{-3} cd m^{-2}, corresponding to a clear moonless night, the threshold illuminance is only about 10^{-2} lm km^{-2}, or that of a sixth-magnitude star. Readers may recall that the scale of stellar magnitudes was originally chosen so that a magnitude of 6 was assigned to a star that was just perceptible under favorable meteorological conditions. The

TABLE 1.7 STELLAR MAGNITUDE AND TERRESTRIAL ILLUMINANCE

Magnitude	Lumens per square kilometer	Log	Magnitude	Lumens per square kilometer	Log
−5	209	2.32	2	0.331	−0.48
−4	83	1.92	3	0.132	−0.88
−3	33.1	1.52	4	0.052	−1.28
−2	13.2	1.12	5	0.0209	−1.68
−1	5.2	0.72	6	0.0083	−2.08
0	2.09	0.32	7	0.00331	−2.48
1	0.83	−0.08	8	0.00132	−2.88

Source: Middleton (1952).

illuminance units of lumens per square kilometer (a rather large area) attest to the small amount of light received from a star.

1.5.4 Visual and Meteorological Ranges

The concepts of visual and meteorological ranges are derived from the ideas of contrast attenuation and visual threshold. Both concepts refer to the wavelength at which the eye has the greatest sensitivity, that is, at 0.55 μm according to Figure 1.3. Huschke (1959) defines visual range as "the distance, under daylight conditions, at which the apparent contrast between a specified type of target and its background (horizon sky) becomes just equal to the threshold contrast of an observer .... The visual range is a function of the atmospheric extinction [scattering, as the terms are used in this book], the albedo [reflectance] and visual angle of the target, and the observer's threshold contrast at the moment of observation."

Values of visual range usually are estimated from the appearance of buildings and special targets at differing distances against the skyline. The formula for the visual range R_v is

$$R_v = \frac{1}{\beta_{sc}} \ln \frac{C}{\epsilon} \tag{1-70}$$

where C is the inherent contrast of the target against the background, and ϵ is the threshold contrast of the observer. The use of β_{sc} rather than β_{ex} in (1-70) implies that absorption by atmospheric particles at visual wavelengths is small enough to ignore. This view is justified except in regions of polluted air. A basis for selecting a value of ϵ can be found, for example, in the Blackwell (1946) data.

The subjective factors and optional target features involved in the strict meaning of visual range are not present in the meteorological range R_m. This is obtained by specifying a black target and letting $\epsilon = 0.02$, a value not far from that of the bottom curve for daylight conditions in Figure 1.10. In practice, the target is made large enough that this value of ϵ is applicable. Since the target is black, its inherent contrast against the sky is unity, and (1-70) becomes

$$R_m = \frac{1}{\beta_{sc}} \ln \frac{1}{0.02} = \frac{3.912}{\beta_{sc}} \tag{1-71}$$

which is a very useful relationship. It is evident from the foregoing relationships that the transmittance for a path length just equal to the meteorological range is 0.02. Duntley (1948a) extended the ideas of contrast reduction and meteorological range from a horizontal to an inclined path by means of a concept called the *optical slant range.* This range "represents that horizontal distance for which the attenuation is the same as that actually encountered along the true inclined path length." The concept is presented as the *equivalent path length* in Section 2.5.3.

Table 1.8 gives values of the meteorological range and scattering coefficient for the indicated meteorological conditions. Also listed are the corresponding numbers of the International Visibility Code. It is noteworthy that R_m, and of course β_{sc}, vary over a range of 1000 to 1 for a gamut of conditions familiar to everyone. The entries on the bottom line are for molecular scattering only, which represents an irreducible restriction on visibility. The corresponding value of β_m refers to air at standard conditions and to a spectrally weighted average wavelength for daylight within the visual spectrum.

In studying visibility problems, values of ϵ other than 0.02 are sometimes used in (1-69). List (1966) gives tabulations of R_m as a function of β_{sc} over a

TABLE 1.8 INTERNATIONAL VISIBILITY CODE, METEOROLOGICAL RANGE, AND SCATTERING COEFFICIENT

Code no.	Weather condition	Meteorological range, R_m		Scattering coefficient, β_{sc} (km^{-1})
		metric	English	
0	Dense fog	< 50 m	< 50 yd	> 78.2
1	Thick fog	50 m	50 yd	78.2
		200 m	219 yd	19.6
2	Moderate fog	200 m	219 yd	19.6
		500 m	547 yd	7.82
3	Light fog	500 m	547 yd	7.82
		1000 m	1095 yd	3.91
4	Thin fog	1 km	1095 yd	3.91
		2 km	1.1 nmi	1.96
5	Haze	2 km	1.1 nmi	1.96
		4 km	2.2 nmi	0.954
6	Light haze	4 km	2.2 nmi	0.954
		10 km	5.4 nmi	0.391
7	Clear	10 km	5.4 nmi	0.391
		20 km	11 nmi	0.196
8	Very clear	20 km	11 nmi	0.196
		50 km	27 nmi	0.078
9	Exceptionally clear	> 50 km	> 27 nmi	0.078
—	Pure air	277 km	149 nmi	0.0141 (β_m)

Source: Hulburt (1941).

wide range of values for threshold contrasts of 0.05 as well as 0.02. From such tabulations, or from the use of (1-71), values of β_{sc} can be determined from measurements of R_m. Such values necessarily include the effects of both molecular and particle scattering according to (1-38). In very clear weather, with the resulting large values of R_m, the coefficient β_p may be comparable in value to β_m. If values of the former are wanted, as for a thin haze aerosol, the value of β_m should be subtracted from β_{sc} when this coefficient has been determined by measuring R_m.

1.5.5 Visibility of Point Sources

The distance at which a given terrestrial point source of light can be seen against a sky of specified luminance and scattering properties can be found from Allard's law (1-36). For this purpose we write the law as

$$E_t = I_0 \frac{\exp(-\beta_{sc} R_t)}{R_t^2} \tag{1-72}$$

where E_t is the illuminance threshold for point source against a specified background, I_0 is the luminous intensity of source, and R_t is the range threshold. In the above form the law is not amenable to direct solution.

Putting the law in logarithmic form we have

$$\ln I_0 - \ln E_t = 2 \ln R_t + \beta_{sc} R_t \tag{1-73}$$

List (1966) provides tabulated values of the left side of this expression as a function of I_0 and E_t, and tabulated values of the right side as a function of R_t and β_{sc}. He then shows how to find the value of any one of the four variables when the other three are given. Middleton (1952) provides a set of nomograms for solving (1-72) for wide ranges of parameter values. When it is desired to enter the problem with E_t as a known factor, the data in Figure 1.11 may be used. Gordon et al. (1975) provide a comprehensive treatment, with nomograms, of the visibility of light sources.

The visibility of stars in the daytime, as well as at night, is an interesting subject and is one of importance to navigators and astronomers. The data in Figure 1.11 and Table 1.7 enable us to predict the magnitude of the star that can just be detected by the unaided eye, with due allowance for atmospheric transmittance. As pointed out in Section 1.5.3, the data argue that a sixth-magnitude star is just perceptible in a very dark night sky when the air is clear. This is confirmed by observation. At the other extreme, for example a bright day sky having a luminance near 10^4 cd m^{-2}, the threshold illuminance corresponds approximately to a stellar magnitude of -3.5. This is the reason why the brightest star, Sirius, which has a magnitude of -1.58, is not visible in a bright sky. Of all the stars and planets, only Venus with a magnitude of about -4 can be seen with the unaided eye in the daytime.

Nevertheless it is an ancient but persisting notion that stars in the daytime sky can be seen from the bottom of a deep canyon or mine shaft. From such a station the visual field appears as a small bright area with a dark surround. The experiments of Hynek (1951) with Vega, whose magnitude is $+0.14$, and Smith (1955) with Pollux, whose magnitude is $+1.21$, showed that these stars could not be seen with the observer sighting through a tall chimney stack. It may be that the ancient notion arose from occasional chance sightings of Venus by observers at subterranean stations. The improvement of star visibility at night by using a telescope is well known of course, but the improvement that can be obtained in the daytime seems less well known. This subject has been investigated theoretically by Koomen (1959), Hardy (1967), and Horman (1967), and both theoretically and experimentally by Tousey and Hulburt (1948). The subject is fascinating and can be easily explored with a simple telescope; a binocular one will suffice. The observer essentials are, in order of increasing importance, visual acuity, knowing where to look, and patience.

1.6 LITERATURE OF ATMOSPHERIC OPTICS

The literature of atmospheric optics, written over the past 200 years, is widely spread among publications in the fields of meteorology, atmospheric physics, astronomy, and geophysical science. The literature exists in books, technical journals, and scientific and engineering reports. A large amount of it is devoted to scattering, and a representative portion is cited in this book.

For convenience in having basic information sources listed in one place, this section identifies several books, journals, and organizational sources of technical reports concerned with atmospheric scattering. This will give the reader an overview of information sources, and perhaps guide his thinking toward library acquisitions that will serve his own interests and needs. A recommended adjunct to this section is the literature survey by Thompson (1971), which covers atmospheric transmission in the radio, microwave, and optical regions of the spectrum. His survey is a very useful guide to the quite extensive literature in these fields. Almost a century ago Lord Rayleigh (1884) observed that "…. it is often forgotten that the rediscovery in the library may be a more difficult and uncertain process than the first discovery in the laboratory." The continuing explosion of technical information in recent years has not weakened the truth of his statement nor lessened the need for guidance.

1.6.1 Textbooks and Monographs

A representative listing of books that treat atmospheric scattering is given in Table 1.9. Although such a listing cannot be really complete, the works listed will enable readers to supplement or extend the treatments of

TABLE 1.9 BOOKS THAT PROVIDE TREATMENTS OF ATMOSPHERIC SCATTERING

	Topics covered				
Reference	1	2	3	4	5
Chandrasekhar (1950)	C	C	—	—	—
Middleton (1952)	A	A	B	C	C
Rozenberg (1966)	C	C	C	C	—
Shifrin (1968)	C	B	C	B	—
Deirmendjian (1969)	B	B	C	C	—
Ivanov et al. (1969)	—	C	—	C	—
Kondratyev (1969)	B	C	B	C	—
Tricker (1970)	B	—	B	—	B
Zuev (1974)	A	C	C	C	B
Humphreys (1940)	B	B	—	—	—
Malone (1951)	A	B	A	B	B
van de Hulst (1952)	A	B	A	B	—
Johnson (1954)	B	B	A	A	C
Sekera (1956)	C	C	—	—	—
van de Hulst (1957)	B	C	C	C	—
Kruse et al. (1963)	—	B	—	A	A
Bullrich (1964)	B	C	B	C	—
Goody (1964)	B	—	B	—	—
Tverskoi (1965)	A	A	A	B	B
Hodkinson (1966b)	—	C	—	B	—
Hudson (1969)	—	A	—	A	—
Kerker (1969)	C	C	—	A	—
Henderson (1970)	A	—	C	C	—

Notes:

1. Rayleigh theory.
2. Mie theory.
3. Skylight.
4. Transmission.
5. Visibility.

A. Brief treatment.
B. Treatment of moderate length.
C. Extended, detailed treatment.

scattering in this book. The topics covered are indicated by the integers at the column headings, according to the legend at the foot of the table. The letters A, B, and C used as column entry marks signify the extent to which a book treats the topic: A means a brief treatment, often a summary presentation of one or more aspects; B means a treatment of moderate length, usually covering several aspects; and C means an extended treatment, usually with derivations of formulas. The books listed in the first group are devoted completely to the topics indicated. The books in the second group are concerned with subjects closely related to atmospheric scattering and treat the indicated topics in one or more chapters.

1.6.2 Society Journals and Technical Magazines

A large part of the literature outputs from the many research and practical measurement programs carried on by industry, universities, and governmental organizations is carried currently in the following periodicals:

Applied Optics (*Appl. Opt.*)
Journal of the Optical Society of America (*J. Opt. Soc. Am.*)
Journal of Quantitative Spectroscopy and Radiative Transfer (*J. Quant. Spectrosc. Radiat. Transfer*)
Journal of the Atmospheric Sciences (*J. Atmos. Sci.*)
Journal of Applied Meteorology (*J. Appl. Meteorol.*)
Quarterly Journal of the Royal Meteorological Society (*Quart. J. Roy. Meteorol. Soc.*)
Journal of Geophysical Research (*J. Geophys. Res.*)
Infrared Physics (*Infrared Phys.*)
Tellus

Back issues of these periodicals, and the *Journal of Meteorology* (*J. Meteorol.*), contain a wealth of information.

Technical literature tends to increase at a seemingly exponential rate. Howard and Garing (1967), in a report to the American Geophysical Union, list 418 papers on atmospheric optics that were published in the preceding 4 years. They point out that 53% of the papers cited were published in the first three periodicals listed above. Additional bibliographies documenting the many contributions of two outstanding workers in atmospheric optics are contained in *J. Opt. Soc. Am.* (1956, 1960). At the end of this book is a list of the literature cited in the individual chapters. All these listings should assist readers who wish to track theories and data back to their sources.

1.6.3 Scientific and Engineering Reports

Several governmental organizations have for many years conducted programs on various aspects of atmospheric optics and have sponsored much additional work by universities and industrial firms. The governmental organizations include:

U.S. Air Force Cambridge Research Laboratories (AFCRL)
U.S. Army Electronics Command (ECOM)
U.S. Naval Research Laboratory (NRL)
National Aeronautics and Space Administration (NASA)
National Oceanic and Atmospheric Administration (NOAA)

Universities that have maintained continuing programs in atmospheric optics include:

University of Alaska
University of Arizona
University of California
Colorado State University
University of Denver
University of Florida
University of Michigan
Massachusetts Institute of Technology
Ohio State University
Texas A & M University
University of Washington
University Corporation for Atmospheric Research

Most of the reports issued by governmental organizations and their contractors in the aerospace field are listed in abstract form in Scientific and Technical Aerospace Reports (STAR). This is an indexing and abstracting journal published semimonthly by NASA. The listings in STAR give the source from which report copies may be ordered and indicate whether microfiche or hard copy is available. A similar service for scientific and trade journals, books, and conference papers is provided by International Aerospace Abstracts (IAA), published semimonthly by the American Institute of Aeronautics and Astronautics. Copies of nonclassified reports issued by governmental agencies and by civilian agencies under government contract usually are available to the general public from the National Technical Information Service (NTIS) or from the Superintendent of Documents, U.S. Government Printing Office (GPO). Guidance to additional sources of information is provided by Thompson (1971).

In the reference list for this book, several organizations are referenced quite often. Some economy of listing is obtained by using the following abbreviations for the organizational names:

AFCRL: U.S. Air Force Cambridge Research Laboratories, Lawrence G. Hanscom Field, Bedford, Mass. 01731.

AMS: American Meteorological Society, 45 Beacon St., Boston, Mass. 02108.

GPO: Superintendent of Documents, U.S. Government Printing Office, Washington, D.C. 20402.

NASA: National Aeronautics and Space Administration, Scientific and Technical Information Division, Washington, D.C. 20546.

NRL: U.S. Naval Research Laboratory, Washington, D.C. 20375.

NTIS: National Technical Information Service, Springfield, Va. 22151.

2

Structure and Composition of the Gas Atmosphere

The sea that covers nearly three-fourths of the globe is only the second biggest thing on earth. Incomparably vaster is the ocean of the atmosphere, which dominates the lives of men and most other creatures as surely as water dominates the lives of fish. Without the atmosphere's oxygen, living things die almost at once. Without rain erosion and the weathering of rocks, there would be no soil for plants to grow in. Without carbon dioxide the plants could not produce carbohydrates, the primary link in the food chain that supports all animal life. Without the high-altitude umbrella of ozone to absorb the lethal ultraviolet rays of the sun, human existence—if any—would be quite different. Yet this is only a fractional list of the free services performed by the atmosphere and taken for granted by the nearly three billion human beings who at this moment are drawing breaths of it. [Beiser (1962)]*

The atmosphere is an enormous theater of diverse but related and incessant activities. All these are involved in one way or another with the environmental conditions collectively known as *weather.* Many of the sheer physical properties entail optical properties, and these are our concern. The gases that maintain themselves (seemingly forever) above the earth's surface, the water vapor endlessly changing to rain and snow and then back again, the airborne particles abounding in all environments and occasionally drifting halfway around the globe—these materials and their changing distributions in time and space are central to the subject of atmospheric optics.

This chapter deals with the structure and composition of the gas atmosphere from the viewpoint of scattering. We start by reviewing certain factors from the kinetic theory of gases; these factors are the essence of any planetary atmosphere. Next, the general features of the earth's atmospheric envelope and their variation with altitude are described. Several analytic models which render the atmospheric vertical structure amenable to calculation are explained, and attention is called to the model U.S. Standard Atmosphere. Finally, several methods are discussed which allow the scattering characteristics of optical paths to be evaluated.

2.1 FACTORS FROM KINETIC THEORY

A planetary atmosphere can exist only by maintaining itself in suspension against the force of gravity, and this property of self-support derives entirely from the thermal energies of the gas molecules. The study of this molecular world belongs to a subject still called the *kinetic theory of gases*, although by now this subject is also based on a large body of fact. Kinetic theory has had a long history, whose sources may be divined in the speculations of Democritus and Lucretius. Scientifically the theory began with Bernoulli (1738) and was developed in the nineteenth century by Dalton, Joule, Clausius, Maxwell, Kelvin, and Boltzmann. Detailed treatments of the theory have been given by Jeans (1925, 1952), Loeb (1934), Kennard (1938), Sommerfeld (1964b), and Vincenti and Kruger (1965). The shorter presentations by Feather (1959) and Christy and Pytte (1965) are recommended to those who wish a relatively quick view of the essentials.

2.1.1 The Molecular Quantities

Several molecular quantities basic to the kinetic theory of gases are reviewed at this point. We recall that a gram molecular weight, or *mole*, of any substance has a mass in grams numerically equal to its molecular weight. Also, a mole of any substance contains the same number of molecules as a mole of any other substance. Thus a mole is a definite quantity of matter called a *molar mass* and denoted here by M_A. Avogadro's law states: "Equal volumes of gas, under like conditions of temperature and pressure, contain equal numbers of molecules." The number of molecules in a mole of any gas is known as *Avogadro's number* N_A and has the value

$$N_A = 6.025 \times 10^{23} \text{ mole}^{-1} \tag{2-1}$$

All numerical values in the following discussion refer to standard air, or to any gas at 0 °C and 1 atm of pressure. The volume of 1 mole of gas is called the *molar volume* V_A, having the value

$$V_A = 2.242 \times 10^4 \text{ cm}^3 = 22.42 \text{ liters} \tag{2-2}$$

The number of molecules in 1 cm^3 of gas is given by Loschmidt's number:

$$N_L = 2.687 \times 10^{19}\ \text{cm}^{-3} \tag{2-3}$$

The number of molecules in any unit volume under consideration under any conditions of temperature and pressure is called the *number density*, denoted in this book by the symbol N.

The mass of a single molecule is equal to the molar mass divided by Avogadro's number. Air is a mixture of several gases, with nitrogen and oxygen predominant and in constant proportions at altitudes below about 90 km. The molecular weight of air, averaged over the constituents, is 28.964 (dimensionless), so that the mass m of an "air molecule" is

$$m_{\text{air}} = \frac{28.964}{N_A} \tag{2-4}$$

Although an air molecule is not a physical entity, the term is convenient and is used herein. The volume of 1 g of air is given by

$$V = \frac{V_A}{M_A} = 7.74 \times 10^2\ \text{cm}^3\ \text{g}^{-1} \tag{2-5}$$

The density ρ, or mass per unit volume, of air is

$$\rho = \frac{M_A}{V_A} = 1.29 \times 10^{-3}\ \text{g cm}^{-3} = 1.29\ \text{kg m}^3 \tag{2-6}$$

or approximately 0.1 lb ft^{-3}.

No review of molecular quantities can proceed far without taking molecular collisions into account. As usual in elementary kinetic theory, we make several simplifying assumptions. The first one is that molecular internal energies can be ignored. This allows the molecule to be treated as a small elastic sphere, not unlike a tiny billiard ball. The property of elasticity derives from the spherically symmetric force field that surrounds the molecule. It is further assumed that the molecules exert no forces on each other except when they collide, and that between collisions they move in straight lines at constant speeds. When two molecules are about to collide, powerful repulsion forces begin to operate as the separation distance becomes small. At still closer approach the repulsion forces alter the directions of the two motions, so that the molecules rebound from each other. In other words, a collision occurs and then each molecule goes its separate way. The collision angle may have any value, from head-on encounter to grazing incidence. The collision is always *elastic*, meaning that the sum of the two translational kinetic energies remains unchanged, although the individual speeds may be greatly altered. *Inelastic collisions*, important in absorption-emission processes, involve exchanges between translational and internal energies and are not treated in this book.

Translational kinetic energy is a direct manifestation of molecular mass and speed. As for a macroscopic body, this energy at any instant is given by

$$E_{tr} = \tfrac{1}{2}mC^2 \tag{2-7}$$

where C is the instantaneous speed, without regard to direction. Because each molecule has three degrees of translational freedom and moves at random, molecular velocities are uniformly distributed over all directions. That is, the velocity space is isotropic. When a gas is confined within a container, the pressure exerted against the walls is due to changes in molecular momenta as the gas molecules rebound elastically from the molecules that constitute the walls. The pressure P is defined by

$$P = \tfrac{1}{3}Nm\overline{C^2} \tag{2-8}$$

where N is the number density, and $\overline{C^2}$ is the mean square speed. The factor $\frac{1}{3}$ accounts for the equipartitioning of momenta along any set of three orthogonal axes no matter now oriented.

The root-mean-square (rms) speed, in terms of the macroscopic bulk properties of the gas, is readily found from (2-8). Gas density is equal to the sum of the molecular masses in a unit volume, that is,

$$\rho = Nm \tag{2-9}$$

Substituting this in (2-8) gives

$$P = \tfrac{1}{3}\rho\overline{C^2} \tag{2-10}$$

from which the rms speed is

$$C_{rms} = (\overline{C^2})^{1/2} = \left(\frac{3P}{\rho}\right)^{1/2} \tag{2-11}$$

which is greater than the average speed by a factor of 1.08. Substituting for P and ρ in (2-11) the values 1.0133×10^5 N m^{-2} and 1.29 kg m^{-3}, respectively, we find that

$$C_{rms} = 485 \text{ m sec}^{-1}$$

This is comparable to the speed of sound, a result to be expected because directed molecular translations are the means for propagating acoustic waves. In fact, the speed of sound in air is given by

$$C = \left(\frac{1.40P}{\rho}\right)^{1/2} \tag{2-12}$$

where the factor 1.40 is the value of the ratio between the specific heat of air at constant pressure and that at constant volume. Hence the theoretical speed of sound in standard air is 332 m sec^{-1}, or 1088 ft sec^{-1}, which agrees with measured values.

The following quantities enable us to appreciate the spatial and time scales of the gas molecular world. An air molecule has an effective diameter

$$d \approx 3.7 \times 10^{-8}\ \text{cm} \approx 3.7\ \text{Å} \qquad (2\text{-}13)$$

Because N_L is the number of molecules per cubic centimeter, the average volume of space available to a molecule is just $1/N_L\ \text{cm}^3$. Hence the average spacing between molecular centers is

$$S = \left(\frac{1}{N_L}\right)^{1/3} \approx 3.3 \times 10^{-7}\ \text{cm} \approx 33\ \text{Å} \qquad (2\text{-}14)$$

or about 9 times the diameter. The mean free path $\bar{l}$, which is the average distance a molecule travels between collisions, is given by

$$\bar{l} = \frac{1}{\sqrt{2}\pi d^2 N_L} \approx 6 \times 10^{-6}\ \text{cm} \approx 600\ \text{Å} \qquad (2\text{-}15)$$

which is about 160 times the diameter. From (2-9) we can write (2-15) as

$$\bar{l} = \frac{m}{\sqrt{2}\pi d^2 \rho} \qquad (2\text{-}16)$$

to show that for a given value of ρ the mean free path depends only on the mass and diameter of the molecules. The average frequency of collisions per molecule, called the *mean collision rate*, is

$$\phi = \frac{(\overline{C^2})^{1/2}}{\bar{l}} \approx 8 \times 10^{9}\ \text{sec}^{-1} \qquad (2\text{-}17)$$

From the relationships given previously, it can be seen that ϕ is proportional to $P^{1/2}$ and to $\rho^{1/2}$.

Thus from (2-12) through (2-14), the diameter, average spacing, and mean free path are in the ratios

$$d : S : \bar{l} = 1 : 9 : 160$$

Because the effective range of intermolecular forces is of the order of the diameter, these ratios justify the assumption that the molecules interact only during the collision process.

The foregoing quantities should be compared with the characteristics of the light waves that engulf the air molecules along atmospheric paths. For example, comparing a reference wavelength of $0.55\ \mu\text{m}$, or $5.5 \times 10^{-5}\ \text{cm}^{-5}$, to the molecular diameter of 3.7×10^{-8} cm, it is clear that the phase of the passing wave is uniform over the molecule. This is an important fact in Rayleigh scattering. The frequency of the reference wave just cited is about 5.5×10^{14} Hz; hence the period is 1.8×10^{-15} sec. During this interval the molecule moves only 8.7×10^{-11} cm, or 0.0023 of its own diameter, which is insignificant. The molecule travels the mean free path in about 1.2×10^{-10} sec. If the molecule scatters light of the reference wavelength, its electronic structure undergoes approximately 6.6×10^4 oscillations during the travel time. The Doppler effect due to molecular velocity is ordinarily ignored in treatments of Rayleigh scattering, but it contributes importantly to the broadening of the spectral lines observed in absorption and emission processes. Finally, random molecular motions destroy any fixed-phase relationships or coherency in the light scattered by a group of molecules, except in the exactly forward direction. In all other directions the scattered light is incoherent.

2.1.2 Pressure, Volume, and Temperature

The pressure, volume, and temperature of a specified mass of gas are related by the equation of state for an ideal gas. The term *ideal* means that the molecules are sufficiently small, compared to their average spacing, to be regarded as point masses. The term also means that the molecules exert no forces on each other except during collisions, a condition that was assumed in the previous section. As the number density increases, however, the average spacing decreases and the gas departs from the ideal condition. A comparison of the molecular diameter and molecular spacing values given in the preceding section indicates that standard air is not actually an ideal gas. However, number density decreases as altitude increases, and the ideal condition is realized at upper altitudes. In elementary kinetic theory, which is adequate for most purposes in atmospheric optics, it is assumed that atmospheric air behaves as an ideal gas. This assumption simplifies the treatment considerably.

Before discussing the equation of state, several units of measurement are noted. Standard temperature of 0 °C on the Celsius scale corresponds to 273.15 K on the Kelvin or absolute scale. At 0 °C a column of mercury 760 mm high exerts a pressure of 1.01325×10^6 dynes cm^{-2}, which is taken as standard atmospheric pressure. This is often referred to as 1 atm of pressure. The *dyne*, a unit in the cgs system, is the amount of force required to accelerate a 1-g mass at 1 cm sec^{-2}. A unit of pressure called the *bar* is equal to 10^6 dynes cm^{-2}. The *millibar* is a customary unit of pressure in meteorology and is equal to 10^{-3} bar. Thus the standard pressure of 1.01325×10^6 dynes cm^{-2} equals 1013.25 mb. In the mks system the unit of force is the *newton*, which is the force required to accelerate a 1-kg mass at 1 m sec^{-2}. This means that 1 N equals 10^5 dynes, and standard pressure can be stated as 1.01325×10^5 N m^{-2}. The foregoing units of pressure are listed in Appendix B.

The equation of state is expressed by

$$PV_A = R_* T \tag{2-18}$$

where V_A is the volume occupied by a molar mass, and R_* is the *universal gas constant* having the value

$$R_* = 8.3143 \text{ J mole}^{-1} \text{ K}^{-1} \tag{2-19}$$

where the presence of the asterisk should be noted. Defined in this way, R_* has the same value for all gases, since it is referred to a molar mass. When the value given by (2-19) is employed, the pressure and volume must be expressed in mks units.

The gas constant can be defined in terms of a unit mass instead of a molar mass. For air, division of the value in (2-19) by the molecular weight 28.964 gives

$$R = 2.8706 \times 10^6 \text{ ergs g}^{-1} \text{ K}^{-1} \tag{2-20}$$

in cgs units, and

$$R = 2.8706 \times 10^2 \text{ J kg}^{-1} \text{ K}^{-1} \tag{2-21}$$

in mks units. When thus defined for a unit mass, the gas constant R has a different value for each species of gas, depending on its molecular weight, and is called the specific gas constant. The asterisk then is not used in the symbol. In the equation of state the symbol V, without the subscript A, then represents the volume of a unit mass of the gas, or the *specific volume* defined by

$$V = \frac{1}{\rho} \tag{2-22}$$

Equation (2-18) then can be written as

$$P = \rho RT \tag{2-23}$$

which is useful in dealing with the atmosphere where the gas is not confined to a particular volume. Parenthetically we note that in routine meteorological work density is not measured directly, because of instrumentation difficulties. Instead, temperature and pressure are measured, and density is calculated as needed from the equation of state.

Basically, the equation of state (2-18) defines the energy inherent in a molar mass of gas at a given temperature. This can be appreciated by reading the right-hand side of (2-18) in light of the units employed for the gas constant R. Looking further in this direction, from (2-6) we can rewrite (2-10) as

$$P = \frac{1}{3}\frac{M_A}{V_A}\overline{C^2} \tag{2-24}$$

Combining this with (2-18) we obtain the equalities

$$3PV_A = M_A\overline{C^2} = 3R_*T \tag{2-25}$$

The total energy inherent in the molar mass M_A resides in the kinetic energy of N_A molecules each having mass m and moving at the rms speed defined by (2-11). Hence the energy of the molar mass is given by

$$\tfrac{1}{2}M_A\overline{C^2} = \tfrac{1}{2}N_A m\overline{C^2} = \tfrac{3}{2}R_*T \tag{2-26}$$

The mean kinetic energy $\bar{E}$ of a single molecule is found by dividing both sides of (2-26) by Avogadro's number N_A, giving

$$\bar{E} = \frac{1}{2}m\overline{C^2} = \frac{3}{2}\frac{R_*T}{N_A} = \frac{3}{2}\kappa T \tag{2-27}$$

where the left-hand group of terms in (2-26) has not been used.

In (2-27) the quotient R_*/N_A, denoted by κ, is known as *Boltzmann's constant*, which has the value

$$\kappa = \frac{R_*}{N_A} = 1.381 \times 10^{-23}\,J\,K^{-1} \tag{2-28}$$

Also, because the specific gas constant R, defined for air by (2-20), refers to unit mass, Boltzmann's constant can be expressed by

$$\kappa = Rm \tag{2-29}$$

where the unit of mass employed for m must be consistent with that for R. It is evident that Boltzmann's constant represents the share of kinetic energy a single molecule can claim as a function of temperature. From the foregoing relationships we may see that, in a given volume of gas at temperature equilibrium, molecules of different masses have equal mean energies whose values are functions only of Boltzmann's constant and the temperature.

2.1.3 Maxwell–Boltzmann Distributions

Kinetic energies and velocities are distributed over the molecular population according to several functions which variously bear the names of Maxwell and Boltzmann. Relevant to our subject is the function that deals with molecular speed, that is, the absolute value of velocity regardless of direction. This function is

$$\frac{dN}{dC} = 4\pi NC^2\left(\frac{m}{2\pi\kappa T}\right)^{3/2}\exp\left(-\frac{mC^2}{2\kappa T}\right) \tag{2-30}$$

where dN is the number of molecules having speeds between C and $C + dC$, N is the number density, and the meaning of κT is seen from (2-27). The ratio dN/N that can be formed from (2-30) gives the fraction of the total population occupying the narrow speed range dC. For a gas in thermal equilibrium, the rate at which molecules enter this speed range is equal to the rate at which they leave it.

A plot of (2-30) for oxygen molecules is shown in Figure 2.1. The ordinate scale has been arranged for easy visualization of the quantity dN/dC, with the differential dC equal to unity. The area under the curve is proportional to N, taken here as equal to 10,000. The area of any strip having height dN/dC and width dC (taken as unity) is proportional to dN. Three concepts of speed should be noted. The most probable speed corresponds to the maximum of the function and is found from the derivative of (2-30). This turns out to be

$$C_{mp} = \left(\frac{2\kappa T}{m}\right)^{1/2} \tag{2-31}$$

The function (2-30) is such that the probability of a speed greater than $4C_{mp}$, for example, is only about 10^{-6}. The average speed $\bar{C}$ equals $1.13C_{mp}$. The rms speed, previously found from (2-11) to be 485 m sec^{-1} for air molecules at 0 °C, is $1.22C_{mp}$. Thus the rms speed is greater than

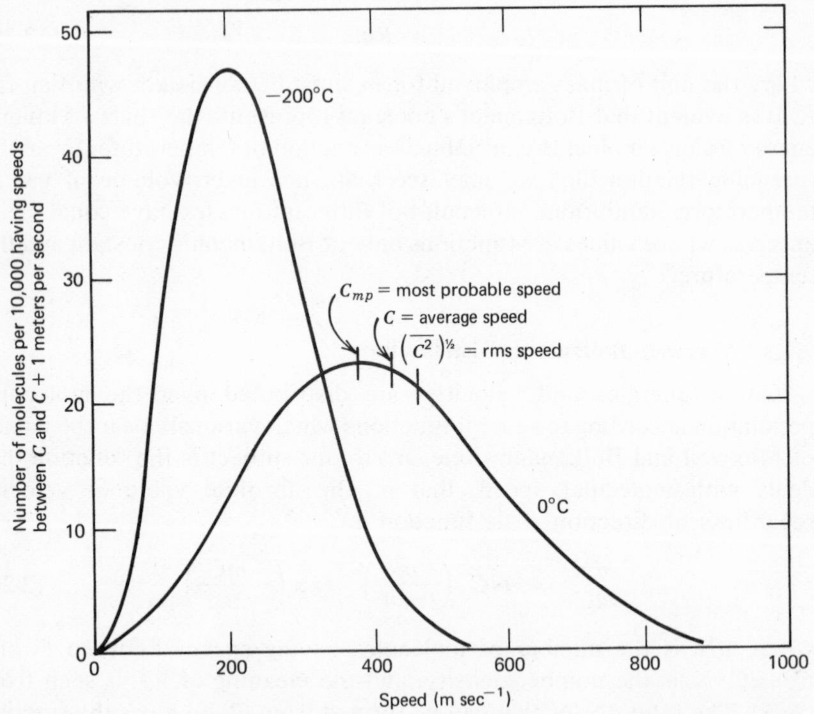

Figure 2.1 Distribution of molecular speeds for oxygen at 0°C and −200°C. From Osgood et al. (1964).

the average speed by a factor of only 1.08. At lower temperatures, the distribution becomes more peaked, and the maximum occurs at lower speeds, as shown by the curve for −200 °C.

The range of speeds shown in the figure helps to explain why the earth can retain an atmosphere while the moon cannot. The escape velocity for an object near the earth's surface is 11.2 km sec^{-1}, or 7.0 mi sec^{-1}. Hence the molecular rms speed of 485 m sec^{-1} at 0 °C is only one twenty-third, and the average speed only one twenty-fifth, of the escape velocity. Jeans (1952) calculated that, if the average speed were one-third of the escape velocity, one-half of the atmosphere would escape into space within a few weeks. For an average speed that is one-fifth of the escape velocity, the escape time is several hundred million years. In great contrast, the escape velocity on the moon is only about 2.4 km sec^{-1}, while its surface temperature exceeds 100 °C during lunar moon. Using this temperature, we find that the average speed for molecules having masses near those of air molecules is approximately 0.225 times the lunar escape velocity. This argues that, if the moon ever had an atmosphere, it would have diffused into space within a period relatively short on the geological time scale.

In the function (2-30) the population distribution is increasingly controlled by the exponential term as the variable C takes on larger values. The numerator of the exponential argument corresponds to kinetic energy, while the denominator is a function of temperature only. With no great rigor but with convenience here, the argument can be regarded as the ratio of a *possible* energy value to the value set by κT. In a simple manner this leads to the essence of a Boltzmann distribution: The probability that a molecule will be at a particular energy level E is given by the term

$$\exp\left(-\frac{E}{\kappa T}\right)$$

The energy may be in either the kinetic form, as in molecular translation, or in potential form, as when the environment is characterized by a force field.

Boltzmann's distribution has many applications in chemistry, physics, and thermodynamics. It is not restricted to molecular entities, but covers all bits of matter from electrons to colloidal and aerosol particles. The force field that creates the possibility of potential energy may be magnetic electrostatic, centrifugal, or gravitational (as in the atmosphere). In considering now two energy states E_1 and E_2 that are accessible to a molecular population, with $E_1 < E_2$, the relative populations of the two states are given by the statistical probability

$$\frac{N_2}{N_1} = \exp\left(-\frac{E_2 - E_1}{\kappa T}\right) \tag{2-32}$$

and so on for any two states. This is Boltzmann's distribution in elementary form and leads to the general conclusion that, when matter has freedom of arrangement or of combination, the most probable resulting state is the one with the lowest total energy. On the microscopic scale this principle governs populations of molecular energy and speed intervals, as we have seen. On the macroscopic scale it governs the vertical distribution of the atmosphere, to which we now turn.

2.1.4 The Law of Atmospheres

The atmosphere can exist only by maintaining itself in kinetic suspension against the force of gravity. In order to examine theoretically the resulting vertical distribution of the atmosphere we assume first that it is in hydrostatic equilibrium. This means that the atmosphere is free to adjust itself in response to applied forces, and that the pressure at any given level is caused only by the downward force (weight) exerted by the overlying column of air. This relationship is expressed by the hydrostatic equation

$$-dP = \rho g\, dZ \tag{2-33}$$

where dP is the increment in pressure due to an increment dZ in altitude, and the minus sign indicates that P decreases as altitude Z increases.

In the real atmosphere, however, ρ is a function of Z, so that the equation in the form (2-33) cannot be integrated to find the total pressure. But from (2-23) we have

$$\rho = \frac{P}{RT} \tag{2-34}$$

Substituting this for ρ in (2-33), we have

$$-dP = \frac{Pg}{RT}\, dZ \tag{2-35}$$

Also, in the real atmosphere, T is a function of Z, as discussed and dealt with in subsequent sections. At this point, however, a useful expression is obtained by assuming that T remains constant as Z changes.

With T thus held constant, integration of (2-35) between the limits P_0, where the altitude is 0, and P_Z, where the altitude is Z, yields

$$P_Z = P_0 \exp\left(-\frac{gZ}{RT}\right) \tag{2-36}$$

Thus the pressure decreases exponentially as the altitude increases, and from (2-23) it is clear that the density decreases in the same manner. Equation (2-36) is the *law of atmospheres*, developed by Laplace. This law can be put in a form that shows the dependence on molecular quantities by substituting for R its value from (2-29), giving

$$P_Z = P_0 \exp\left(-\frac{gmZ}{\kappa T}\right) \tag{2-37}$$

This has the form of the Boltzmann distribution (2-32), but the numerator of the argument of the exponential is now identifiable as the potential energy of a molecule due to its height above the earth's surface, while the denominator represents the kinetic energy of that molecule. The law expressed by (2-36) is the basis for the isothermal model of the atmosphere, which is examined further in Section 2.3.2.

2.2 THE ATMOSPHERIC ENVELOPE

This section presents an overall view of the atmospheric envelope that encloses the earth, thus providing a context for later descriptions of the constituents and their changing distributions in time and space. These changing distributions are governed by the physical processes embraced by the science of *meteorology*. As developed over many years, meteorology has become an interdiscipline having various specialized fields and is an astonishing totality of diversified but related subjects. A general familiarity with atmospheric physical processes, whose workings are often called *the weather*, is helpful to an understanding of atmospheric optics. Since meteorology itself is beyond the scope of this book, readers who wish to make or renew an acquaintance with that subject are referred to the many tutorial treatments available. Several of those that we have found helpful are cited below; many others of equal value could have been consulted.

Ludlum (1971) provides an extensive bibliography on all aspects of meteorology and weather, with attention to the needs of beginners. Sutton (1961) describes the science of meteorology and discusses the problems facing the atmospheric scientist. Thompson and O'Brien (1965) give a broad treatment of weather-making processes for the laymen, with many outstanding illustrations. AW (1965) treats weather from a very practical standpoint for aircraft pilots, employing a wealth of diagrams and photographs. Battan (1974) gives a readable, nonmathematical survey of established principles and recent developments in meteorology.

Definitive meteorological textbooks, which also describe measurement instrumentation to varying extents, are provided by Hess (1959), Pogosyan (1965), Petterssen (1969), Cole (1975), Longley (1970), and Byers (1974)—to name a few. Mirtov (1964) reviews the gaseous composition of the atmosphere and describes many methods of measurement. Huschke (1959) gives detailed definitions of all meteorological terms. For works that emphasize the physics of the atmosphere, readers are referred to the classic by Humphreys (1940), Johnson (1954), Kuiper (1954), Craig (1965), Fleagle and Businger (1963), Roll (1965), Tverskoi (1965), Webb (1966), and Goody and Walker (1972). Valley (1965) provides a comprehensive compilation of meteorological and physical data, along with guidance for their use. Practically all aspects of atmospheric physics, meteorology, and climatology are treated in Malone (1951) and WMO (1973a, 1973b).

2.2.1 The Troposphere

Many of the varied phenomena occurring in the total atmospheric envelope can be generally grouped by type and altitude region. Strong correlations also exist between physical (and optical) properties and altitude. This natural stratification of the atmosphere is illustrated in Figure 2.2, where the strata are actually spherical shells which surround the earth. It must be understood that the interfaces, designated by the suffix *pause,* separating the strata are not sharply defined, particularly those at higher altitudes. Throughout much of the envelope, pressure, mass density, and number density decrease exponentially (or nearly so) as altitude increases, in general conformity with (2-36). Temperature, however, does not obey so simple a rule.

In considering the role of temperature in the atmosphere, it is convenient to introduce here the idea of temperature gradient with altitude Z. The gradient is known as the *lapse rate* and is defined by

$$\gamma = -\frac{dT}{dZ} \tag{2-38}$$

The minus sign makes the lapse rate positive for the usual case of temperature decrease with altitude, up to about 11 km. Customarily the value $\gamma = 0.65$ °C per 100 m is taken as an average for this altitude range. This *environmental* or normal lapse rate, which applies to a free atmosphere in equilibrium, should be distinguished from the *adiabatic* lapse rate, which applies to a rising parcel of air. In the latter case, the air is cooled by expansion during its ascent, without gaining heat from the surrounding air. The adiabatic lapse rate therefore is greater than the normal lapse rate and has the nominal value 0.98 °C per 100 m. In this book we are concerned only with the environmental lapse rate.

The *troposphere* extends from the earth's surface to the *tropopause,* which is the altitude where the lapse rate as defined by (2-38) becomes zero, or nearly so. The altitude of the summer tropopause is between 15 and 18 km over the equator, and 8 to 10 km over the polar regions. In winter the tropopause is often difficult to locate. A nominal altitude of about 11 km, or 36,000 ft, may be assumed for the tropopause. Because

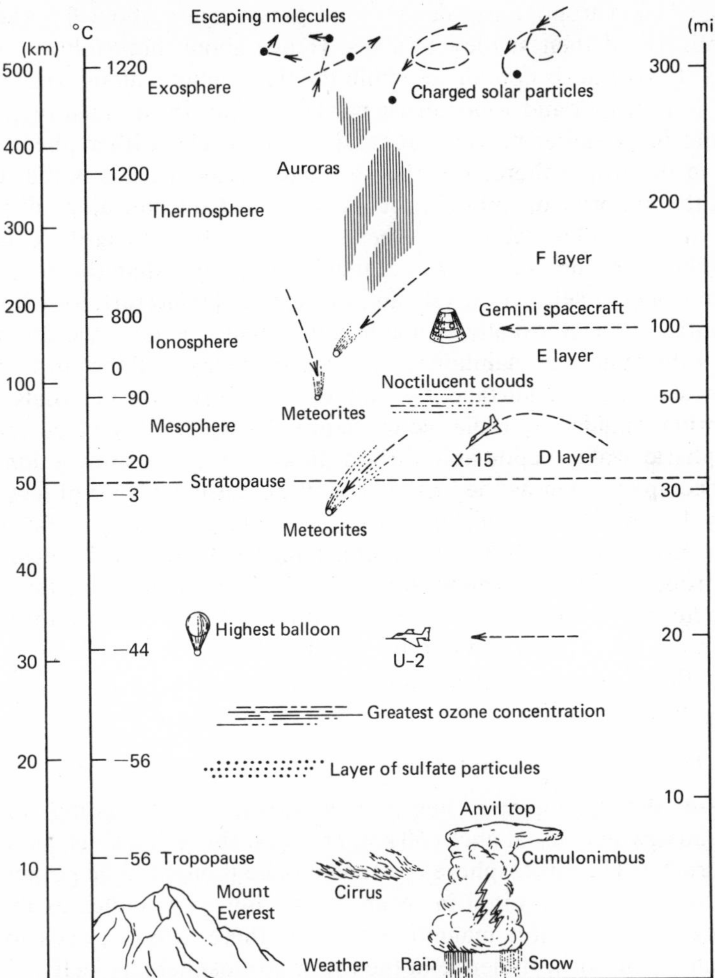

Figure 2.2 Stratification of the atmospheric envelope.

the environmental or normal lapse rate in the troposphere is 6.5 °C km^{-1}, the temperature at an 11-km tropopause is near −56 °C when the sea-level temperature is 15 °C. During periods of calm weather, temperature frequently increases with altitude in the lower troposphere, a condition known as *temperature inversion.* The surface air then is capped with higher-temperature air, so that there is little tendency for convection currents to develop. The lowest air layers thus are very stable, and pollutants may become trapped over their sources, as many city dwellers can attest.

At an 11-km tropopause, density and pressure are about 0.3 and 0.2, respectively, of their sea-level values. Hence about three-fourths of the earth's store of air is contained within the troposphere, along with nearly all the water vapor and atmospheric particles. Since these constituents are essential for weather-making processes, nearly all weather phenomena occur in the troposphere. For the very same reasons, this is the region where the majority of optical processes occur. Concentrations of water vapor and particles decrease rather exponentially with altitude in the troposphere, on the average, but at much faster rates than the density of the atmosphere itself. At the cruising altitudes of jet aircraft, for example, horizontal visibility usually is much better than it is near the ground.

The troposphere is maintained in average energy equilibrium with the solar-heated earth's surface by convection currents and winds. The equilibrium is aided by radiative exchanges between the surface and the atmospheric water vapor and carbon dioxide. Normal convection currents become weaker as they rise to the tropopause, because of lesser air density, but mixing with the air above still continues. Sharply meandering jet streams with speeds up to several hundred knots are found in the upper troposphere and somewhat above. The highest cirrus clouds, often the precursors of storms, float at these altitudes. When thunderstorms are developing below, the anvil tops of cumulonimbus clouds may rise quickly through the tropopause and spread horizontally across the lower stratosphere for miles.

2.2.2 The Stratosphere and Above

Above the tropopause lies the stratosphere, extending to the stratopause which lies at about 50 km, or 30 mi, above sea level, as shown in Figure 2.2. The stratosphere is a region once thought to be perpetually calm. Information acquired in recent years, however, indicates that air currents of considerable magnitude do exist there. Aircraft occasionally encounter clear air turbulence in the lower stratosphere as well as in the troposphere. Such encounters are without visible warning, as the name implies, and may be destructively violent. Radar tracking of balloons discloses frequent winds of high velocity and erratic vertical currents to a considerable altitude. All these mixing motions mean that there is little difference in the composition of upper tropospheric and that of lower stratospheric air. The tropopause is not a physical boundary but, as noted previously, is the altitude range where the lapse rate becomes virtually zero.

Air pressure and density continue to decrease exponentially with altitude in the stratosphere, and at the stratopause are only one-thousandth of their sea-level values. Temperature remains fairly constant, at about −56 °C, from the tropopause to an altitude of about 20 km. This is an isothermal region, where simple expressions such as (2-36) are quite valid. Temperature then increases with altitude above 20 km, until it reaches a value near 0 °C at the stratopause. Ozone is both produced and destroyed by photochemical processes in the stratosphere, and its greatest concentrations are found in an altitude range near 22 km. The concentrations of water vapor in the stratosphere are small, but relative humidity is still a few percent because of the generally low temperatures. Atmospheric particles are few, but several layers where they are more numerous have been revealed by ground-based lidar probes and balloon soundings.

Although air density in the middle stratosphere at about 32 km, or 20 mi, is only one-hundredth of its sea-level value, manned balloons have ascended to this altitude, which is the usual ceiling for such vehicles. Specialized surveillance aircraft such as the U-2 can just about reach this altitude. Instrumented balloons regularly reach altitudes near 90,000 ft, floating for hours while they collect a wide variety of data. In the upper stratosphere, convection currents can scarcely be maintained, because of the small values of air density, and energy exchanges are effected principally by radiative absorption and emission. The ozone layer is particularly active in this respect and is probably responsible for the temperature increase in the upper stratosphere.

Above is the mesosphere, extending from the stratopause up to 80 km, or 50 mi. Air density and pressure continue to decrease with altitude until their values near the top of the mesosphere are only about 10^{-5} of their sea-level values. Temperature decreases steadily from near 0 °C at the stratopause to approximately −90 °C at about 80 km. Data on the structure and composition of the mesosphere are understandably difficult to acquire, and much of our knowledge concerning this region is inferential. Instrumented sounding rockets are the only feasible means for sampling the mesosphere and the layers above. Most meteors that survive swift passage through the outer atmosphere are completely vaporized in this region. Optical observations of their visible trails yielded the early data on air density in the mesosphere. Filmy noctilucent clouds, occasionally visible from high latitudes during summer dusk, float in the upper mesosphere.

Next is the ionosphere, often considered to overlap the upper mesosphere, extending to perhaps 500 km, or about 300 mi. The ionosphere has been known since the early days of "wireless telegraphy." Lacking the protection of a sensible layer overhead, the ionosphere is strongly irradiated by solar energy at very short ultraviolet wavelengths and by electrons and protons in the streaming solar winds. Many ionospheric molecules are thereby ionized and dissociated, so that free electrons and atomic oxygen and nitrogen are relatively plentiful. Spectacular auroral displays, or northern lights, occur when bombardment by solar particles is severe. The ionosphere is marked by several identifiable layers of electrons which provide mirror-like reflections for radio waves of certain wavelengths. Values of atmospheric density are so small that molecular mean free paths are measured in thousands of feet. Temperatures in the upper ionosphere are 1000 °C and greater, although our everyday concept of temperature hardly fits so rarified a medium. Above is the exosphere, where electrons and fragmented molecules travel at speeds of about 1 mi sec^{-1}, seldom colliding and frequently escaping into space. The upper exosphere, gradually merging into the interplanetary environment, is the inexact but ultimate limit of our atmosphere.

2.2.3 Permanent Gas Constituents

The gases of the atmosphere are usually classified as *permanent* or *variable* with respect to amount, but these classifications are mostly a matter of degree. On a sufficiently long time scale, although not necessarily a geological one, perhaps any species of atmospheric gas is variable. Those in the permanent class, however, have not exhibited significant variations with geographical location and time since measurement data first became acceptable. By contrast, those in the variable class exhibit variations—sometimes large—with location and time, where the scale for the latter may be seasonal or even diurnal. Glueckauf (1951) gives a succinct account of atmospheric gaseous composition, covering both permanent and variable gases. Hutchinson (1954) discusses the gases in detail and examines the processes that diminish and replenish their amounts. Junge (1958, 1963) deals with all the atmospheric gases except nitrogen, oxygen, and the noble gases in his extensive treatments of atmospheric chemistry. Additional information on the permanent gases can be found in Roll (1965) and Butcher and Charlson (1972).

The permanent gases are listed in Table 2.1, where their amounts are expressed in three different ways. The first two, commonly used in chemistry, are self-explanatory. The measurement unit *atmosphere-centimeter* (atm-cm) is often used to define the total amount of gas distributed along a path of interest. Consider that the entire amount of a particular species in a given length of path is reduced to a layer at standard conditions of 0 °C and 1013.3 mb, or 1 atm, and that all other species are excluded. The resulting thickness of the layer is then expressed in atmosphere-centimeters. For a path length of 1 km, the amount is stated as atm-cm km^{-1}. It is evident that the cross-sectional area of the path is not a factor, but a unit area is implied. The values in

TABLE 2.1 ATMOSPHERIC GASES PRESENT IN PERMANENT AMOUNTS

Constituent	Volume ratio (%)	Parts per million	Total amount in vertical column (atm-cm)
Nitrogen, N_2	78.084	—	6.244×10^5
Oxygen, O_2	20.948	—	1.675×10^5
Argon, Ar	0.934	—	7.47×10^3
Carbon dioxide, CO_2	0.314	—	2.51×10^2
Neon, Ne	—	18.18	14.5
Helium, He		5.2	4.2
Krypton, Kr		1.1	0.9
Xenon, Xe		0.09	0.07
Hydrogen, H_2		0.5	0.4
Methane, CH_4		2.0	1.6
Nitrous oxide, N_2O*		0.5	0.4
Carbon monoxide, CO*		1.1	0.9

*Has varying concentration in polluted air.
Source: USSA (1962).

atmosphere-centimeters in the table refer to a vertical column extending from sea level to the top of the atmosphere. In subsequent discussions the term *total amount* is used to designate this measurement concept.

The gases listed in Table 2.1 are equally effective in scattering, although minor differences appear in their angular patterns because of differing molecular anisotropy. The first four entries, with argon about three times as plentiful as carbon dioxide, account for nearly all the permanent composition. Argon, neon, krypton, and xenon are monatomic and chemically inert. Their resistance to combining with other elements earned for them the designation *noble gases.* Methane, nitrous oxide, and carbon monoxide are present in what may be called *trace amounts,* but they are important chemically and have measurable infrared absorptions over long paths. In fact, these species were first identified as atmospheric gases by observing their absorption bands in the solar spectrum with the sun low in the sky. Despite their differing molecular weight, all the gases listed in the table have constant volume ratios up to altitudes of about 60 km, because of thorough mixing by winds and convection currents. At greater altitudes the volume ratios of the lighter gases, hydrogen and helium, increase somewhat as they diffuse upward and tend to "evaporate" from the tenuous top of the atmosphere. Information on this subject is given by Hunten (1973).

2.2.4 Variable Gas Constituents

The variable gases are listed in Table 2.2. These constituents, along with several trace gases not listed, enter into many types of reactions with the other gases and particles of the atmosphere. Junge (1958, 1963), SRI (1961), and Butcher and Charlson (1972) should be consulted for tutorial treatments of the variable gases and their reactions. Recent studies and measurements are reported by Ehhalt et al. (1975), Stewart and Hoffert (1975), and Wilkniss et al. (1975). The volume ratios of the variable gases fluctuate between wide limits, so that local information is usually needed to determine their amounts along a given path. Both ozone and water are important in atmospheric optics, because of their strong absorption of ultraviolet and infrared, respectively. In fact, the absorption of lethal ultraviolet energy from the sun by ozone is positively necessary for the existence of life on earth. Additional importance is attached to water, because of the influence this gas (water vapor) has on the growth behavior of atmospheric particles. This gas therefore is considered in greater detail in the following section.

Nitric acid vapor was discovered at altitudes above 19 km by balloon-borne observations of the solar absorption spectrum, as reported by Murcray et al. (1969, 1973). Further information on this gas and its distribution have been obtained from atmospheric emission spectra (Williams et al., 1972) and from direct sampling (Lazrus and Gandrud, 1974). A set of reactions for its production, involving nitrogen dioxide and the hydroxyl radical, is discussed by Rhine et al. (1969). Many reactions take place in the stratosphere among the several oxides of nitrogen and

TABLE 2.2 ATMOSPHERIC GASES PRESENT IN VARIABLE AMOUNTS

Constituent	Volume ratio (%)	Parts per million
Ozone, O_3	—	0–0.07 (ground)
	—	1–3 (20–30 km)
Water vapor, H_2O	0–2	—
Nitric acid vapor, HNO_3	—	$(0–10) \times 10^{-3}$
Ammonia, NH_3	—	Trace
Hydrogen sulfide, H_2S	—	$(2–20) \times 10^{-3}$
Sulfur dioxide, SO_2	—	$(0–20) \times 10^{-3}$
Nitrogen dioxide, NO_2	—	Trace
Nitric oxide, NO	—	Trace

Source: Data from USSA (1962) and Junge (1963).

water, ozone, and atomic oxygen, as presented by Cunnold et al. (1975). Several of these are of such a nature as to indicate the possibility that large-scale emission of nitrogen oxides by supersonic aircraft in the stratosphere could diminish the ozone concentration. Readers wishing information on this complex but important matter should consult Crutzen (1970, 1972), Johnston (1971), and McElroy et al. (1974).

A trace amount of ammonia apparently is produced by the decay of organic matter; this gas was discovered to be an atmospheric constituent by analysis of rainwater. Both hydrogen sulfide and sulfur dioxide are injected into the atmosphere by volcanos and hot springs, and sulfur dioxide is also produced in relatively large quantities by the burning of fossil fuels. In the presence of water vapor, ammonia, hydrogen sulfide, and sulfur dioxide enter into a series of photochemical reactions which culminate in a family of minute aerosol particles, as summarized in Section 3.1.2. The oxides of nitrogen, nitrogen dioxide and nitrous oxide, are produced by bacterial action in the soil and by internal combustion engines. Both these gases, and probably traces of ozone, are involved in the production of photochemical smog, an especially irritating variety occurring in metropolitan regions.

Ozone occurs principally in the stratosphere, where it is both produced and destroyed by photochemical reactions. It is the most unstable of all the atmospheric gases and dissociates readily into atomic and molecular oxygen, whence its vigorous oxidizing property. It enters eagerly into many chemical reactions, particularly those that involve easily oxidizable materials such as sulfur dioxide and various hydrocarbons. An ozone concentration of only 0.1 ppm is irritating to the nose and throat, while 1 ppm is harmful to plants. Readers will recall that small amounts of this gas are produced by electric discharges in air, and some persons have been able to report smelling it in the vicinity of lightning strikes. In fact, ozone was discovered by Schonbein near electric motors about 1845. Most ozone in the troposphere, however, is due to the slow overturning of stratospheric air.

The history of atmospheric ozone investigations has been recounted by Dobson (1968). The presence of this gas in the upper atmosphere was first verified by Fabry and Buisson about 1920, when they matched its ultraviolet absorption spectrum to the observed solar spectrum. Long before airborne methods of measurement were generally feasible, ground-based optical methods were developed by Dobson (1930), Götz et al. (1934), and Strong (1941). The ensuing measurement programs over a period of years established both the total amount of ozone and the altitude region of its greatest concentration. These optical methods are explained by Götz (1951), Johnson (1954), Craig (1965), and Griggs (1966).

Accurate measurements of ozone profiles have been made during recent decades by balloon-borne optical and chemical methods, as summarized by Griggs (1966). Determinations of the vertical and geographical distribution of ozone have been made by Prabhakara et al. (1970, 1971) from data gathered by an infrared spectrometer on the polar orbiting Nimbus 3 satellite. Additional measurement work is referenced in Section 2.4.2.

Chapman (1930), and others of his era, postulated that atmospheric ozone is produced by a series of photochemical reactions at high altitudes where solar ultraviolet energy is abundant. Such processes are difficult to duplicate in the laboratory, primarily because of wall effects in the required chambers. The important reactions are

$$\begin{aligned} &(a)\quad O_2 + h\nu \longrightarrow O + O \qquad \lambda < 2420\ \text{Å} \\ &(b)\quad O_2 + O + M \longrightarrow O_3 + M \\ &(c)\quad O_3 + h\nu \longrightarrow O_2 + O \qquad \lambda < 11{,}100\ \text{Å} \\ &(d)\quad O + O_3 \longrightarrow 2O_2 \\ &(e)\quad O + O + M \longrightarrow O_2 + M \end{aligned} \tag{2-39}$$

In reaction *a* atomic oxygen is produced by molecular dissociation due to absorption of a photon having a wavelength less than 2420 Å. Reaction *b* produces ozone by collisional recombination, with M representing any third body such as molecular nitrogen or oxygen. The third body is needed to carry away the energy liberated by the recombination, thus preventing immediate dissociation of the just-created ozone molecule. Above altitudes of about 80 km, quanta of solar radiation are so plentiful that reaction *a* proceeds at a great rate. At such altitudes, however, three-body collisions are relatively rare because of the small values of number density. Hence in the uppermost stratosphere and in the ionosphere atomic oxygen (and also atomic nitrogen) are predominant.

At midstratosphere altitudes, which are still penetrated by some photons having short wavelengths, the probability of three-body collisions is much greater. These two requirements are sufficiently met at altitudes between 15 and 30 km that the highest rates of ozone production occur in this range. At lower altitudes high-energy photons are rare, because of the very effective shielding above, and the rate of ozone production again becomes small. At all altitudes, ozone is destroyed by reactions *c* and *d* in (2-39), while atomic oxygen is destroyed by reactions *b*, *d*, and *e*. Thus qualitatively we may see that varying amounts of ozone are in equilibrium, of a sort, at different altitudes. There are diurnal variations of course, due to the nightly interruption of solar energy. Circumstances combine to yield the greatest concentrations between altitudes of 15 and 30 km, with a maximum near 22 km.

2.2.5 Properties and Measures of Water Vapor

The potential abundance of water vapor can be appreciated from the fact that more than three-fourths of the earth's surface is covered with water. Diffusing into the air by evaporation and transported by winds and convections, atmospheric water is a mobile reservoir of energy for meteorological processes. The unique ability of water to store and release energy is indicated by its thermal characteristics:

$$\text{Specific heat} = 1\ \text{cal g}^{-1}\ ^\circ\text{C}^{-1}$$
$$\text{Heat of fusion} = 80\ \text{cal g}^{-1}$$
$$\text{Heat of vaporization} = 540\ \text{cal g}^{-1}$$

Except for large-scale synoptic and geographical patterns, the horizontal distributions of water vapor tend to be as variable as the weather itself. Although the total amount aloft is also quite variable, an average vertical distribution is definable, as discussed in Section 2.4.2.

Atmospheric water exists in gas, liquid, and solid phases. The amount of water that can evaporate into a given parcel of air depends only on the air temperature. The *saturation vapor pressure* is the maximum vapor pressure that can be produced, at a given temperature, by evaporation from the liquid or solid phase. When this pressure is reached, the rate of molecular entry into the gas phase is just balanced by the rate of return to the other phase. The two phases then are in equilibrium, and the space containing the vapor is *saturated.* Saturation pressure e_s and saturation density ρ_s are listed in Tables 2.3 and 2.4 with respect to plane surfaces of pure water and ice. The stipulation of a plane surface is important in analyzing the growth of small droplets, as discussed in Section 3.1.3. The rapidly increasing capacity of air to contain water vapor as the temperature is increased should be noted in the tables. If the air is cooled until the saturation pressure is less than the actual vapor pressure, the air is *supersaturated.* Vapor pressure and density are related by the equation of state which, in the notation for water vapor, reads

$$e_v = \rho_v R_v T \tag{2-40}$$

The gas constant R_v has the value $4.615 \times 10^2\ \text{J kg}^{-1}\ \text{K}^{-1}$, which should be compared with that given by (2-21) for dry air.

The matter of condensation is important. Water vapor becomes liquid water only when wettable surfaces, or small centers of substance, are present to retain the impinging water vapor molecules. That is, molecules do not ordinarily aggregate spontaneously to initiate condensation. When the air is saturated, or supersaturated, condensation occurs at ordinary wettable surfaces, such as on a glass of ice water. Substances that are

TABLE 2.3 SATURATION PRESSURE e_s (mb) OF WATER VAPOR OVER PLANE SURFACES OF PURE WATER AND ICE

Temperature (°C)	0	1	2	3	4	5	6	7	8	9
Over water										
−20	1.25	1.15	1.05	0.96	0.88	0.81	0.74	0.67	0.61	0.56
−10	2.86	2.64	2.44	2.25	2.08	1.91	1.76	1.62	1.49	1.37
−0	6.11	5.68	5.28	4.90	4.55	4.21	3.91	3.62	3.35	3.10
0	6.11	6.57	7.05	7.58	8.13	8.72	9.35	10.01	10.72	11.47
10	12.27	13.12	14.02	14.97	15.98	17.04	18.17	19.37	20.63	21.96
20	23.37	24.86	26.43	28.09	29.83	31.67	33.61	35.65	37.80	40.06
30	42.43	44.93	47.55	50.31	53.20	56.24	59.42	62.76	66.26	69.93
100	1013.25									
Over ice										
−50	0.039	0.035	0.031	0.027	0.024	0.021	0.018	0.016	0.014	0.012
−40	0.128	0.115	0.102	0.091	0.081	0.072	0.064	0.057	0.050	0.044
−30	0.380	0.342	0.308	0.277	0.249	0.223	0.200	0.179	0.161	0.144
−20	1.032	0.937	0.850	0.771	0.699	0.632	0.572	0.517	0.467	0.421
−10	2.597	2.376	2.172	1.984	1.811	1.652	1.506	1.371	1.248	1.135
0	6.107	5.623	5.173	4.757	4.372	4.015	3.685	3.379	3.097	2.837

Source: List (1966).

TABLE 2.4 WATER VAPOR DENSITY ρ_s (g m^{-3}) AT SATURATION OVER PLANE SURFACE OF PURE WATER

Temperature (°C)	0	1	2	3	4	5	6	7	8	9
−20	1.07	0.99	0.91	0.84	0.77	0.71	0.65	0.59	0.54	0.50
−10	2.36	2.19	2.03	1.88	1.74	1.61	1.48	1.37	1.26	1.17
−0	4.85	4.52	4.22	3.93	3.66	3.41	3.17	2.95	2.74	2.54
0	4.85	5.19	5.56	5.95	6.36	6.80	7.26	7.75	8.27	8.82
10	9.40	10.01	10.66	11.35	12.07	12.83	13.63	14.48	15.37	16.31
20	17.30	18.34	19.43	20.58	21.78	23.05	24.38	25.78	27.24	28.78
30	30.38	32.07	33.83	35.68	37.61	39.63	41.75	43.96	46.26	48.67

Source: List (1966).

hygroscopic, such as table salt, are effective agents for condensation and become moist when the air is not even saturated. Small centers of substance deserve special mention. At large supersaturations gas ions are such centers for condensation, as in the Wilson cloud chamber. More relevant here, several types of small atmospheric particles are sufficiently hygroscopic that they are basic in all condensation processes. These particles, known as condensation nuclei, are discussed in Section 3.1.2.

The amount of water vapor in a parcel of air is expressed in several ways. *Absolute humidity* is defined either by the actual vapor pressure in millibars, usually called the partial pressure, or by the actual vapor density in grams per cubic meter. Hence the values in Tables 2.3 and 2.4 are saturation absolute humidities. *Relative humidity* U is the ratio of actual vapor pressure, at a stated temperature, to the saturation value at that temperature:

$$U = \frac{e_v}{e_s} \times 100 = \frac{\rho_v}{\rho_s} \times 100 \tag{2-41}$$

in percent. Thus the absolute humidity corresponding to a stated temperature and relative humidity is found by applying that ratio to the appropriate saturation value in Table 2.3 or 2.4. The *mixing ratio* w_v is the mass of vapor contained in a unit mass of dry air, hence is equal to the ratio of the two densities when both are in the same units. Very often, however, the mixing ratio is expressed in grams per kilogram. The saturation mixing ratio, as the name implies, is the value of the ratio that exists when the air is saturated. Hence

$$U = \frac{w_v}{w_s} \tag{2-42}$$

Dewpoint T_d and *frostpoint* T_f temperatures are frequently employed as measures of moisture. Visualize that a parcel of moist air at a given temperature and vapor pressure is cooled at constant pressure, that is, without changing the amount of vapor per unit volume. As the cooling proceeds, a saturated condition is reached, and condensation starts. Dewpoint temperature then is defined as the temperature to which air must be cooled in order that the saturation pressure, with respect to water, will equal the initial vapor pressure. Dewpoint temperature thus is a saturation temperature. Frostpoint temperature is defined in the same way, except that the saturation is with respect to ice. Instrumentation for measuring these temperatures is widely used in balloon soundings of the atmosphere.

The absolute humidity of a parcel of air at a given dewpoint or frostpoint temperature is determined by finding that temperature in Table 2.3 and reading out the associated value of e_s. This is really the vapor pressure e_v for the parcel regardless of temperature; hence it is the absolute humidity of the parcel. To find the relative humidity of the parcel, its actual temperature must be known. The ratio of the value of e_v, as just determined, to the value of e_s, as read from the table for the actual temperature of the parcel, is the relative humidity U, as in (2-41). Numerical relations between these measures of moisture are given in Table 2.5. The concept of dewpoint is retained even for saturation temperatures down to about −40 °C. This is because clouds contain many undercooled droplets which remain in liquid form at such temperatures. If the air in the cloud is saturated, which is the usual condition, the saturation is with respect to these liquid droplets. The air thus is supersaturated with respect to any ice crystals that may be present, as may be seen by comparing corresponding entries in the two parts of Table 2.3. For clouds in which precipitation is forming, this may lead to the growth of ice crystals at the expense of water droplets.

The amount of water vapor along a path is defined by the depth of liquid water that would be produced if all the vapor were precipitated or condensed. Such a depth is independent of the cross-sectional area of the

TABLE 2.5 CONVERSION OF DEWPOINT TO OTHER MEASURES OF MOISTURE

Dewpoint (°C)	Frostpoint (°C)	Vapor pressure (mb)	Vapor density (g m^{-3})
−40	−36.5	0.189	0.176
−35	−31.8	0.314	0.286
−30	−27.2	0.509	0.453
−25	−22.5	0.807	0.705
−20	−18.0	1.254	1.074
−15	−13.4	1.912	1.605
−10	−8.9	2.863	2.358
−5	−4.5	4.215	3.407
0	0	6.108	4.847
5	—	8.719	6.797
10	—	12.272	9.399
15	—	17.044	12.83
20	—	23.373	17.30
25	—	31.671	23.05
30	—	42.430	30.38
35	—	56.236	39.63
40	—	73.777	51.19

Source: Gringorten et al. (1966).

path, but a unit area is implied. Precipitable water in centimeters (pw-cm) thus is the counterpart of the unit atmosphere-centimeter employed in Table 2.1 for gases that do not condense in the atmosphere. Since the density of water can be taken as $1\ \text{g cm}^{-3}$, the precipitable water per kilometer along a path evidently is

$$\text{pw-cm} = 0.1\bar{\rho}_v\ (\text{km}^{-1}) \tag{2-43}$$

where $\bar{\rho}_v$ is the average vapor density in grams per cubic meter, and the factor 0.1 reconciles the numerical value for the different measurement units. Figure 2.3 shows precipitable water-centimeters per kilometer and per mile as a function of temperature and relative humidity, which usually are either known or are readily measured. The chart is applicable to a homogeneous path to which a mean temperature and relative humidity can be assigned. Usually this can be done when the path is horizontal or nearly so. Truly slant paths involve the vertical distribution of water vapor; this distribution is considered in Section 2.4.2.

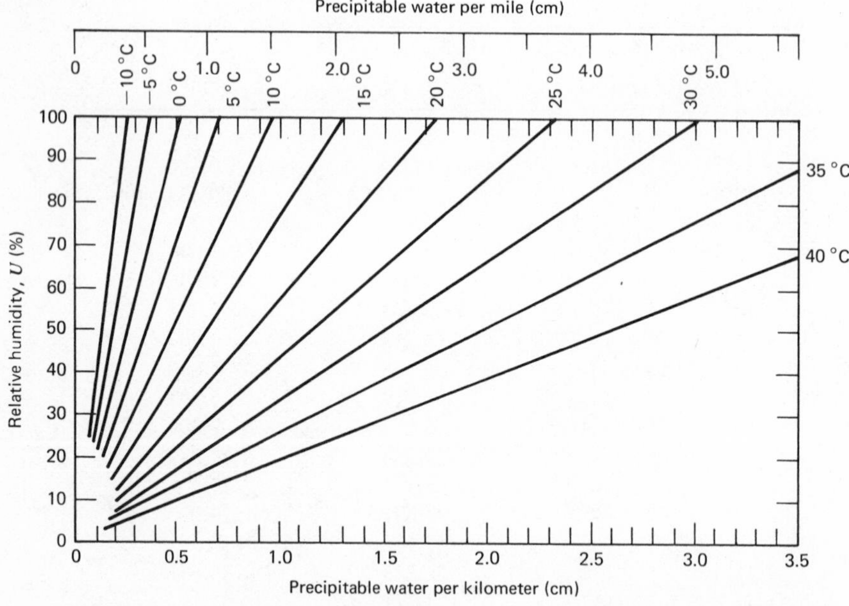

Figure 2.3 Precipitable water as a function of temperature and relative humidity.

2.3 MODELS OF THE ATMOSPHERE

The gas atmosphere properties of greatest importance to scattering are the *surface values* and *altitude profiles* of pressure, temperature, lapse rate, and density. These properties, like the weather with which they are associated, are extremely variable from place to place and time to time, often in a seemingly unpredictable fashion. Although the properties are strongly interrelated, the dependences in the real world are complex and may not be readily expressible in closed form. Of necessity, atmospheric physics has devised several analytic models which represent in a simple manner certain relationships of interest. Also, there is need for a reference model of the real atmosphere—a model that portrays the mean values of properties about which the actual values fluctuate. In this section we review three analytic models that find wide application, and the standard reference model. Each of these models is useful in determining atmospheric optical properties.

2.3.1 Constant-Density Model

An interesting model is obtained by assuming that atmospheric density does not decrease as altitude increases, but remains at the sea-level value all the way to the top. Such a concept is called a *homogeneous atmosphere.* Seen from the viewpoint of the law of atmospheres (2-36), this assumption has the corollary that air is an incompressible fluid. Although such assumptions are not of the real world, they are the basis of an analytic model of the atmosphere—the constant-density model.

In this model the variation in pressure with altitude is found from the hydrostatic equation (2-33). As noted there, ordinarily ρ is a function of Z, so that the equation cannot be directly integrated. With the assumption of constant density, however, this difficulty is removed. Integration of (2-33) between the altitude limits 0 and Z gives

$$P_Z = P_0 - \rho_0 g Z \qquad (2\text{-}44)$$

showing, not unexpectedly, that pressure decreases linearly with altitude in a constant-density fluid. A familiar example is seen in the relation between pressure and depth in a body of water, which is almost incompressible. At the very top of the constant-density atmosphere, the condition $P_Z = 0$ obtains, and (2-44) becomes

$$Z_{\text{top}} = \frac{P_0}{\rho_0 g} \qquad (2\text{-}45)$$

where the subscript 0 indicates a reference value. For standard conditions at sea level, (2-45) yields the value $Z = 8.00$ km, or about 26,200 ft, so that several Himalayan peaks would still stand above the top of a constant-density atmosphere.

In all atmospheres, whether model or real, the rate of temperature change with altitude is the lapse rate γ, defined by (2-38). In the constant-density model, γ is found by differentiating the equation of state (2-23), which gives

$$\frac{dP}{dT} = \rho R \tag{2-46}$$

where R is the gas constant for a unit mass, as defined in (2-21). Substituting the value of dP from (2-33) into (2-46) we find

$$-\frac{dT}{dZ} = \gamma = \frac{g}{R} = 3.42\ °\text{C per } 100\ \text{m} \tag{2-47}$$

This is known as the *autoconvective lapse rate*, which contrasts strongly with the normal or environmental lapse rate of 0.65 °C per 100 m in the real atmosphere. It is seen from (2-47) that the lapse rate in the constant-density model is determined by the ratio of two *physical constants*. The temperature at any altitude is given by

$$T_Z = T_0 - Z\gamma \tag{2-48}$$

which applies to all models and the real atmosphere as well. It may be verified that the temperature at the upper limit of the constant density model is 0 K, equivalent to a cessation of molecular motion.

2.3.2 Isothermal Model

The isothermal or constant-temperature model is obtained by letting $\gamma = 0$, so that $T_Z = T_0 = \text{constant}$. This isothermal condition is fairly representative of the lower and middle stratospheric regions and is the assumption used in deriving the law of atmospheres (2-36). Analytic expressions for this model are developed easily. Taking (2-33) and substituting therein the value of ρ from (2-23) gives

$$-dP = \frac{P}{RT} g\, dZ \tag{2-49}$$

Since temperature is constant, by definition, the pressure at any altitude is found by integrating (2-49) between the limits 0 and Z, yielding

$$P_Z = P_0 \exp\left(-\frac{gZ}{RT}\right) \tag{2-50}$$

which is the same as (2-36) but is now to be viewed in a broader context.

The density at any altitude is found by realizing from (2-23) that, in an isothermal atmosphere,

$$\frac{\rho_Z}{\rho_0} = \frac{P_Z}{P_0} \tag{2-51}$$

Hence (2-50) may be rewritten as

$$\rho_Z = \rho_0 \exp\left(-\frac{gZ}{RT}\right) \tag{2-52}$$

Since molecular number density N is proportional to mass density ρ, we have

$$N_Z = N_0 \exp\left(-\frac{gZ}{RT}\right) \tag{2-53}$$

Thus an isothermal atmosphere has no definite upper boundary, but tapers off exponentially toward infinity.

The above expressions containing the exponential term are Boltzmann distributions of the form (2-32). Here the numerator of the exponential argument represents the potential energy of a unit mass of air at altitude Z above sea level. The denominator represents the average thermal or kinetic energy of this mass. When $Z = RT/g$, the value of P_Z, ρ_Z, or N_Z is just $\exp(-1)$ of the value at $Z = 0$. Hence a *scale height* H for the exponentially distributed molecular atmosphere can be defined as

$$H = \frac{RT}{g} = \frac{\kappa T}{mg} \tag{2-54}$$

where the last part of the expression is based on the relationship $\kappa = Rm$ from (2-29). When $T = 0°$ C, as for standard air, $H = 8.00$ km. From the meaning of scale height, it follows that

$$P_Z, \rho_Z, \text{ or } N_Z = (P_0, \rho_0, \text{ or } N_0) \exp\left(-\frac{Z}{H}\right) \tag{2-55}$$

Scale height, or *scale factor* generally, is a useful parameter of exponential distributions. By reading (2-54) in the light of relationship (2-23), it is seen that (2-54) can be made identical to (2-45). Thus the isothermal scale height is equal to the height of the constant-density atmosphere when the sea-level pressure and density of the two models are made the same. The isothermal scale height is constant with altitude, varies directly with temperature, as shown by (2-54), and is the same for both pressure and density, as shown by (2-55).

2.3.3 Polytropic Model

Characteristics of this model, often called the constant–lapse rate model, are intermediate between those of the constant-density and isothermal models. Whereas in the first the lapse rate of 3.42 °C per 100 m is fixed by the ratio g/R and in the second the lapse rate is zero by definition, in the polytropic model the lapse rate is selected arbitrarily. Such freedom allows this model to represent each altitude region of the real atmosphere except the isothermal regions. The model is developed by substituting the value of dZ from (2-47) into (2-49) to give

$$\frac{dP}{P} = \frac{g}{R\gamma}\frac{dT}{T} \tag{2-56}$$

Integrating between the limits P_0, P_Z and T_0, T_Z, and eliminating the logarithmic form we have

$$P_Z = P_0\left(\frac{T_Z}{T_0}\right)^{g/R\gamma} \tag{2-57}$$

When altitude rather than upper air temperature T_Z is the data quantity given, an alternative form of (2-57) is useful in finding P_Z. Substituting for T_Z in (2-57) its value from (2-48) gives

$$P_Z = P_0\left(1 - \frac{Z\gamma}{T_0}\right)^{g/R\gamma} \tag{2-58}$$

Expressions for density as a function of altitude and temperature analagous to (2-57) and (2-58) are

$$\rho_Z = \rho_0\left(\frac{T_Z}{T_0}\right)^{(g/R\gamma)-1} = \rho_0\left(1 - \frac{Z\gamma}{T_0}\right)^{(g/R\gamma)-1} \tag{2-59}$$

which are readily developed from several of the above relationships. The altitude corresponding to a given pressure P_Z is found by solving (2-58) for Z, thus

$$Z = \frac{T_0}{\gamma}\left[1 - \left(\frac{P_Z}{P_0}\right)^{R\gamma/g}\right] \tag{2-60}$$

Since P_Z, as well as ρ_Z and T_Z, is zero at the top of a polytropic atmosphere, the height of such an atmosphere is, from (2-60), just

$$Z_{\text{top}} = \frac{T_0}{\gamma} \tag{2-61}$$

The versatility of the polytropic model lies in the freedom of selecting the lapse rate. When this rate is 3.42 °C per 100 m, the polytropic upper limit is the same as that of the constant-density model. With very small lapse rates, the polytropic model approaches the isothermal, but the value $\gamma = 0$ is inadmissible of course. Since scale height is a function of temperature, scale height itself is a function of altitude in a polytropic model, rather than being constant with altitude as in the isothermal case. The dependence of pressure and density on altitude, however, is very similar in these two models. This can be seen, for example, by comparing the first few terms of the series expansions of (2-50) and (2-58). Exponential treatments can be applied effectively to the polytropic distributions of pressure and density that occur in several altitude regions of the U.S. Standard Atmosphere described in the following section.

The pressure scale height H_P of the polytropic model depends on the same factors as the scale height of the isothermal model and is defined, analogously to (2-54), by

$$H_P = \frac{RT}{g} = \frac{kT}{mg} \tag{2-62}$$

Because the lapse rate is nonzero in the polytropic model, temperature, hence pressure scale height, is a function of altitude. The density scale height H_ρ in the polytropic model is related to H_P by

$$H_\rho = \frac{H_P}{1 - R\gamma/g} \tag{2-63}$$

which likewise varies with altitude.

2.3.4 U.S. Standard Atmosphere

Various "standard atmospheres" have been established from time to time as references for general scientific purposes and for designing and testing aerospace vehicles. Each of these references incorporated the best data available from the real atmosphere at the time. Where data were sparse, interpolations and extrapolations from theory necessarily were made. An early standard atmosphere was devised in 1925 by the National Advisory Committee for Aeronautics (NACA), the predecessor of the National Aeronautics and Space Administration (NASA). This early NACA standard was based on a constant lapse rate of 0.65 °C per 100 m from the surface to the tropopause at 11 km, and an isothermal region from that altitude to 20 km. Properties of this NACA atmosphere, which is still adequate for many purposes, are tabulated by List (1966). As aviation developed and aircraft performance factors were improved, several additional standard atmospheres were devised by aviation organizations.

The advent of ballistic missiles and satellites created many needs for an understanding of the upper air, while rocket probes made possible new techniques of measurement. In 1956 the Air Research and Development Command of the U.S. Air Force established an ARDC Standard Atmosphere, and this model was continually reviewed as space programs proceeded. The cooperative efforts of many nations during the International Geophysical Year of 1957–1958 provided much new information. Discrepancies began to appear between the assigned density values for the upper atmosphere and the observed perturbations of satellite orbits as a result of atmospheric drag. From several such studies and the evaluation of new measurement data, the 1956 model was revised into the ARDC model of 1959, presented in detail by Campen et al. (1960).

The U.S. Standard Atmosphere of 1962 (USSA-1962) was established under joint sponsorship of the U.S. Air Force, U.S. Weather Bureau, and NASA. This model generally embodies science's best knowledge of atmospheric structure, from sea level up to 700 km, and is listed in the references as USSA (1962). The tabulations comprise mean annual values of the principal physical properties at altitude intervals of 50 m in the altitude range 0 to 11 km, and at greater intervals thereafter, for the midlatitude belt. Valley (1965) provides an abridgement of these very detailed tabulations. This model, like the earlier ones, deals only with the permanent gas atmosphere whose constituents are listed in Table 2.1. From the optical standpoint, the important properties of such an atmosphere are temperature, pressure, and density. Values of these properties for several altitudes are listed in Table 2.6, and the complete ranges are graphed in Appendix E.

The real atmosphere at a given time and location necessarily differs from an annual model constructed for a single latitude belt. Weather, on both local and synoptic scales, produces large fluctuations in the properties in the troposphere, and the fluctuations frequently extend into the stratosphere. In addition, there are systematic variations due to latitude and season, and solar factors become important at high altitudes. Several supplementary model atmospheres have been established to take latitudinal and seasonal variations into account. These are embodied in the U.S. Standard Atmosphere Supplements of 1966 (USSAS-1966), listed in the references as USSAS (1966). The tabulations are similar to those in USSA-1962 but cover summer and winter values at latitudes 30, 45, 60, and 75 deg north. From these standard atmospheres and supplements, several workers have constructed models for expressing the attenuation along optical paths, taking account of variations due to season, latitude, and visibility conditions at the earth's surface. Perhaps the most widely used of such models are those by Elterman (1968, 1970a) and McClatchey et al. (1971).

Although detailed values of T_Z, P_Z, and ρ_Z are listed in USSA-1962 and

TABLE 2.6 USSA-1962: TEMPERATURE, PRESSURE, AND DENSITY AT SELECTED GEOMETRIC ALTITUDES

Altitude (m)	Temperature (K)	Pressure (mb)	Density (kg m^{-3})	Density ratio (ρ/ρ_0)	Number density (N m^{-3})
−100	288.80	1.025×10^{3}	1.24×10^{0}	1.01×10^{0}	2.57×10^{25}
0	288.15	1.013×10^{3}	1.23×10^{0}	1.00×10^{0}	2.55×10^{25}
1,000	281.65	8.99×10^{2}	1.11×10^{0}	9.07×10^{-1}	2.31×10^{25}
2,000	275.15	7.95×10^{2}	1.01×10^{0}	8.22×10^{-1}	2.09×10^{25}
3,000	268.66	7.01×10^{2}	9.09×10^{-1}	7.42×10^{-1}	1.89×10^{25}
4,000	262.17	6.17×10^{2}	8.19×10^{-1}	6.69×10^{-1}	1.70×10^{25}
5,000	255.68	5.40×10^{2}	7.36×10^{-1}	6.01×10^{-1}	1.53×10^{25}
8,000	236.22	3.57×10^{2}	5.26×10^{-1}	4.29×10^{-1}	1.09×10^{25}
10,000	223.25	2.65×10^{2}	4.14×10^{-1}	3.38×10^{-1}	8.60×10^{24}
15,000	216.65	1.21×10^{2}	1.95×10^{-1}	1.59×10^{-1}	4.05×10^{24}
20,000	216.65	5.53×10	8.89×10^{-2}	7.26×10^{-2}	1.85×10^{24}
25,000	221.55	2.55×10	4.01×10^{-2}	3.27×10^{-2}	8.33×10^{23}
30,000	226.51	1.20×10	1.84×10^{-2}	1.50×10^{-2}	3.83×10^{23}
40,000	250.35	2.87×10^{0}	4.00×10^{-3}	3.26×10^{-3}	8.31×10^{22}
50,000	270.65	7.98×10^{-1}	1.03×10^{-3}	8.38×10^{-4}	2.14×10^{22}
60,000	255.77	2.25×10^{-1}	3.06×10^{-4}	2.50×10^{-4}	6.36×10^{21}
70,000	219.70	5.52×10^{-2}	8.75×10^{-5}	7.15×10^{-5}	1.82×10^{21}
100,000	210.02	3.01×10^{-4}	4.07×10^{-7}	4.06×10^{-7}	2.62×10^{16}
200,000	1236	1.33×10^{-6}	3.32×10^{-10}	2.71×10^{-10}	7.82×10^{15}
300,000	1432	1.88×10^{-7}	3.59×10^{-11}	2.93×10^{-11}	9.53×10^{14}
400,000	1488	4.03×10^{-8}	6.50×10^{-12}	5.31×10^{-12}	1.96×10^{14}
700,000	1508	1.19×10^{-9}	1.54×10^{-13}	1.26×10^{-13}	5.73×10^{12}

Source: USSA (1962).

supplements, it is convenient to have numerical expressions for these properties as functions of altitude. In this regard, the relevant defining parameters for the several altitude regions of USSA-1962 are given in Table 2.7. It is seen that the entire model encompasses five polytropic and three isothermal regions. The desired expressions then can be obtained by substituting the parameter values, as appropriate, into the equations for the polytropic and isothermal models described in the two preceding sections. By taking the troposphere as an example and employing the meter as the altitude unit, the temperature at any altitude is found from (2-48) to be

$$T_Z = 288.2 - 6.5\times10^{-3}Z \quad \text{K} \tag{2-64}$$

TABLE 2.7 USSA-1962: VALUES OF THE DEFINING PARAMETERS FOR VARIOUS ALTITUDE REGIONS*

Altitude (km)	Temperature (K)	Lapse rate (K/km)	Pressure (mb)	Density ($kg\ m^{-3}$)
0	288.15		1013.25	1.225
		−6.5		
11	216.65		227.00	3.648×10^{-1}
		0.0		
20	216.65		55.29	8.891×10^{-2}
		1.0		
32	228.65		8.89	1.356×10^{-2}
		2.8		
47	270.65		1.16	1.497×10^{-3}
		0.0		
52	270.65		6.22×10^{-1}	8.01×10^{-4}
		−2.0		
61	252.65		1.97×10^{-1}	2.703×10^{-4}
		−4.0		
79	180.65		1.24×10^{-2}	2.349×10^{-5}
		0.0		
88.74	180.65		2.07×10^{-3}	4.0×10^{-6}

*These values may be used, where appropriate, in the polytropic and isothermal regions.
Source: USSA (1962).

From (2-57) the pressure is

$$P_Z = 1013.3(1 - 2.255 \times 10^{-5} Z)^{5.256} \quad \text{mb} \tag{2-65}$$

and from (2-58) the density is

$$\rho_Z = 1.2250(1 - 2.255 \times 10^{-5} Z)^{4.256} \quad \text{kg m}^{-3} \tag{2-66}$$

Numerical expressions for the other polytropic regions are found in a similar manner, with due regard for proper numerical substitutions.

As noted in the preceding section, pressure and density in a polytropic region have an approximately exponential dependence on altitude. This dependence in the troposphere may be seen in the semilog plots of pressure and density data from USSA-1962 in Figure 2.4. The two plots are nearly straight lines from which are deduced an average pressure scale height $\overline{H_P}$ of 7.7 km and an average density scale height $\overline{H_\rho}$ of 9.3 km. The same values can be obtained by solving (2-65) and (2-66) for the conditions

$$\frac{P_Z}{1013.3} = \exp(-1)$$

$$\frac{\rho_Z}{1.2255} = \exp(-1)$$

Such values of scale heights can be used in the exponential function (2-55), which is more convenient for ordinary calculations than the polytropic functions (2-58) and (2-59).

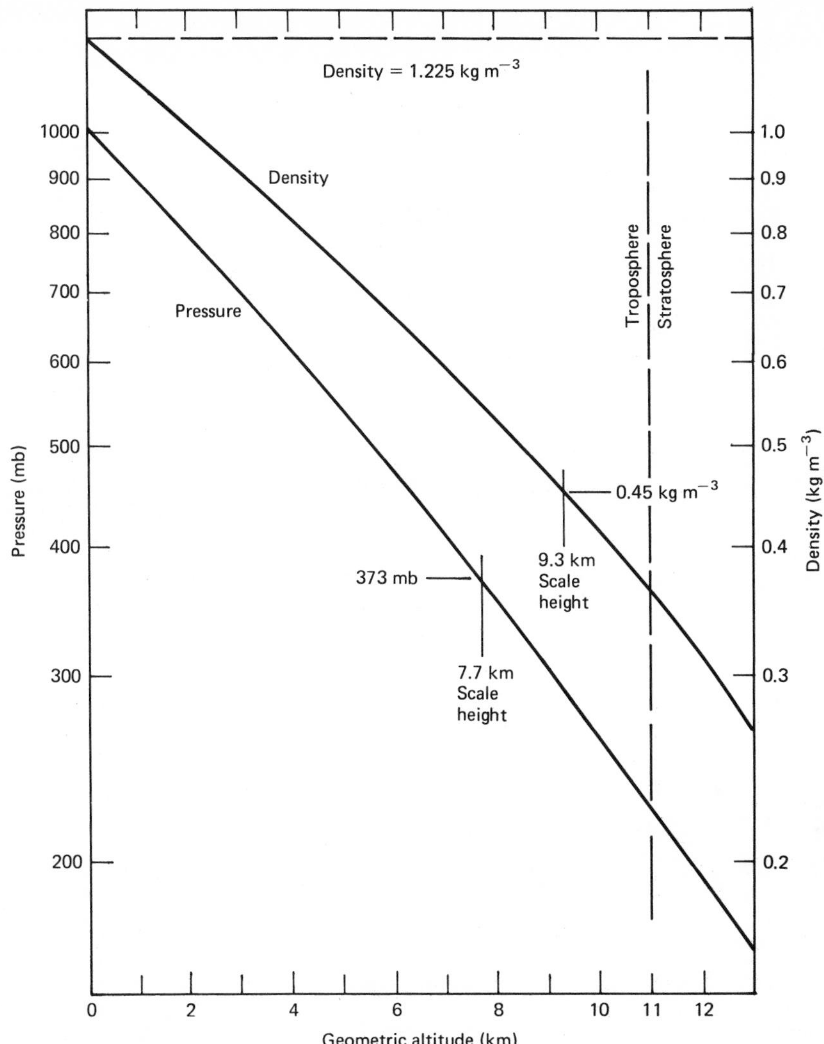

Figure 2.4 United States Standard Atmosphere—1962: Pressure, density, and average scale height in the troposphere.

It is emphasized that the foregoing values of scale height for the troposphere are average values, as may be seen from the manner of their derivation. The exact scale heights, as defined by (2-62) and (2-63), are functions of temperature and lapse rate. Therefore they decrease continuously in the troposphere as altitude increases, and the effects of the decrease are evident in the slight curvature of the plots in Figure 2.4. The exact values of pressure and density scale height at sea level are found, by substituting in (2-62) and (2-63), to be

$$H_P = 8.44\ \text{km}$$
$$H_\rho = 10.4\ \text{km} \qquad (2\text{-}67)$$

2.4 MEASURES OF THE OVERHEAD ATMOSPHERE

In studying atmospheric physical processes and forecasting the weather, meteorology is concerned with both vertical and horizontal distribution of the temperature, pressure, and density of the permanent and variable constituents. Tomorrow's weather is just over the horizon—almost. In atmospheric optics, emphasis is placed on the vertical distribution of properties. This is because the properties usually do not vary greatly in horizontal directions for distances ordinarily associated with lines of sight. The variations that do occur are really fluctuations about mean values whose permanent gradients are weak, that is, the gradients are characterized by distance scales that are long.

In the vertical direction, however, fluctuations are superposed on the strong, permanent gradients that exist in the several altitude regions of the real atmosphere. These gradients, whose theoretical bases are explained in Sections 2.1 and 2.3, define the atmospheric vertical structure and are called *vertical profiles*. Such profiles are indispensable for predicting and evaluating the characteristics of optical paths. For these purposes they are usefully complemented by information regarding the total amount, that is, the total mass, of atmosphere directly above a point at a specified reference altitude. This information is provided by the concept of *reduced height*. Thus, in defining spatial distribution and quantity, the profiles and reduced height are the measures of the overhead atmosphere. This section describes and presents data on these measures.

2.4.1 Vertical Profiles of Permanent Gases

Vertical profies of temperature, pressure, and density for any of the model atmospheres can be computed by means of the expressions given previously. The results are applicable only to the model, however, and depend on the values assumed for the constants. Profiles closer to reality can be constructed from the data in USSA-1962 and USSAS-1966 for the standard and supplemental atmospheres. Nonetheless it is desirable to have reference profiles of the real atmosphere in which some but not all of the variations have been removed by averaging. For example, annual or even semiannual mean values do not show seasonal variations which may be important. At the other extreme, daily values do not correlate with those from other years. The problem is one of compressing data without losing significant information.

Valley (1965) provides and discusses data on atmospheric properties for several locations in North America. A good view of atmospheric structure is provided in the mean monthly profiles developed by Kantor and Cole (1965) following earlier work by Cole and Kantor (1964). These profiles extend from sea level to 80 km and are based on data gathered by balloon and rocket probes over a nearly 8-year period. Numerical values of these profiles for latitudes 30, 45, and 60 deg north are listed in the tables of Appendix F.

Additional information on pressure between the 30- and 90-km levels is provided by Kantor (1966). Tabulations of mean properties between 30 and 300 km, including composition, are given by Champion (1965). From recent studies of the 80-to-120-km region, Adams (1970) provides tabulated values of the nitrogen, molecular oxygen, and atomic oxygen constituents. Groves (1971) presents a review of observational data on temperature, pressure, density, and winds from 25 to 110 km, along with information on their seasonal and latitudinal dependence.

Our knowledge of atmospheric properties at very high altitudes is still largely inferential, although it has grown enormously since rocket probes and space vehicles came into being Details of the properties and the causes of their variations are difficult to unravel, because there are many sources and sinks of energy in the tenuous upper atmosphere immediately brushing the space environment. Important inputs to the physical and chemical processes are the changing fluxes of solar particles and ultraviolet energy, changing amounts of atomic oxygen and nitrogen, solar and lunar tides, and variations in geomagnetism that affect the motions of charged particles. Gerson (1952), Bates (1953), Nicolet (1953), Whipple (1953), Hulburt (1957), and Khvostikov (1965) examine these upper-atmosphere processes and provide many references.

Malone (1951) has brought a group of experts to bear on just about every aspect of the upper atmosphere as it was known at the time. The book by Mitra (1952) has long been regarded as a standard reference. More recent works have been published by Chamberlain (1961), Craig (1965), and Rishbeth and Garriott (1969).

2.4.2 Vertical Profiles of Variable Gases

Except for ozone and water, the variable gases listed in Table 2.2 usually are not present in amounts greater than 20 ppb. These amounts may be exceeded of course in regions near natural and industrial sources. At a given location the amounts depend on source proximities, winds, and diffusion factors and may be difficult to establish in an overall sense unless the weather is very calm. Although these gases are very minor constituents with respect to quantity, they are active chemically among themselves and with other gases. Usually they lose their identity by such activities. Except for nitric acid they do not appear to have long-term vertical profiles that are significant in atmospheric optics.

A profile of nitric acid has been established by the work of Murcray et al. (1973), Williams et al. (1972), and Lazrus and Gandrud (1974). They showed that this gas occurs at altitudes between 14 and 28 km, with concentration maxima near 16 and 22 km. This altitude region coincides approximately with the region of greatest ozone concentration. The reactions that culminate in nitric acid, as well as others that also involve the oxides of nitrogen, have ramifications that may affect the amount of ozone in the stratosphere. Hence concern has been expressed that the injection of large amounts of nitrogen oxides into the stratosphere by supersonic aircraft may reduce the ozone concentration, as referenced in Section 2.2.4.

The vertical distribution of ozone is closely linked to the photochemical reactions that create and destroy this gas in the stratosphere. These reactions are outlined in Section 2.2.4, along with an indication of the altitude region in which each one tends to be the most effective. Geographical distributions of total ozone, however, vary widely. Presumably this is due, at least partially, to horizontal advection currents at different stratospheric altitudes, to vertical movements of stratospheric air, and to varying rates of ozone production and destruction. The resulting geographical and vertical distributions have been investigated so thoroughly that upper-air research is almost synonomous with ozone research. Many data are reviewed by Junge (1963), Craig (1965), and Griggs (1966). Data from a network of ozonesonde stations extending from the Panama Canal Zone to Thule, Greenland, are presented by Hering (1964), Hering and Borden (1964, 1965), and Hering and Dütsch (1965).

The average amounts of total ozone for the northern hemisphere, as found by London (1962), are illustrated in Figure 2.5. Variations according to latitude and season are relatively large. In middle latitudes the increase from fall to spring is about 20% while in the polar regions it is nearer 40%. A representative vertical profile has been established by Elterman (1968, 1970a) for his model of a clear standard atmosphere, as shown in Figure 2.6. This profile, representing 0.35 atm-cm of total ozone, has been widely used in exploratory studies. Dowling and Green (1966), in their studies of earth-atmosphere radiance in the middle ultraviolet due to backscattering, devised analytic functions to fit the upper and lower portions of ozone profiles. The function for the upper portion of the profile in Figure 2.6 is, in their notation,

$$W_3 = W_p \exp\left(-\frac{y - y_p}{h}\right) \tag{2-68}$$

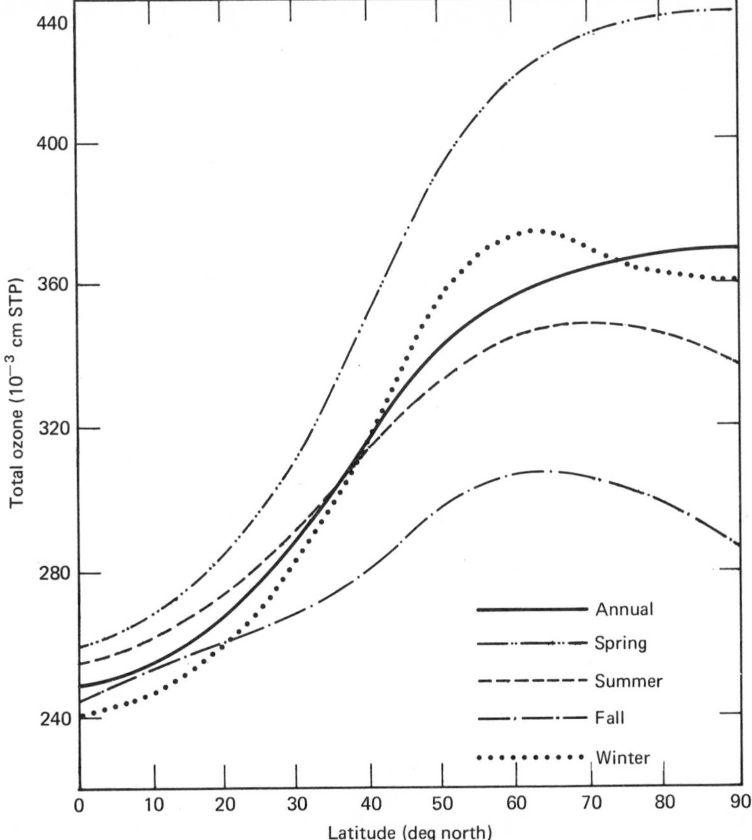

Figure 2.5 Variation in total ozone with latitude and season. From London et al. (1962) and Craig (1965).

where W_p, y_p, and h are adjusted constants, determined from Elterman's data to be, respectively, 0.160 atm-cm, 23.25 km, and 6.92 km. This function is plotted as the dashed line in the figure, where the fit is extremely close.

Because of its basic importance to meteorology, the vertical distribution of water vapor has been investigated thoroughly. Gutnick (1962a,b) constructed a mean annual profile for middle latitudes from sea level to 31 km by selective use of data from a diversity of measurement programs. Not all the data on stratospheric moisture were consistent, however, and Gutnick (1961) had previously reviewed the differing interpretations current at the time. Since then many balloon soundings have been made with highly sophisticated instrumentation, as described by Mastenbrook (1968, 1971) and Sissenwine et al. (1968a,b), for example. In particular, the

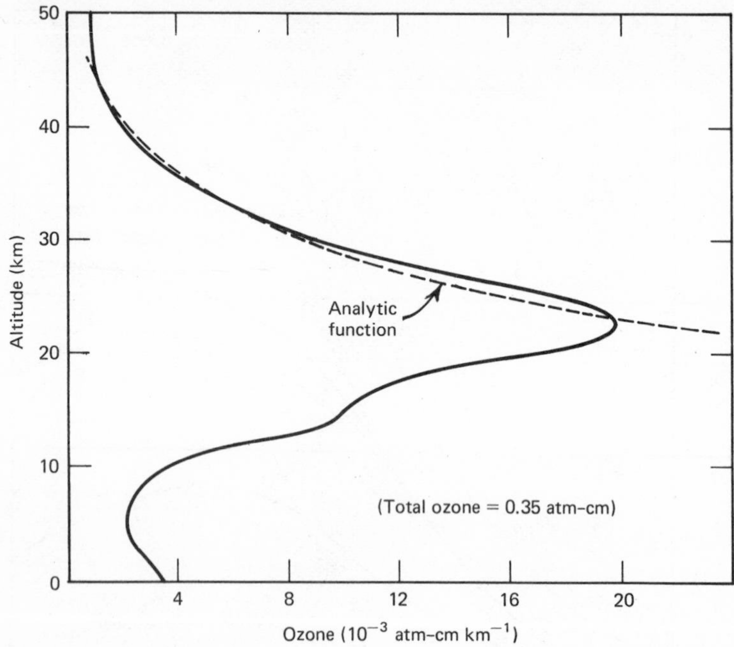

Figure 2.6 Representative vertical profile of ozone concentration. From Ellerman (1964).

series of frostpoint measurements made at Chico, California, by AFCRL over a 15-month period provided consistent data up to 27 km. From these and other investigations, Sissenwine et al. (1968b) established a representative moisture profile from the surface to 80 km. Some of the profile quantities to 50 km are listed in Appendix G, and the values of vapor density up to this altitude are plotted in Figure 2.7.

Several comments regarding this profile are noted. The relative humidity value at the surface is a weighted average for the United States; values at all other altitudes were computed from the frostpoint data. For frostpoint temperatures greater than 40 °C, saturation with respect to water was assumed for the reason given in Section 2.2.5. Mixing ratios were calculated with respect to the air densities of USSA-1962. The value of precipitable water at each altitude is the amount contained in a 2-km vertical column above that altitude. It is seen that the total amount is near 1.5 cm, or 0.6 in. The amount in centimeters of precipitable water per kilometer for a horizontal path at any of the altitudes listed is found by substituting that value of vapor density into (2-43). Linear interpolation within an altitude interval leads to little error in determining precipitable water for either vertical or horizontal paths. The entire profile represents a

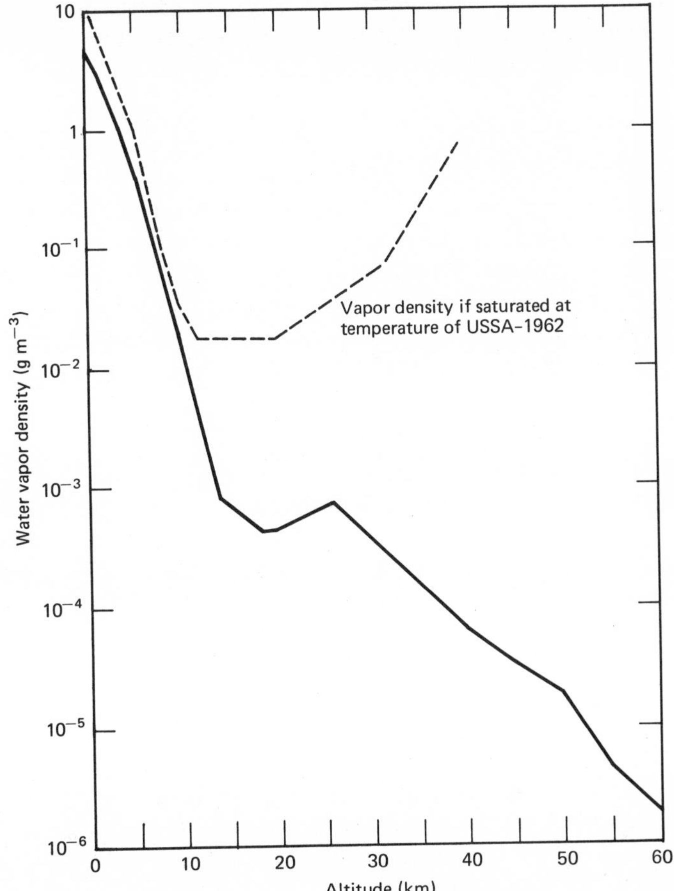

Figure 2.7 Representative vertical profie of water vapor density. From Sissenwine et al. (1968b).

very consistent set of data for the mean vertical distribution of water vapor and is recommended as an engineering reference.

Gringorten et al. (1966) have compiled an atlas showing mean seasonal distributions of water vapor over the northern hemisphere. The data are 4-year averages and cover ocean as well as continental areas. The atlas presents isopleths of mixing ratios at the surface and isopleths of dewpoint temperatures at the surface and at the 850-, 700-, and 400-mb pressure levels. Since the probabilities of high and low humidity do not follow a gaussian distribution, separate seasonal charts are provided for the percentile groups: 5, 25, 50, 75, and 95%. Seasonal moisture profiles, up to the 400-mb level, for any location may be constructed from the dewpoint values given. The tropospheric portion of the reference profile described above employs data from this atlas, which has great value in atmospheric optics as well as in meteorology.

2.4.3 Atmospheric Reduced Height

Despite the great vertical extent of the atmosphere, a column of air having a cross section of 1 in.^{2} and reaching from sea level to the uppermost limit weighs only about 14.7 lb. This effect can equally well be regarded as due to a column having a uniform density equal to that of sea-level air and a height sufficient to produce the observed force or pressure. Such a height, known as the *reduced height* and denoted here by Z'_0, is given by

$$Z'_0 = \frac{P_0}{\rho_0 g} \tag{2-69}$$

where P_0 and ρ_0 are the pressure and density of sea-level air. Since (2-69) is identical to (2-45), the concepts of constant-density atmosphere and reduced height are equivalent. Alternatively, Z'_0 can be defined in terms of sea-level temperature by substituting the value of P/ρ from (2-23) into (2-69), giving

$$Z'_0 = \frac{RT_0}{g} \tag{2-70}$$

This expression is identical to (2-54), which means that the sea-level value of the reduced height is equal to the scale height of an isothermal atmosphere when the temperatures of the two are equal.

The reduced height is useful when dealing with molecular scattering and absorption by the overhead atmosphere. The value of Z'_0 for sea-level air at standard conditions is obtained by substituting values from Appendix B into (2-69), which results in

$$Z'_0 = 7.997 \times 10^3 \text{ m} \approx 8.00 \text{ km} \tag{2-71}$$

The reduced height of each permanent gas constituent is equal to the product of Z'_0 and the volume ratio of that constituent. In this manner the atmosphere-centimeter values in Table 2.1 were determined for the vertical path from sea-level to the upper atmospheric limit. Thus *reduced height* is seen to be an overall application of the method described in connection with that table for specifying the amount of gas along a given path and leads directly to finding the total number of molecules along a vertical path. Since Loschmidt's number $N_L = 2.687 \times 10^{19}$ cm^{-3} for any gas under standard conditions, the number of molecules from sea level to the top of the atmosphere is 2.15×10^{25} cm^{-2}. Likewise, the total mass of any of the gases in such a column is equal to the product of the sea-level density of that gas and the reduced height.

The concept of reduced height is not restricted to sea-level location nor to air under standard conditions. Defined generally, reduced height Z' is the quotient of the total mass of air in a vertical column of a unit cross section, extending from the observer's location altitude to the top of the atmosphere, divided by the density of sea-level air. Reduced heights for several observer altitudes are listed in Table 2.8 and plotted in Figure 2.8,

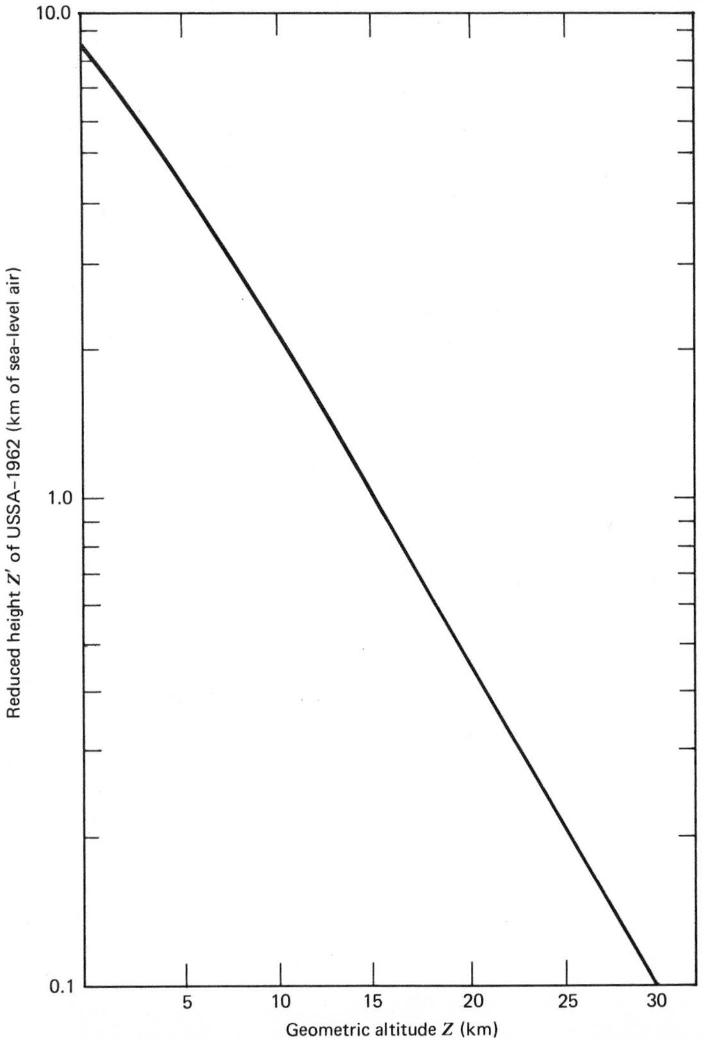

Figure 2.8 Atmospheric reduced height as a function of altitude.

TABLE 2.8 REDUCED HEIGHT AS FUNCTION OF OBSERVER ALTITUDE IN USSA-1962

Altitude (km)	Reduced height (km)	Altitude (km)	Reduced height (km)
0	8.44	8	2.97
1	7.48	9	2.56
2	6.62	10	2.21
3	5.84	11	1.89
4	5.06	13	1.38
5	4.50	15	1.01
6	3.93	20	0.46
7	3.42	30	0.10

Source: USSA (1962).

calculated by means of (2-69) from pressure and density values of USSA-1962. It is seen that the value of Z_0' is greater than the value given by (2-71). This is because the sea-level temperature of USSA-1962 is 15 °C, whereas (2-71) refers to 0 °C. The amount of reduced atmosphere for a vertical path between any two altitudes in Table 2.8 and Figure 2.8 is equal to the difference between the corresponding reduced heights. This amount can be considered as the equivalent amount of sea-level atmosphere between the two altitudes.

2.5 QUANTITATIVE TREATMENT OF OPTICAL PATHS

The previous sections have dealt with atmospheric constituents and their physical properties, emphasizing their interrelations and vertical distributions. With such structure and composition as a background, this section presents several methods of evaluating the physical parameters of optical paths. In a sense, all the preceding material has been directed toward this section. Here the geometric relationships between the path direction and the earth's surface and atmosphere are described first. Next, the concept of equivalent path in terms of a homogeneous or known atmosphere is developed for the flat earth approximation and then for the spherical earth. Finally, attention is directed to published tabulations of equivalent paths.

2.5.1 Earth Coordinates, Refraction, Horizon Geometry

Two coordinate systems are commonly used to specify the directions of optical paths. The first, the earth coordinate system, is anchored to the direction of the gravity vector at the observer's location and so employs the natural orientation induced in the observer by the force of gravity. The second, the celestial coordinate system, is tied to the earth's polar axis and equator extended to the celestial sphere. This allows the basic positional data of celestial objects to be defined in terms that are analogous to terrestrial latitude and longitude. This section describes the earth coordinate system, often called the *altitude-azimuth* system. The celestial coordinate system and conversion relationships between the two systems are described by Nassau (1948), Dunlap and Shufeldt (1969), and McNally (1975).

Earth coordinates provide a convenient framework for specifying directions and distances from an observer's particular location to points appearing on or above the earth's surface. The direction is specified by the *zenith* (or the *elevation*) and the *azimuth* angles measured from reference axes particularized for each location. Basic to this coordinate system is a horizontal plane and a reference direction (usually north) in this plane, as shown in Figure 2.9. By definition the horizontal plane is normal to the local vertical, or line to the zenith, which is established by the direction of the gravity vector at that location. Hence the horizontal plane is tangent to the earth's mean surface, which is assumed to be spherical, and is normal to the radius of the sphere. An unlimited number of such planes over the earth's surface can be conceived. The intersection of the horizontal plane with the meridian plane, which contains the local vertical and the earth's axis, establishes a reference north-south line.

As indicated in the figure, the zenith angle ζ and the elevation angle h are measured in the vertical plane that contains the line of sight and the local vertical. This vertical plane passes through the earth's center, hence its trace on the earth's spherical surface is a great circle (not shown in the figure). The elevation angle is positive when the line of sight is above the horizontal, and negative (depression angle) when it is below. In astronomy and celestial navigation, the elevation and zenith angles are called *altitude* and *zenith distance*. The azimuth *az* is measured in the horizontal plane and usually is reckoned clockwise from north. Any spatial direction is defined unambiguously by its elevation and azimuth angles.

We note briefly the matter of refraction and its effect on the direction of an optical path. Because air density varies with altitude, a density gradient exists *across*, that is, normal to the direction of, any path except the one directed to the zenith. Also, the refractive index of a gas increases with the density, so that a gradient of refractive index exists across the path. Since the velocity of a light wave varies inversely with the refractive index, the propagation path is bent in the direction of the gradient.

In the atmosphere this means that propagation paths are concave toward

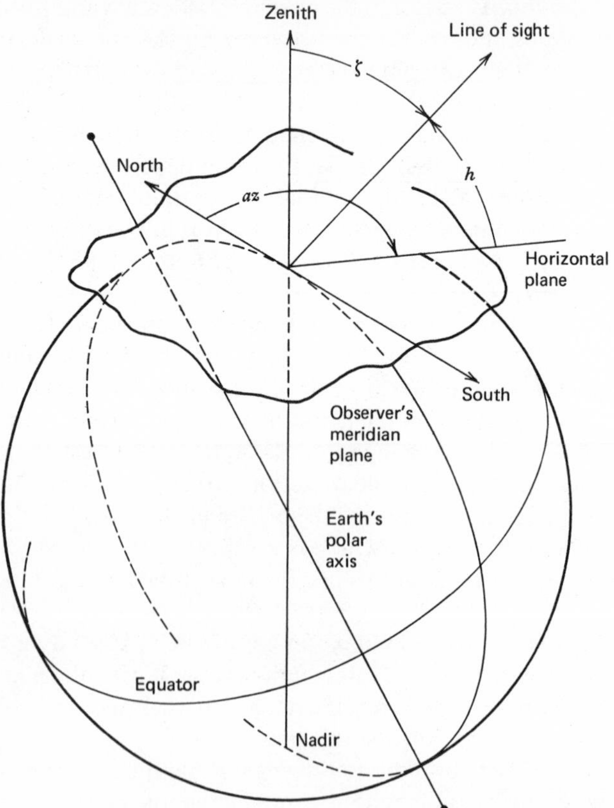

Figure 2.9 Geometry of earth coordinate system.

the earth, except when a density inversion is present. Because of this, the zenith angle of the optical line to a distant object is less than the zenith angle of the geometric line, as shown for the line of sight to a star in Figure 2.10. The difference between the two angles is the *refraction correction*, which varies from zero at the zenith to about 35 arc-min at the apparent or optical horizon. As a result, when the setting sun appears tangent to this horizon, it is already below the geometric horizon. Values of the refraction correction to be applied to *apparent elevation angles* are listed in Table 2.9. These corrections are applicable only to a line of sight having one end at sea-level and the other end at a point essentially beyond the atmosphere.

The line at which the earth's curved surface apparently meets the sky is called the horizon. It is a matter of easy observation that the distance to the horizon increases with the height of the observer's eye above the earth's surface. This can be strikingly demonstrated by slowly wading into still

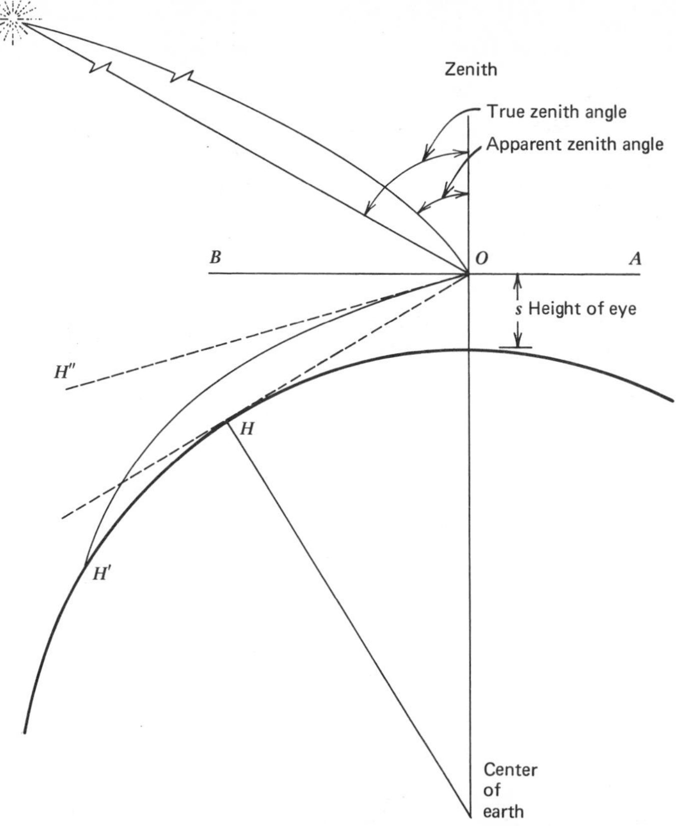

Figure 2.10 Horizon geometry and the effects of refraction.

water having an unbroken expanse of several miles until the eye is at the same level as the surface. Such a dependence of sighting distance on the height of eye is evidence in itself that the earth's surface is curved. As shown in Figure 2.10, two types of horizons are distinguished conceptually: the geometric horizon H and the optical or visible horizon H'. The geometric horizon is the locus of tangency points for all straight lines that graze the earth from a given observer station. Because of refraction, however, the observer sees a horizon apparently higher and more distant than the true geometric horizon. This displaced horizon, which appears in the direction OH'', is the only one that can be seen or sensed and is the *optical horizon.* Thus refraction enables the observer to see over and beyond the "hump" presented by the earth's surface at H.

The depression angle between the line of sight to the optical horizon and the horizontal plane, whose trace on Figure 2.10 is the line AOB, is the *dip*

TABLE 2.9 SEA-LEVEL REFRACTION CORRECTIONS FOR ELEVATION ANGLES OF ASTRONOMICAL LINES OF SIGHT*

Observed altitude (deg)	(min)	Refraction correction (min)	Observed altitude (deg)	Refraction correction (min)
0	00	34.5	11	4.9
	15	31.4	12	4.5
	30	28.7	13	4.1
	45	26.4	14	3.8
1	00	24.3	15	3.6
	15	22.5	16	3.3
	30	20.9	17	3.1
	45	19.5	18	2.9
2	00	18.3	19	2.8
	15	17.2	20	2.6
	30	16.1	25	2.1
	45	15.2	30	1.7
3	00	14.4	35	1.4
4	30	10.7	50	0.8
5	00	9.9	55	0.7
6	00	8.5	60	0.6
7	00	7.4	65	0.5
8	00	6.6	70	0.4
9	00	5.9	80	0.2
10	00	5.3	90	0.0

*The correction is to be subtracted from the observed (apparent) angle.

Source: NA (annual)

angle. The value of the dip angle depends on the height of the eye and the amount of refraction, and uncertainties in this latter quantity can be troublesome when good accuracy is wanted. The uncertainties arise from the fact that the line of sight to the horizon lies close to the earth's surface for much or all of its length. Depending on differences in air and land or water temperature, abnormal temperature gradients and even local temperature inversions may exist in the lowest few feet of the atmosphere. In the free atmosphere, a temperature gradient is equivalent to a density gradient. Thus the amount by which the line of sight is bent, per unit of length, often is quite variable and difficult to predict. For average refractive conditions the formula for the dip angle, represented by angle BOH'' in the figure, is given by Bowditch (1958) as

$$\text{Dip angle} = 0.97 s^{1/2} \quad \text{arc-min} \tag{2-72}$$

where s is the height of the eye in feet. Dip angles for several eye heights are listed in Table 2.10.

It is clear from Figure 2.10 that refraction increases the distance at which the horizon can be observed. The distance to the optical horizon for average values of refraction is given by Bowditch (1958) as

$$\text{Horizon distance} = 1.14 s^{1/2} \quad \text{nmi} \tag{2-73}$$

TABLE 2.10 HORIZON DIP ANGLES AND DISTANCES

Altitude (ft)	Dip angle (arc-min)	Horizon distance (nmi)	Altitude (1000 ft)	Dip angle (arc-min)	Horizon distance (nmi)
0	0.0	0.0	2	43	52
1	1.0	1.1	4	61	73
5	2.2	2.6	6	75	89
10	3.1	3.6	8	87	103
15	3.8	4.4	10	98	116
20	4.4	5.1	12	107	127
25	4.9	5.7	14	116	137
30	5.4	6.3	16	125	146
35	5.8	6.8	18	133	155
40	6.2	7.2	20	140	163
45	6.6	7.7	22	148	172
50	7.0	8.1	24	155	180
60	7.6	8.9	26	161	187
70	8.2	9.6	28	168	193
80	8.8	10.2	30	174	200
90	9.3	10.9	40	202	231
100	9.8	11.4	50	226	257
150	12.0	14.0	60	247	280
200	13.9	16.2	70	267	300
300	17.0	19.8	80	287	320
400	19.6	22.9	90	306	339
500	21.9	25.6	100	326	359
1000	31.0	36.2			

Source: At altitudes ≤1000 ft, dip angles from Nassau (1948) and horizon distances from Bowditch (1958). At altitudes ≥1000 ft, data from Sweer (1938).

where s is the height of the eye in feet. From this we have

$$\text{Horizon distance} = 3.839s^{1/2} \quad \text{km} \tag{2-74}$$

with s in meters. Horizon distances for several eye heights are listed in Table 2.10. As may be understood from the figure, no part of the earth's surface whose elevation angle is less than that of the optical horizon can be seen beyond this horizon. When the visibility is good, however, an elevated object such as the mast of a ship or a lighthouse can be seen beyond the curve of the earth. Such an elevated object has a horizon distance of its own, and the limiting distance at which it can be seen is equal to the sum of its horizon distance and the observer's horizon distance. During periods of abnormal refraction, often brought about by strong inversions, objects far beyond and below the horizon may become visible. This is a form of mirage known as *looming*.

In the foregoing discussion it is understood that the term *refraction* refers to an average steady-state, or nonturbulent, condition of the air. On any scale of measurement the air is never really motionless, and some degree of turbulence is associated with all fluid flow in nature. Even small fluctuations in air density, on microscales measured in centimeters of length, produce corresponding fluctuations in the refractive index. These give rise to varying small-scale deviations of the wavefronts of light waves and are manifested in the shimmer of distant landscapes on a hot day and in the scintillation of stars even when the air near the earth's surface appears quiet.

2.5.2 Geometry of Atmospheric Scattering

The outstanding features of an unclouded day sky, apart from the sun itself, are the radiance, color, and polarization of the scattered light that reaches an observer from a hemisphere of directions. Figure 2.11 illustrates the geometry applicable to such scattering. Although for illustration the sun is assumed to be the source of light, the relationships are valid for any extraterrestrial source. In the figure the observer is at point O, and the hemispherical outline represents one-half of the atmospheric envelope surrounding the earth. The direction of the sun is given by the principal ray OS, and all other solar rays, such as PP'', incident on the atmosphere are taken parallel to OS. The solar altitude and zenith distance are denoted by h_0 and ζ_0 and the solar azimuth by az_0.

The direction of observation to any point P in the sky is OP, whose elevation angle, zenith angle, and azimuth angle are, respectively, h, ζ, and az. As described in Section 1.4.1 and illustrated by Figure 1.7, angular scattering measurements are made in a plane of observation formed by the propagation direction of the incident light and the direction of observation.

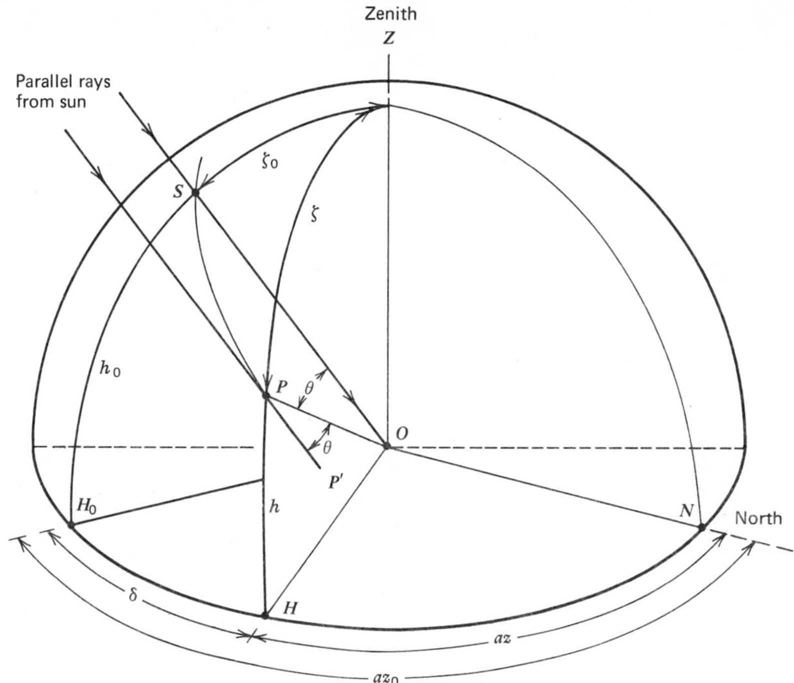

Figure 2.11 Geometry of atmospheric scattering of sunlight.

Further, the observation angle θ lies in this plane and is measured from the forward direction of the incident light. Hence in Figure 2.11 the plane of observation is the plane defined by *OS* and *OP*; only a portion of this plane is shown. The observation angle θ is the angle $P'PO$, which is equal to the angle *SOP*. The projection of *SOP* on the horizontal plane is the angle H_0OH, or δ, which is the bearing of the direction of observation from the sun. Angle θ can be found by solving the spherical triangle *SZP*, formed by the three great circle arcs, from the expression

$$\cos\theta = \cos\zeta_0 \cos\zeta + \sin\zeta_0 \sin\zeta \cos\delta \tag{2-75}$$

It can be visualized from Figure 2.11 that the line *OP* can be pointed to any direction in the sky by suitably choosing the observation angle θ and the orientation angle of the observation plane about the observer-sun line *OS*. All the atmosphere along any path, such as that illustrated by the line *OP*, is exposed to the solar rays that are considered here to be parallel. The solar flux is attenuated by total scattering of course as it penetrates to lower altitudes. Hence the sky radiance observed from point *O* is the aggregate result of (1) the angular scattering of attenuated solar flux from all the elementary volumes along OP to the outer limit of the atmosphere, and (2) the subsequent attenuation of this scattered flux as it proceeds toward point O.

Several additional quantities are noted in the following discussion. The vertical plane ZNO is the observer's meridian plane. The vertical plane ZSH_0O contains the direction of the sun and is frequently called the *sun's vertical.* This plane is used in polarization studies of skylight as a reference plane for specifying the perpendicular and parallel components of scattered light. Directions in this plane in the same quadrant as OS are said to be on the solar side; those on the opposite side (not shown in the figure) of OZ are on the antisolar side. Not shown but readily visualizable is the *almucantar,* which is a small circle of equal elevation angle on the hemisphere. Such a circle would be formed by rotating line OP about the zenith direction OZ. Sky radiance values sometimes are tabulated with respect to almucantars.

2.5.3 Equivalent Paths: Flat Earth Approximation

In general we can regard an optical path in the atmosphere as a *slant path,* with horizontal and vertical paths being special cases. The concept of *equivalent path* was developed by Duntley (1948a) from considerations of optical contrast reduction by scattering and was called the *optical slant range* (see also Middleton 1952). It is defined as the horizontal distance that has the same attenuation as that of the true (slant) path length. The concept permits evaluation of the amount of optical materials, that is, gas molecules and particles, and the resulting scattering along a slant path in terms of a horizontal path length at a specified reference altitude. This altitude is selected so that the concentration of the material, in amount per unit volume or per unit path length, is either known or can be estimated reasonably well. The equivalent path is denoted by the symbol $\bar{R}$. When the reference altitude is taken as sea level, which is often done, we have the equivalent sea-level path $\bar{R}_{sl}$.

When the zenith angle of the path is not greater than about 75 deg, it is usually permissible to assume that the earth is flat and that the atmospheric strata are parallel to the earth's surface. In this simplification it is also assumed that the effects of refraction can be ignored. These assumptions result in the plane-parallel approximation illustrated in Figure 2.12 for several general cases. In Figure 2.12*a*, the lower end of the path is at sea level, while the upper end either extends beyond the atmosphere toward S or terminates at P within the atmosphere. In Figure 2.12*b*, the lower end is above sea level but not necessarily above the ground, while the upper end either extends beyond the atmosphere toward S' or terminates at P'. The

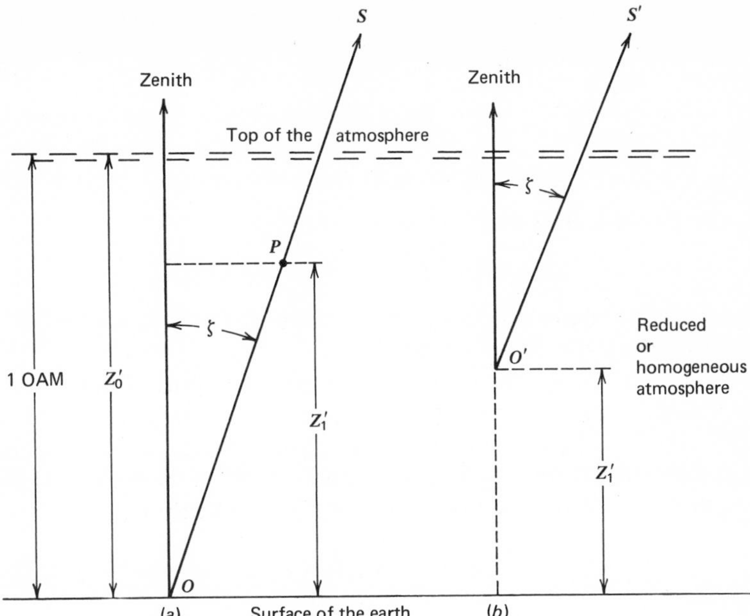

Figure 2.12 Geometry of slant paths in the plane-parallel approximation of the reduced atmosphere.

path having one end beyond the atmosphere is called an *astronomical* path, and the path having both ends within the atmosphere is a *terrestrial* path.

The amount of atmosphere between any two altitude points is a minimum when the path is vertical, as can be seen from Figure 2.12. The amount along the vertical path from sea level to the top of the atmosphere is termed one *optical air mass* (OAM), or just *air mass.* This use of the term must not be confused with its meteorological sense. There it means a large parcel of air, usually many miles in horizontal extent, which is different from the neighboring air with respect to temperature, pressure, and humidity. It is evident that the "total amount" of atmosphere, or of any constituent, in one OAM is equal to its reduced height for sea level. The concept of OAM applies strictly only to an astronomical path having its lower end at sea level.

For a sea-level station, the reduced height Z'_0 has the value 8.44 km when the parameters are those of USSA-1962, as shown in Table 2.8 and Figure 2.8. Consequently, this is the total amount of sea-level atmosphere in one OAM, and it is the value of the equivalent sea-level path $\bar{R}_{sl}$. As explained in Section 2.4.3, the equivalent amount of sea-level atmosphere in a vertical path between any two altitudes is equal to the difference between their reduced heights. In symbols, and referring to Figure 2.12,

$$\bar{R}_{sl} \text{ (vertical)} = Z_0' - Z_1' \tag{2-76}$$

where Z_1' has the value 0 at the top of the atmosphere. For a slant path we have

$$\bar{R}_{sl} \text{ (slant)} = (Z_0' - Z_1') \sec \zeta \tag{2-77}$$

When the lower end of the path is at sea level,

$$\text{Number of OAM} = 1 \times \sec \zeta \tag{2-78}$$

for the flat earth approximation. The concept of OAM is applied to the spherical earth in the following section.

We next develop equivalent-path expressions in terms of density scale height, assumed to be constant. This condition exists in all isothermal regions, and it is approximated in the troposphere, as discussed in connection with Figure 2.4. The total number N_t of molecules along a slant path of unit cross section and length R evidently is

$$N_t = \int_0^R N(R)\, dR \tag{2-79}$$

where $N(R)$ is the number density as a function of position along the path. Because of the flat earth approximation, altitude is given by

$$Z = R \cos \zeta$$

From this relationship and (2-55), it is clear that N is a function of R along the slant path according to

$$N(R) = N_0 \exp\left(-\frac{R \cos \zeta}{H_\rho}\right) \tag{2-80}$$

where N_0 is the number density at sea level (or a selected reference altitude), and H_ρ is the density scale height. For a slant path between two altitudes Z_1 and Z_2 we can combine (2-79) and (2-80) and write

$$N_t = N_0 \int_{R_1}^{R_2} \exp\left(-\frac{R \cos \zeta}{H_\rho}\right) \tag{2-81}$$

By definition, the equivalent path lies in a conceptually homogeneous atmosphere, and the total number of molecules in a length $\bar{R}$ is

$$N_t = N_0 \bar{R} \tag{2-82}$$

Substituting this into (2-81) and integrating, we find

$$\bar{R}_{sl} = H_\rho \sec \zeta \left[\exp\left(-\frac{R_1 \cos \zeta}{H_\rho}\right) - \exp\left(-\frac{R_2 \cos \zeta}{H_\rho}\right)\right] \tag{2-83}$$

This expression is applicable to a slant path between altitudes Z_1 and Z_2 at distances R_1 and R_2 from the sea-level end of the path extended downward as needed to reach that reference datum.

We consider three limiting cases, assuming for each one that the associated altitude region is isothermal so that the scale height has a constant value. When the lower end of the path is at sea level, both Z_1 and R_1 are equal to 0, and (2-83) becomes

$$\bar{R}_{sl} = H_\rho \sec \zeta \left[1 - \exp\left(-\frac{R_2 \cos \zeta}{H_\rho}\right)\right] \tag{2-84}$$

When the lower end lies at altitude Z_1, at distance R_1 from the sea-level end of the extension, and the upper end is at the top of the atmosphere, with Z_2 and R_2 effectively equal to infinity, (2-83) becomes

$$\bar{R}_{sl} = H_\rho \sec \zeta \exp\left(-\frac{R_1 \cos \zeta}{H_\rho}\right) \tag{2-85}$$

When the path encompasses the entire height of the atmosphere, so that Z_1 is equal to 0 and Z_2 is equal to infinity, (2-83) reduces to

$$\bar{R}_{sl} = H_\rho \sec \zeta \tag{2-86}$$

For this last condition, (2-86) is equivalent to (2-77) where, correspondingly, Z_1' becomes equal to 0. The basis of the equivalence may be seen by comparing (2-54) and (2-70).

The above relationships can be applied to several atmospheric constituents. For permanent gases having $H_\rho \approx 9.3$ km in the troposphere, as in Figure 2.4, we can write (2-84) as

$$\bar{R}_{sl}\,(\text{air}) = 9.3 \sec \zeta\,[1 - \exp(-0.11 R_2 \cos \zeta)] \tag{2-87}$$

to find the equivalent length of a path whose lower end is at sea level. Values calculated from (2-87) for several zenith angles are plotted in Figure 2.13. Such values can be used as the multiplier of the total scattering coefficient β_m, in expressions such as (1-35), to determine the attenuation due to molecular scattering along the path. When $\bar{R}_{sl}$ and N_0 employ the same unit of length, their product is the total number of molecules along the path. Likewise, the product of $\bar{R}_{sl}$ and ρ_0 is the total mass of gas. For any gas species listed in Table 2.1, multiplication of its volume ratio by $\bar{R}_{sl}$ (in centimeters) gives the amount of that gas (in atmosphere-centimeters) along the path.

2.5.4 Equivalent Paths: Spherical Earth

As the zenith angle ζ increases beyond 75 deg, the flat earth approximation becomes unrealistic, as may be appreciated from Figure 2.14. In

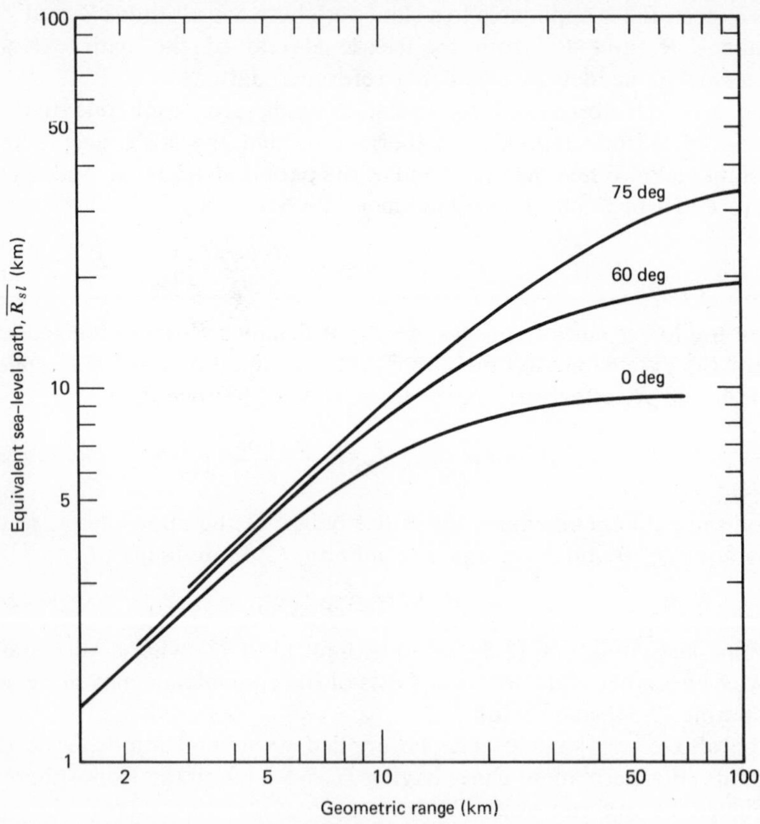

Figure 2.13 Equivalent sea-level path lengths in the plane-parallel approximation of the exponential atmosphere for the indicated zenith angles.

addition, the effects of refraction are significant at large values of ζ. The resulting errors in the simple approximation are inadmissably large, and it is necessary to take into account the sphericity of the earth and its atmospheric envelope. These factors are included in the formula by Laplace (see Kondratyev 1969) which he developed for an isothermal atmosphere. The formula is

$$\text{OAM} = \frac{\text{refraction in arc-seconds}}{58.36\ \text{arc-sec} \times \sin \zeta} \tag{2-88}$$

Here the quantity OAM is the number of optical air masses along an astronomical path lying at angle ζ and having its lower end at sea level. The numerator represents the refraction associated with the angle ζ. Equation (2-88) was refined by Bemporad (1907) who took into account the

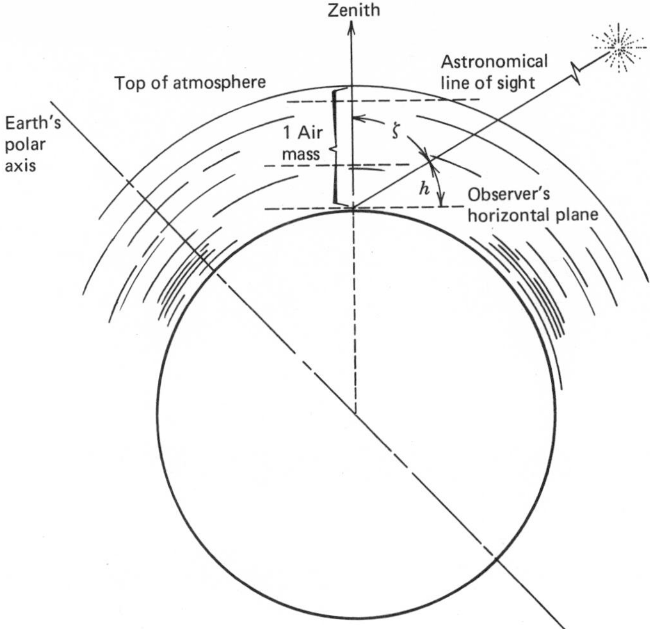

Figure 2.14 Geometry of the optical air mass.

environmental lapse rate of 6.5 °C km^{-1} discussed in Section 2.2.1. Values of OAM from his work are listed in Table 2.11. In earth-atmosphere geometry a suitable value from the table can be used instead of sec ζ at any value of ζ. The extent to which the flat earth approximation is in error as ζ becomes large can be seen by comparing the values in the table with corresponding values of sec ζ. When the lower end of the path is not at sea level, or when the pressure is other than standard, the correction factor at the foot of the table should be used. At high altitudes the Chapman function described later in this section should be used in lieu of the air mass.

The equivalent sea-level length of a slant astronomical path at any value of ζ is readily found by combining the reduced height with a tabulated value of OAM. According to the preceding paragraph, the equivalent length is given by

$$\bar{R}_{sl} = Z_0' \times \text{OAM} = 8.44 \times \text{OAM} \qquad \text{km} \tag{2-89}$$

where 8.44 km is the value of Z_0' at sea level in USSA-1962. For example, when the sun's apparent zenith angle is 89 deg, the equivalent path for the solar rays is found by substitution from Table 2.11 into (2-89), giving

$$\bar{R}_{sl} = 228 \text{ km} \approx 123 \text{ nmi}$$

TABLE 2.11 VALUES OF THE OPTICAL AIR MASS*

Sun's zenith distance	Optical air mass									
	0°	1°	2°	3°	4°	5°	6°	7°	8°	9°
0°	1.00	—	—	—	—	—	—	—	—	—
10	1.02	—	—	—	—	1.04	—	—	—	—
20	1.06	1.07	1.08	1.09	1.09	1.10	1.11	1.12	1.13	1.14
30	1.15	1.17	1.18	1.19	1.20	1.22	1.23	1.25	1.27	1.28
40	1.30	1.32	1.34	1.37	1.39	1.41	1.44	1.46	1.49	1.52
50	1.55	1.59	1.62	1.66	1.70	1.74	1.78	1.83	1.88	1.94
60	2.00	2.06	2.12	2.19	2.27	2.36	2.45	2.55	2.65	2.77
70	2.90	3.05	3.21	3.39	3.59	3.82	4.07	4.37	4.72	5.12
80	5.60	6.18	6.88	7.77	8.90	10.39	12.44	15.36	19.79	26.96

*If the pressure P at the ground level is different from the standard pressure $P_0 = 1013.3$ mb, the above values are to be multiplied by P/P_0.
Source: List (1966).

of sea-level air at 15 °C and 1013.3 mb. All things considered, this is a long sighting path. The resulting attenuation, even during clear weather, suggests why the rising or setting sun can be viewed by the unprotected eye, and why stars are seldom visible near the horizon. A numerical example is revealing. We note from Table 1.8 that the scattering coefficient β_m for pure air has the value $0.014\ \mathrm{km}^{-1}$. Substituting this into (1-49), along with $\bar{R}_{sl} = 228$ km as determined above, we find that the transmittance is

$$\tau = 4.1 \times 10^{-2}$$

for molecular scattering only. The transmittance due to scattering by haze is frequently several orders of magnitude less than this.

On occasion it is necessary to evaluate terrestrial paths at high altitudes and at zenith angles greater than about 75 deg. In general, the use of air mass values for such cases becomes unreliable, and recourse to other techniques is advisable. Graphical methods have been devised by Carpenter et al. (1957) and Altshuler (1961); the method of Carpenter is reviewed by Plass and Yates (1965). Graphical methods, however, may be tedious to use, and they lack the nicety of analytic solutions whose computational results can be tabulated for direct use at any time.

Such a solution is provided by the function Chapman (1931) developed in his studies of the absorption of solar radiant flux by atmospheric gases at high altitudes. He assumed an isothermal atmosphere of arbitrary temperature, a constant value of density scale height according to (2-54), and ignored the effects of atmospheric refraction on path length. The function, values of which have been tabulated, directly replaces sec ζ at all values of ζ in the calculation of equivalent paths. Despite this utility, surprisingly few references to the function appear in the literature of atmospheric optics other than treatises on upper atmospheric physics. Green and Griggs (1963) employ it in a very useful method for calculating infrared transmission, but most workers seem unaware of it. Craig (1965) gives a succinct derivation of the function; here we are concerned with its application.

The geometry of the Chapman function is illustrated by Figure 2.15. The original problem was to compute the absorption of solar radiant flux at point P; hence the attenuation characteristics of the path PP' were to be determined. The given altitude of P is Z_P. Denoting the earth's radius by r_e, the distance from P to the earth's center is $Z_P + r_e$. The zenith angle of PP' at P is ζ, and at P' is ψ. As point P' moves in from an infinite distance to P, angle ψ varies continuously from 0 to ζ. Angle ψ therefore can be regarded as a running variable which specifies the position of P' on the path, and therefore the altitude Z. A parameter χ is defined such that

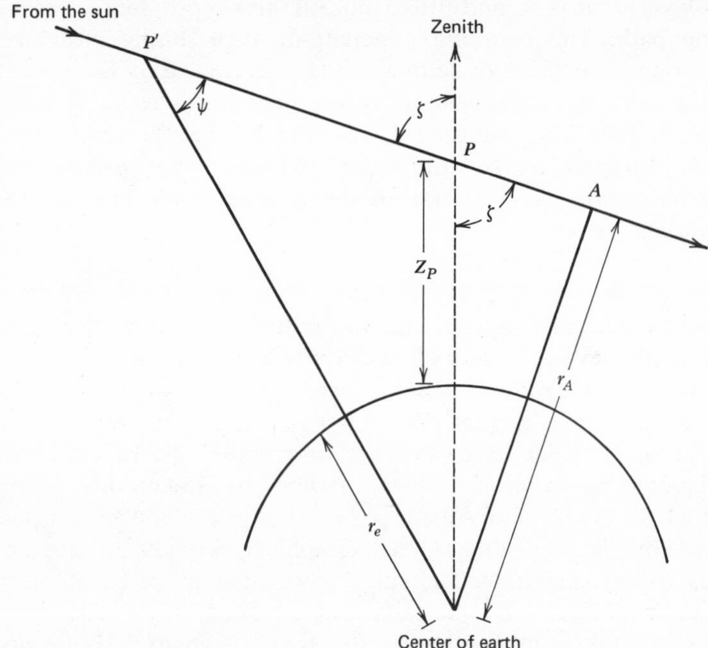

Figure 2.15 Geometry for the Chapman function. The altitude of P above the earth's surface relative to the earth's radius has been exaggerated somewhat for illustrative purposes. From Craig (1965).

$$\chi = \frac{Z_P + r_e}{H_\rho} \tag{2-90}$$

where H_ρ is the density scale height, as in previous discussions. Equation (2-90) thus defines the distance from P to the earth's center in units of scale height.

The Chapman function Ch (χ, ζ) is defined by

$$\text{Ch}\,(\chi, \zeta) = \chi \sin \zeta \int_0^\zeta \exp\,(\chi - \chi \sin \zeta \csc \psi) \csc^2 \psi \, d\psi \tag{2-91}$$

in which all the quantities have been defined in the previous paragraph. In the original solution, the solar irradiance at point P, or altitude Z_P, was found to be

$$E_Z = E_0 \exp\,[-\kappa H_\rho \rho_Z \,\text{Ch}\,(\chi, \zeta) \tag{2-92}$$

where E_0 is the irradiance outside the atmosphere, and κ is the *mass* absorption coefficient of the particular gas being considered. From a study of (2-92) and Figure 2.15 it can be appreciated that the product

$$H_\rho \, \mathrm{Ch}\,(\chi, \zeta)$$

is the equivalent length of path from the outer edge of the atmosphere to point P *in terms of* the density ρ_Z at the corresponding altitude Z_P. Multiplication of this equivalent length by ρ_Z, as in (2-92), gives the total mass of gas in the path, and the multiplication by κ gives the optical thickness of the path in the manner of (1-51).

A study of these relationships, along with the role of sec ζ in slant path expressions such as (2-77) for the flat earth approximation, leads to an important conclusion pointed out by Craig (1965). Such expressions, valid as written for $\zeta < 75$ deg, can be made valid for any value of ζ by substituting Ch(χ, ζ) for sec ζ. This is because the function has been derived explicitly for a spherical earth, a concentric atmospheric en-

TABLE 2.12 VALUES OF THE CHAPMAN FUNCTION

ζ	$\chi = 600$	$\chi = 700$	$\chi = 800$	$\chi = 900$	$\chi = 1000$
70	2.889	2.893	2.897	2.900	2.902
71	3.030	3.036	3.040	3.044	3.046
72	3.188	3.194	3.199	3.203	3.206
73	3.363	3.370	3.376	3.381	3.385
74	3.559	3.568	3.575	3.581	3.586
75	3.780	3.791	3.800	3.807	3.812
76	4.031	4.045	4.055	4.064	4.070
77	4.318	4.335	4.348	4.358	4.366
78	4.650	4.671	4.687	4.700	4.710
79	5.036	5.062	5.083	5.099	5.112
80	5.491	5.525	5.551	5.572	5.590
81	6.034	6.079	6.113	6.141	6.164
82	6.692	6.752	6.799	6.836	6.867
83	7.503	7.585	7.650	7.702	7.745
84	8.523	8.639	8.732	8.807	8.870
85	9.838	10.008	10.144	10.257	10.352
86	11.580	11.839	12.051	12.228	12.378
87	13.970	14.384	14.730	15.024	15.277
88	17.388	18.087	18.686	19.209	19.670
89	22.523	23.784	24.905	25.914	26.830
90	30.719	33.177	35.466	37.615	39.648
91	44.789	50.030	55.211	60.362	65.505
92	71.132	83.522	96.753	—	—

Source: Wilkes (1954).

velope, and a station (such as point P in Figure 2.15) at an arbitrary altitude Z_P.

Values of Ch (χ, ζ) have been tabulated by Wilkes (1954) for wide ranges of the two variables. The earth's radius r_e has the value 6.371×10^3 km, and suitable density scale heights for the troposphere and stratosphere are in the approximate range 6 to 10 km, although they are much greater for the upper atmosphere. Hence the values of χ, defined by (2-90), most applicable to the lower atmosphere lie between 600 and 1000. Values of Ch (χ, ζ) are listed in Table 2.12 for $\chi = 600$ (100) 1000 and $\zeta = 70$ (1) 92 deg. The values for $\chi = 800$ probably have the greatest applicability. The limiting value of ζ (beyond 90 deg) at which the path PP' touches the horizon depends on the altitude of P, as may be seen from Figure 2.15. When the path line lies at or near such a grazing angle, the effect of refraction in lengthening the path may be significant.

The utility of the Chapman function lies in the freedom of selecting an altitude, scale height, and zenith angle that fit a variety of problems. The function has unique importance in studies of horizons from high altitudes,

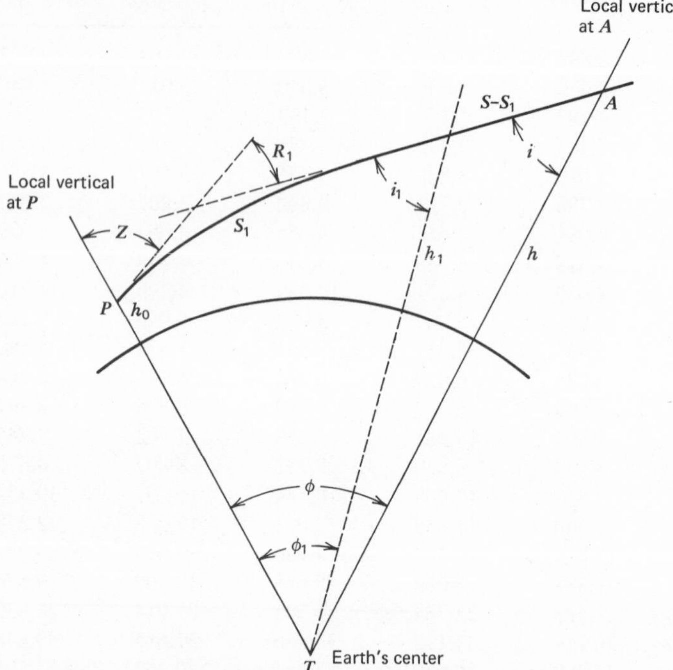

Figure 2.16 Geometry for the light trajectory tabulations of Link and Neuzil (1969). The path PA is the trajectory of interest. From Link and Neuzil (1969) *Tables of Light Trajectories in the Terrestrial Atmosphere.* Hermann, Paris.

where zenith angles exceed 90 deg and atmospheric path lengths can reach maximum possible values. As can be visualized from the figure, a horizon observed from a very high altitude involves a path that traverses the atmospheric envelope *twice.* Swider (1964) extended the function to take account of scale heights that exhibit gradients instead of remaining constant as assumed here. Green and Martin (1966) generalized the function for application to absorbing species such as ozone, which is not distributed with altitude according to a single exponential.

Very often the labor involved in calculating the parameters of an optical path can be eased by employing tabulated values of certain elements. Link and Neuzil (1969) have compiled a set of such tables sufficiently extensive that anyone concerned with optical slant paths should have access to a copy. These tables are a modernization of similar but less extensive tables compiled by Link and Sekera (1940). The present tables, which number more than 600, define the light trajectory or path for altitudes between 0 and 50 km, and for zenith angles from 75 to 90 deg, in terms of the following elements which are illustrated in Figure 2.16, where the notation is that of Link and Neuzil.

1. The geocentric angle ϕ between the radius vectors of points P and A.
2. *The refraction angle* R, which is the angle of the tangents at P and A.
3. The angle i between the tangent at point A and its radius vector TA or the local vertical at A.
4. The length s of the trajectory between P and A.
5. The air mass M, or the equivalent path between P and A, expressed in kilometers of homogeneous air having the density appropriate to the altitude at the lower end of the path. (Note that this use of the term *air mass* differs from that in this book.)

The tables cover 14 climatic situations and atmospheric structures for various north latitudes and seasons, which correspond to those of the USSA-1962 and USSAS-1966.

3

Particles in the Atmosphere

Atmospheric air is never free from particles having a variety of origins, sizes, and chemical compositions. Remaining suspended for varying and indefinite periods of time, they are transported by vertical and horizontal currents, frequently to great distances. These particles, all of which are good scatterers, are the subject of this chapter. We review first the dust grains that are the rather passive constituents of haze. The hygroscopic particles of haze, whose size depends on the relative humidity, are then considered. Size distribution parameters are explained, and two distribution functions widely used in haze investigations are discussed. The altitude dependence of particle concentration is explained next. Typical size distributions of fog droplets and the relationship to liquid water content receive detailed attention. In conclusion, representative size distributions of cloud droplets are presented.

3.1 ORIGINS AND PROPERTIES OF HAZE AEROSOLS

An aerosol is a dispersed system of small particles suspended in a gas; the term *haze aerosol* emphasizes the particle nature of haze. From the optical standpoint, haze is a condition wherein the scattering property of the atmosphere is greater than that attributable to the gas molecules but is less than that of fog. Haze scattering imparts a distinctive gray hue to the sky, which would otherwise appear a deep blue, and is usually the determining factor of visibility. The atmosphere is never free from haze, and the particles constituting the haze aerosol occur in greatest concentration near the earth's surface. On the average the concentration decreases rapidly with altitude, but significant numbers of particles are found well into the stratosphere and indeed throughout the atmosphere. The particles tell a story of diverse origin and global travel. Cosmic dust, volcanic ash, foliage exudations, combustion products, bits of sea salt—

all these are found to varying degrees in haze. For our purposes the size range of haze particles extends from about 0.01 to 10 μm.

We note in advance that certain types of particles, such as ordinary dust grains, are nonhygroscopic, while other types, such as bits of salt, are highly hygroscopic. The first type are relatively passive components of the aerosol although, if their surfaces are wettable, they accumulate a film of water under some conditions. The hygroscopic type act as centers or *nuclei* for condensation of water vapor, and these are the really active members. Their size depends on the ambient relative humidity and available supply of water vapor, and on the extent to which they coagulate by collisions. Because of this potential for particle growth, haze leads to consequences extending far beyond its immediate optical effects. In a broad sense, clouds and precipitation are the meteorological results of nuclei that began life as members of a haze aerosol. A haze particle from a random sampling can be a dry solid, a solid coated with water, or a water droplet carrying its hygroscopic essence in solution. The origin, composition, and growth behavior of condensation nuclei are discussed thoroughly by Houghton (1951), Junge (1951, 1958, 1963), Butcher and Charlson (1972) and Hidy (1972).

The literature on the origin and properties of haze is voluminous. In this chapter we treat the factors related to its functioning as an active optical medium spread widely throughout the atmosphere. This section deals with (1) the origin and nature of dust particles and condensation nuclei, (2) the growth behavior of nuclei as the relative humidity changes, and (3) the dynamic processes that remove small particles from the atmosphere. Readers wishing to go beyond this material should consult the studies cited herein. In addition to the authors listed in the preceding paragraph, much information on several aspects of haze (and smog) can be found in Green and Lane (1967), Cadle (1965, 1966, 1975), the review by Germogenova (1970), the proceedings edited by Blifford (1971), and the compilation edited by Hidy (1972).

3.1.1 Dust Particles and Sources

Large amounts of dust are continually placed in the atmosphere by several natural processes. Consider first the interplanetary debris swept up by the earth as it moves through 3×10^{16} m^3 of space each year. Petterssen (1960) estimated that approximately 10^7 tons of this material may be collected in a single year. This accretion, however, increases the earth's mass by a factor of only 4×10^{-6} in a billion years. The total debris covers an astonishing range of sizes, from objects of many tons to particles of 10^{-16} g. Many aspects of this subject are covered by Cadle (1966) and Hawkins (1964, 1967). Data on the size distribution of interplanetary particles are given also by Dubin and McCracken (1962) and Southworth (1964). Such particles, distributed throughout the immense volume of the solar system, are effective in scattering sunlight and are partially responsible for zodiacal light.

All interplanetary objects in the earth's vicinity are known as *meteoroids.* The term *micrometeoroid* refers to particles smaller than a few hundred micrometers. Meteroids become *meteors* when they enter the atmosphere. Entry speeds are quite high, typically 10 to 20 km sec^{-1}, so that all meteors quickly become incandescent as a result of aerodynamic heating. Their courses are marked by streaks of light and trails of ionization, allowing both optical and radar measurements of their speed and trajectory. These observing techniques and resulting data are described by Whipple (1951, 1952), Hawkins (1962), Millman (1962), and Eshelman and Gallagher (1962). Small micrometeors are decelerated in the ionosphere without being destroyed, and samples are obtained with sounding rockets, as described by Hemenway and Soberman (1962). Meteors having masses between 10^{-12} and 10^{2} g are destroyed in the upper atmosphere, but leave residues. Meteors having still greater initial masses are able to reach the earth's surface, although undergoing vaporization and disintegration along the way, and are called *meteorites.* Thus the atmospheric dust load at all altitudes is continuously augmented by micrometeors that are not destroyed and by condensed vapors and other residues of disintegrated meteors.

All volcanic activities inject huge quantities of fine ash into the atmosphere. The famous eruption of Krakatoa in 1883 changed an island into a sea cavity, raised clouds of ash to 90,000 ft, and turned day into night at Batavia 100 mi away. Within a few weeks the smaller particles, remaining high in the stratosphere, had spread across a broad belt of latitudes and completely around the globe. For several years this impalpable dust produced spectacular lighting effects at sunrise and sunset. In 1963 the volcano Gunung Agung in Bali erupted with similar optical aftereffects. These are described by Cadle (1966) and Volz (1969b; 1970a,b). According to *Bull. Am. Meteorol. Soc.* (1971) and Ellis and Pueschel (1971), as a result of the Agung eruption the average *insolation* (contraction of *incoming solar radiation*) decreased 1.5% at the high-altitude observatory at Mauna Loa, Hawaii, and 5.3% at the South Pole. Seven years were required for the insolation at Mauna Loa to return to the average preeruption level.

More recently, new stratospheric dust layers at altitudes of about 16 and 20 km, investigated by McCormick and Fuller (1975) with ground-based lidar, have been attributed to eruptions of Fuego Volcano in Guatemala starting about mid-October 1974. Fortunately, such large-scale eruptions as those noted above are rare, but continual volcanic activity in several regions steadily injects dust and gases into the atmosphere. Newell (1964, 1971) and Corby (1970) describe the great circulation systems that transport such materials to all parts of the globe.

Surface winds of more than a few miles per hour are effective in loosening, lifting, and transporting the soil dust that is characteristic of all lands during dry seasons. The enormous deposits of loess, which is a soft rocklike material composed of wind-deposited, compacted dust, in many parts of the world are indications of the great winds that blew in early times. In modern times the winds continue to create and transport large quantities of surface dust. Even a moderate breeze across a dry, tilled field produces a cloud of dust. The dust storms that occasionally occur in the plains and desert regions of North America are extreme examples of such lifting and transporting. A strong wind, charged with grains of sand, scours new grains into being from every soil and rock surface in its path. The fantastic sandstone formations in southern Utah, for example, attest to the sculpturing powers of wind and water. In less spectacular fashion, dust-transporting processes are steadily at work in many parts of the world. To cite but one instance, Prospero (1968) describes dust studies at Barbados, in the Lesser Antilles, where much of the dust has been carried from the Sahara Desert by the prevailing trade winds. Optical effects attributable to this dust have been measured by Volz (1970a,c).

Many industrial operations and most construction activities are prolific sources of dust particles. Iron and steel mills, ore smelters, and cement plants, without which our technologically based society could not exist, inject many forms of siliceous and mineral dusts into the atmosphere. Such particles are mostly nonhygroscopic and cover a broad range of sizes. The larger ones settle from the stack plumes within minutes, coating the countryside for miles around. Those smaller than about 1 μm move with the winds and mingle with the hygroscopic particles described in the following section. They all remain suspended until washout by rain or snow, or until agglomeration with other particles, followed by washout, finally carries them to earth.

3.1.2 Hygroscopic Particles and Sources

In contrast to the aridity implied by a dust cloud, the green world of plants and trees creates its own organic type of aerosol, described by Went (1955) as summer *heat haze.* Investigations by Rasmussen and Went (1965) have disclosed that aromatic hydrocarbon vapors exuded by foliage, and known generally as terpenes, are responsible. Under the influence of sunlight and ozone the terpenes oxidize and condense or nucleate into minute droplets of complex tars and resins. The reactions probably are similar to those occurring in some types of photochemical smog. It appears that these tiny particles, whose average radius is estimated to be about 0.15 μm, are somewhat hygroscopic; hence they act as condensation nuclei. These organic aerosols perpetually overlie the tropical rain forests of Africa and South America, creating a dense haze to altitudes of a mile or more. So prevalent is this condition that aerial surveys of these regions commonly employ infrared photography to provide adequate detail. Large forested regions elsewhere usually are crowned by these unique aerial canopies, in less extreme form, which often may be seen from a distance. The blue-gray veil over the timbered slopes of the Great Smoky Mountains is a familiar example.

All combustion processes create many small particles, both directly and through subsequent reactions with other materials in the atmosphere. The burning of coal, for example, produces fine mineral residues known as fly ash and unburned bits of carbon and tar known as soot, as well as carbon monoxide, carbon dioxide, and sulfur dioxide. All these materials except fly ash are also produced by the burning of petroleum derivatives, whether in furnaces or in internal combustion engines. The latter device is also a source of several nitrogen oxides formed by recombinations of atomic nitrogen and oxygen produced by molecular dissociation at high combustion temperatures. Several of the gases noted above undergo photochemical reactions, yielding additional types of gases and small hygroscopic particles. The result is a polluted atmosphere which bears a visible sign. The air above and around industrial and metropolitan regions is marked by a pall of dark haze which can be seen from aircraft 100 mi or more away. Some recent measurements of an urban pall are reported by Breeding et al. (1975).

Not all smoke particles are due to industry. According to Cadle (1966), in the country an ordinary grass fire covering 1 acre yields an estimated 2×10^{22} particles ranging from inorganic ash to complex hydrocarbons. If all these were uniformly distributed to a height of 10,000 ft over the acre, the resulting concentration would be about 2×10^9 cm^{-3}. All such particles are extremely small, usually having a radius of less than 0.1 μm. The number of particles from a forest fire is best handled by imagination. Identifiable smokes from large forest fires sometimes travel great distances, as in 1950 from western Canada to Scotland, where a blue sun was observed through the smoke. The Scottish moon also was blue at that time, illustrating that an event of classic improbability can indeed happen. These rare phenomena were reported by Wilson (1952) and Penndorf (1953).

Several types of particles are produced by photochemical reactions and nucleations between combustion gases and atmospheric trace gases, as in the following two examples. Sulfur dioxide, resulting from the combustion of coal and petroleum derivatives and also injected into the atmosphere by volcanos and fumaroles, is readily oxidized to sulfur trioxide by ozone and ultraviolet light. In the presence of atmospheric moisture, minute droplets of sulfuric acid are then formed. The hygroscopic as well as the corrosive properties of this acid are well known. Another series of reactions, described by Manson et al. (1961) and Mason (1961), produces ammonium sulfate in the following manner. Nitrogen dioxide is formed by lightning discharges and is also a trace component of automobile exhaust gases. Ammonia is produced by decaying organic substances and by the widespread use of nitrogenous fertilizers. Hydrogen sulfide is produced in all littoral areas by the decomposition of stranded sea organisms. The presence of other sulfur gases from the burning of fossil fuels has been noted. Photochemical reactions between all these materials, perhaps aided by ozone, then result in minute particles of ammonium sulfate. Scott et al. (1969) and Friend et al. (1973) investigated and duplicated some of these reactions in the laboratory. A veritable compendium of information on atmospheric chemical processes which produce condensation nuclei, and ultimately metropolitan smog, is provided by Hidy (1972).

Junge (1960a) and Manson et al. (1961) determined that sulfate particles are a major constituent of both urban and rural aerosols. Even in regions as remote from industry as the Greenland ice cap, the concentration of sulfate ions is an order of magnitude greater than that of any other ion. Using airborne sampling instruments, Junge repeatedly found a stable layer of these particles at altitudes near 20 km. This appears to be one of the tenuous aerosol layers whose existence can also be verified from the ground by lidar techniques. The altitudes of greatest sulfate concentration are approximately those of greatest ozone concentration, as may be seen from Figure 2.6. Until recently it was thought that the sulfate particles were ammonium sulfate. However, investigations by Rosen (1971), as well as those by Bigg et al. (1970) and Cadle et al. (1970), indicate that they are primarily droplets of concentrated water solutions of sulfuric acid.

Sea salt particles, also very hygroscopic, are important components of all aerosols. The salt is injected into the air as small droplets of seawater by the bursting of the innumerable bubbles formed by cresting waves at all ocean surfaces. Woodcock (1953), Junge (1958), Mason (1975), Roll (1965), Cadle (1966), and Blanchard (1967) describe these processes. On the average each bubble produces a few large droplets and 100 or more small droplets. The largest are of such a size that the salt residue, when the water is evaporated, has an effective diameter of 2 to 3 μm. The small droplets tend to give residues less than 0.3 μm in diameter. In addition to sodium chloride, the droplets contain proportionate amounts of the other materials carried in seawater. Most of the droplets are caught up by the wind, carried aloft, and transported great distances. Thus entrained in atmospheric motion, each bit of sea salt is a condensation nucleus, alternating between the crystalline and aqueous states as the relative humidity decreases and increases. Sea salt nuclei are not restricted to maritime aerosols but are also found well inland.

Investigators have differed with respect to the importance of salt particles as nuclei. Wright (1936, 1939) studied haze aerosols of mixed composition at several sites in England for many years. In particular he investigated the relationship between relative humidity and visibility and concluded that sea salt rather than combustion nuclei has the predominant role in both haze and fog. Later observations (see Wright, 1940) were conducted at Valentia on the southwestern tip of Eire during periods when winds blew from the ocean. These precautions avoided the contamination from industrial regions of the British Isles. Thus presumably dealing with maritime aerosols, he found striking correlations between relative humidity and meteorological range, the latter decreasing sharply at relative humidities greater than about 65%.

Simpson (1941a) did not dispute this relationship but advanced strong arguments against the salt nucleus theory. From his calculations of rainfall, the theory apparently called for each square centimeter of ocean surface to produce more than 1000 droplets per second—a quite impossible feat. In his view (also see Simpson, 1941b), the principal nuclei are minute droplets of nitrous acid, which also is very hygroscopic. Middleton (1952) has discussed these opposing viewpoints. A current view, after many intervening experiments, is that sea salt nuclei predominate in strictly maritime aerosols, where the wind has a long, uninterrupted track over the ocean. In continental aerosols, where the wind has a long sweep over land, the nuclei have a diversity of sources—organic, combustion, photochemical. Whatever their origin and composition, nuclei are hygroscopic, and the consequences of this property are considered next.

3.1.3 Size Behavior of Condensation Nuclei

The world of small particles is a world of very small masses but exaggerated surfaces. The ratio of surface area to mass becomes enormous as particle size decreases, as may be seen in Table 3.1. For each size class it is assumed that 1 cm^3 of water has been dispersed to form droplets all of that radius; the resulting number of droplets is listed in the second column. This number of particles would be required to make only 1 cm^3 of substance, or 1 g in the case of water. The large surface/mass ratios mean that the material of a small particle is almost entirely at the surface, where it is exposed completely to environmental influences. Hence small hygroscopic particles, or condensation nuclei, respond quickly to changes in relative humidity, absorbing and releasing moisture as fast as the relative

TABLE 3.1 SURFACE- TO MASS RATIOS FOR SMALL DROPLETS*

Radius (μ)	N (cm^{-3})	Total surface (cm^2)	Surface of one droplet (cm^2)	Mass of one droplet (g)	Surface/mass of one droplet ($cm^2\,g^{-1}$)
10	2.4×10^8	3.0×10^3	1.3×10^{-5}	4.2×10^{-9}	3.0×10^3
1.0	2.4×10^{11}	3.0×10^4	1.3×10^{-7}	4.2×10^{-12}	3.0×10^4
0.1	2.4×10^{14}	3.0×10^5	1.3×10^{-9}	4.2×10^{-15}	3.0×10^5
0.01	2.4×10^{17}	3.0×10^6	1.3×10^{-11}	4.2×10^{-18}	3.0×10^6

*Calculated for 1 cm^3 of water.

humidity changes. Condensation nuclei thus are the more active components of haze aerosols, as contrasted with particles that are not hygroscopic. In this section we examine the growth behavior of these nuclei.

Condensation nuclei are classified according to size: *Aitken*, having radii less than 0.1 μm; large, having radii between 0.1 and 1 μm; and *giant*, having radii greater than 1 μm. Smaller nuclei are abundant in the lower atmosphere but are difficult to capture and study. Their sizes are below the resolution limit of an optical microscope, usually taken as about 0.3 μm. Under an electron microscope their substance partially evaporates in the required hard vacuum, and the electron beam tends to shatter the structure. Except for data obtained by spectrochemical analysis, much of our information on the smallest nuclei is inferential. Even the smallest ones, however, can be detected with the instrument originated by Aitken (1888) and known as the *Aitken counter.* Stated in simple terms, the nuclei are made to grow into droplets visible under a microscope by condensation of water vapor in a supersaturated environment. Modern versions of the instrument are described by Junge (1961), Nesti (1970), Hogan et al. (1975), Liu et al. (1975), and Cadle et al. (1975). An outgrowth of this basic technique is the Wilson cloud chamber, in which the condensation occurs on gas ions.

Dessens (1949) was the first to observe the behavior of large sea salt nuclei under varying humidity. Employing experimental techniques as simple as those of Lord Rayleigh, he captured the particles by exposing very fine spider webs to a gentle breeze. These webs were spun by tiny spiders, gathered in the field, walking to and fro on coarser webs spun by larger spiders. Measurement of the finest threads with electron microscopy indicated that their diameters were about 100 Å. Observing the particles with an optical microscope, he changed the nucleus from crystal to liquid, and back again, by varying the relative humidity—with his breath.

Occasionally the nucleus shattered in returning to the crystalline state, thereby creating additional nuclei. In most cases he found that a relative humidity near 70% marked the onset of rapid condensation. The original account by Dessens, or as reviewed by Middleton (1952), is a story of skill, insight, and patience.

Theoretical reasons for the growth behavior now are reviewed. Wright (1939), Houghton (1951), Keith and Arons (1954), Johnson (1954), Fleagle and Businger (1963), Byers (1965), List (1966), Amelin (1967), and Katz and Kocmond (1973) may be consulted for details. First, the vapor pressure required for equilibrium (saturation) over the strongly curved surface of a pure water droplet is greater than that required for saturation over a plane surface. The ratio of the two saturation pressures is given by Kelvin's expression:

$$\ln \frac{e'_s}{e_s} = \frac{2\gamma}{\rho R_v T r} \tag{3-1}$$

where

e'_s = saturation pressure over a curved surface
e_s = saturation pressure over a plane surface
γ = surface tension of water (75.6 dynes cm^{-1})
ρ = density of water (1 g cm^{-3})
R_v = gas constant for water vapor (4.615×10^6 ergs g^{-1} K^{-1})
T = absolute temperature K
r = radius of the water surface in cm

When these substitutions are made, including $T = 273$ K and $e_s = 6.108 \times 10^3$ dynes cm^{-2} or 6.108 mb, the ratio e'_s/e_s as a function of droplet radius is shown by the dashed curve in Figure 3.1. If the actual vapor pressure in the adjacent air is denoted by e_v, the size of the droplet remains constant when

$$\frac{e_v}{e_s} = \frac{e'_s}{e_s} \qquad \text{or} \qquad e_v = e'_s$$

If we recall that percent relative humidity is defined by

$$U = \frac{e_v}{e_s} \times 100$$

it is clear from the ordinate scale that the vapor and liquid phases can be in equilibrium only when the air is supersaturated. The equilibrium is unstable, however, because, if U becomes even slightly less than the supersaturation required at a given value of r, the droplet loses water. This reduction in size makes the disparity between actual and required values of U even greater, so that the droplet loses water at an increasing rate. Thus a

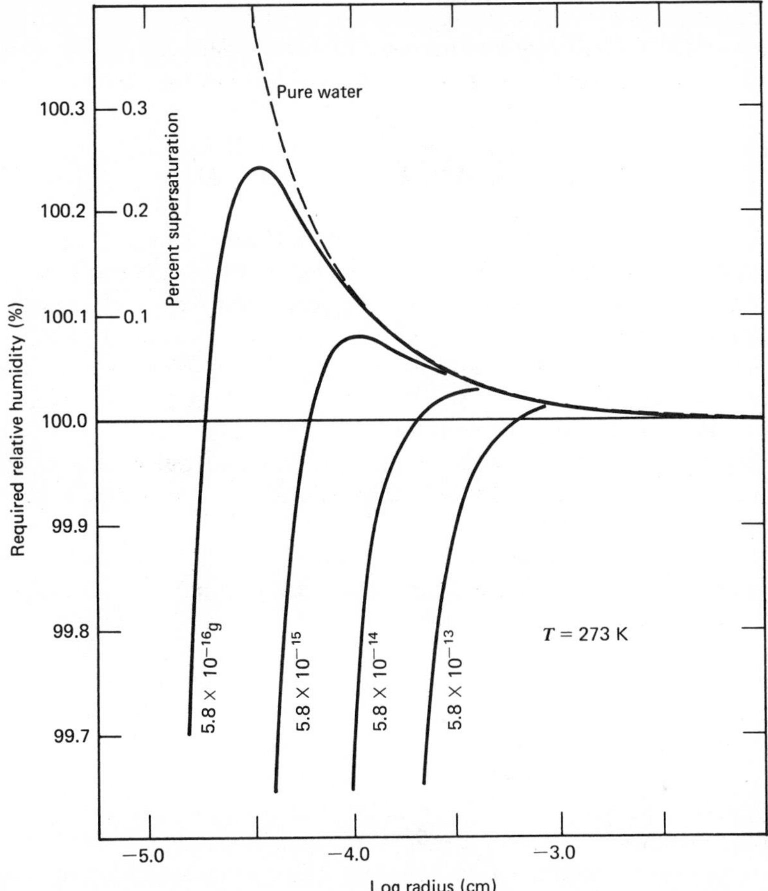

Figure 3.1 Relative humidity required for size equilibrium of condensation nucleus containing stated amount of sodium chloride. Data from List (1966).

runaway condition ensues, and the small droplet goes out of existence. Should U become slightly greater than the required supersaturation, an oppositely directed sequence of events takes place. The droplet now gains water at an increasing rate and grows rapidly. It continues to grow as long as the supersaturation of the adjacent air is greater than the decreasing equilibrium value. When the droplet has grown to a few micrometers in radius, the curvature effect practically disappears, so that pure water droplets of this size and larger are in equilibrium at ordinary saturation. According to the ordinate scale, the required supersaturation becomes very great as r becomes very small, from which we conclude that very small droplets of pure water are a rarity.

Second, the vapor pressure required for saturation over a plane surface of salt solution is less than that required for saturation over a plane surface of pure water. The relationship is given by Raoult's law for dilute solutions:

$$\frac{e_s'' - e_s}{e_s} = -\frac{M'}{M+M'} \quad \frac{e_s''}{e_s} = 1 - \frac{M'}{M+M'} \tag{3-2}$$

where e_s'' is the equilibrium pressure for solution, M' is moles of solute, and M is moles of solvent. The number of moles of either solute or solvent is equal to its mass in grams divided by its molecular weight. The mass of the solute is the mass of the undiluted condensation nucleus, while the mass of the solvent is given by $4\pi r^3\rho/3$. Apart from the curvature effect given by (3-1), evaporation, equilibrium, or condensation would occur accordingly as the actual vapor pressure e_v is less than, equal to, or greater than the equilibrium value e_s''. It may be seen from (3-2) that, when M becomes large, as from condensation, the effect of the solute becomes negligible.

Hence the size of a small, hygroscopic droplet is governed by two contrary factors: (1) raising of the equilibrium vapor pressure by the curvature effect, and (2) lowering of this pressure by the dissolved salt. Figure 3.1 shows the composite result for typical nuclei masses. Each solid curve is the reversible growth characteristic for a droplet having the solute mass of sodium chloride as indicated. If the relative humidity is less than the ordinate value corresponding to a particular radius, the droplet shrinks. If it is greater, the droplet grows. The reader may demonstrate to himself, by an argument similar to that for a pure water droplet, that the size of a hygroscopic droplet is either stable or unstable, depending on whether its size lies on the ascending or descending branch of the equilibrium curve. Chen (1974), in his study of vapor pressure over small droplets, shows equilibrium curves for nuclei composed of sodium chloride, ammonium sulfate, and a mixture of the two salts.

A droplet can grow only by collecting water molecules from the adjacent vapor, so that growth itself depletes the supply. Conversely, droplet shrinkage replenishes the supply. Because a given parcel of air contains a fixed amount of moisture, relative humidity and droplet size and number are closely interrelated. Each droplet is a reservoir giving water molecules to the air when the temperature rises, with a consequent increase in the saturation value, and taking them back when the temperature falls. The temperature itself is somewhat affected by these processes, because water has a large heat of vaporization. Very small droplets, requiring a higher value of supersaturation for equilibrium, are less effective than larger droplets in capturing and retaining water molecules. In this competition for water, the larger entities grow at the expense of the smaller, a process not unfamiliar in other domains.

The effects of this continual balancing between relative humidity and droplet size can be observed during periods of calm weather. A blue-gray haze, veiling the landscape all day, thickens to a *radiation fog* at night according to these natural influences. Such fogs are common in the valleys of a rolling countryside during a temperature inversion. After sunset the ground loses heat by radiation, thus chilling a shallow layer of overlying air, with a consequent reduction in its saturation pressure. If the absolute humidity is great enough, the air becomes slightly supersaturated, and some nuclei grow into droplets according to Figure 3.1. The resulting fog persists until the rising sun warms the ground and air, with consequent increase in saturation pressure. The fog droplets then shrink in size about an order of magnitude, and haze reappears, very like that of the previous day. Colloquially it is said that "the sun has burned away the fog." Such cycles between haze and fog are repeated until winds bring in air having fewer nuclei. The strong correlation between relative humidity and droplet size, as indicated by values of meteorological range, is shown in Figure 3.2. The curve is an average of many observations at the Los Angeles airport, a region notably rich in nuclei.

Similar processes of droplet growth are at work aloft. Although the physics of clouds and precipitation is quite involved, the evolution of a simple convective cloud is analogous to the evolution of fog. Consider the fair-weather cumulus cloud. During the morning hours a thermal updraft, or convective cell, develops over the warming ground. Warm, moist air

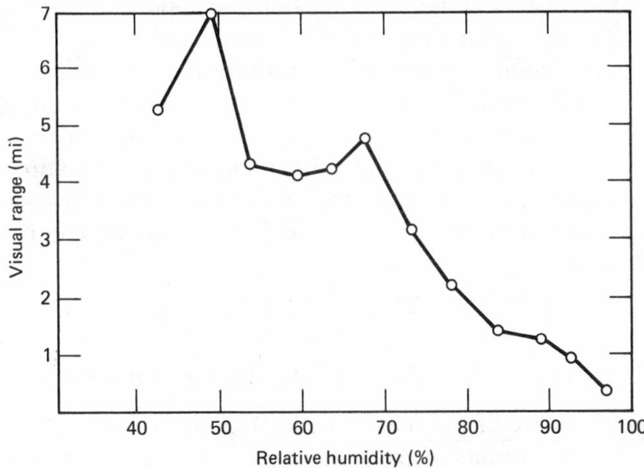

Figure 3.2 Variation in visual range with relative humidity at Los Angeles airport. From Neiburger and Wurtele (1949) and George (1951).

containing many nuclei thereby is lifted several thousand feet, with consequent cooling by adiabatic expansion and development of a saturation condition. Some of the nuclei then grow into droplets, forming a billowing cloud in the clear sky. The cloud is the visible mark of a saturated parcel of air at a lower temperature. Even casual observation reveals that the wispy edges vanish and form anew as surrounding air mixes by turbulent diffusion with the parcel dominated by the cloud.

3.1.4 Particle Removal Processes

The particles continually placed in the atmosphere by the processes described earlier are continually removed by indispensable processes known as *coagulation*, *fallout*, and *washout*. Particles in the atmosphere are in transit between birth and demise, and it is appropriate to think of a "particle residence time." The aggregate result of all emplacement and removal processes is reflected in the measured concentrations and size distributions, because the atmosphere continually adjusts its particle load to some kind of dynamic equilibrium. The removal processes are part of a subject aptly called the *mechanics of aerosols*. Fuchs (1964) and Davies (1966) treat this subject thoroughly, but with comparatively few applications to natural aerosols. Green and Lane (1957) give informative, highly readable treatments of the three removal processes, along with other aspects of aerosol science. Junge (1963) treats the removal processes in a completely atmospheric context, as does Byers (1965). The broad subject of precipitation scavenging is covered in detail by Engelmann and Slinn (1970).

Coagulation occurs when two particles collide and then coalesce, resulting in fewer but larger particles. Collisions of Aitken nuclei in still air are brought about by brownian motion alone. Usually in the real atmosphere, and especially for larger particles, collisions are produced by small-scale, turbulent diffusion. Principally, collisions affect the large population of small particles more than the small population of large particles. Consider now a unit volume of air containing N particles in two size groups having radii r_1 and r_2. The coagulation rate is given in simplified form by

$$\frac{dN}{dt} = -\frac{\kappa T s}{6\eta}\frac{(r_1 + r_2)^2}{r_1 r_2}\left(1 + \frac{a\bar{l}}{r}\right)N^2 \qquad (3\text{-}3)$$

where r is the mean value of r_1 and r_2, and the constants are:

$s = \dfrac{\text{radius of sphere of particle influence (capture)}}{\text{radius of particle}} : 2$

η = viscosity of air (1.8×10^{-4} g cm^{-1} sec^{-1})

a = empirical constant (0.9 cm)

$\bar{l}$ = molecular mean free path [6×10^{-6} cm, as in (2-15)]

The value $s = 2$ implies that any two colliding particles will coalesce on touching at any contact angle from head-on to grazing incidence.

Looking at (3-3), letting $T = 288$ K, assuming that $r_1 = r_2$ (a monodispersion), and substituting values from the above listing, the first group of terms becomes 7.4×10^{-11} cm^3 sec^{-1} and the second group becomes 4. If $r = 5.4 \times 10^{-7}$ cm, which is in the size range of very small nuclei, the parenthetical factor is 11. If N is taken as 10^4 cm^{-3}, the coagulation rate is 0.33 sec^{-1}. If $r = 5.4 \times 10^{-6}$ cm, the parenthetical factor is 2, and the coagulation rate is only 0.059 sec^{-1}. Thus the smaller the particles and the greater the concentration, the greater the rate of coagulation.

In (3-3), the term

$$\frac{(r_1 + r_2)^2}{r_1 r_2}$$

has its minimum value, which is 4, when $r_1 = r_2$; that is, when all particles are the same size. The value increases rapidly as the size difference increases. This means that an aerosol coagulates faster when its particles are of nonuniform size than when they are nearly of the same size. According to Byers (1965), particles of radius 10^{-6} cm coagulate 3 times as fast with particles of radius 10^{-5} cm as with those of radius 2×10^{-6} cm, and 30 times as fast with particles of radius 10^{-4} cm. Thus small aerosol particles ($r < 10^{-5}$ cm) tend to coagulate rapidly with cloud droplets ($r > 10^{-4}$ cm).

Using a more complete theory than indicated above, Junge (1963) calculated the effects of coagulation and aging time on a model haze aerosol. The results are shown in Figure 3.3. The concentration of the smaller Aitken particles decreases at a rapid rate compared to the usual time scale of meteorological processes. Thus the maximum of the distribution shifts steadily toward larger radii, but the resulting increase in concentration of the large particles is so slight that it is barely discernible. As may be inferred from the figure, in natural aerosols the effects of coagulation are such that the maximum of the distribution is found between radii of 0.01 and 0.1 μm. This is because any maximum below 0.01 μm disappears rapidly, while the shift beyond 0.1 μm is very slow. Particles smaller than 5×10^{-3} μm can exist only briefly. The small-size limit of the distribution curve tends to be set by the coagulation rate and by the rate at which new nuclei come into being.

Fallout or sedimentation is the tendency of any particle, whatever its size, to fall by gravity through the atmosphere. The settling tendencies of smaller particles are usually masked by winds and convection currents. In falling through undisturbed air, the particle attains a terminal or equilibrium velocity such that the gravitational force (negative buoyancy) is balanced by the opposing force due to viscous drag. The terminal velocity

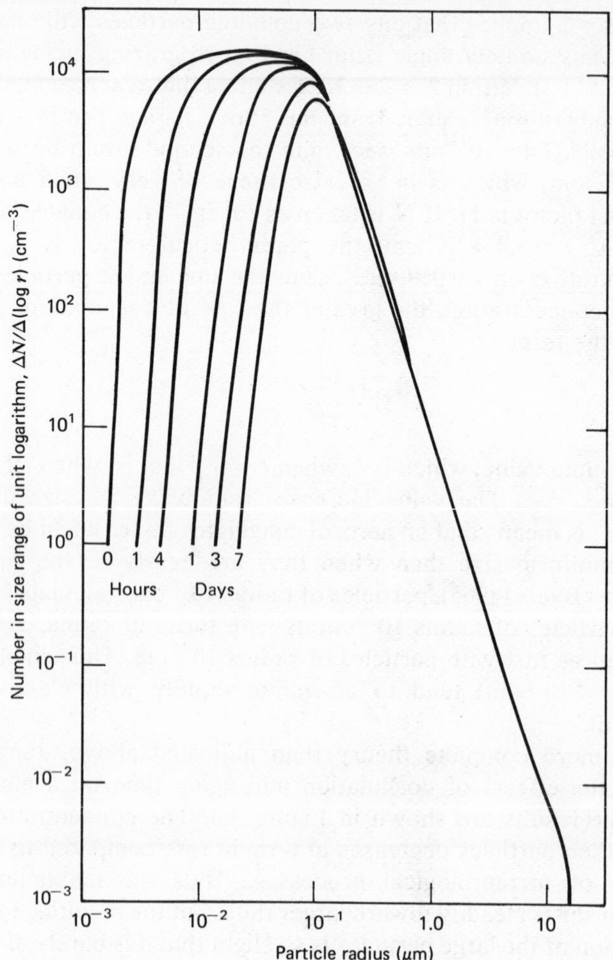

Figure 3.3 Size distribution of aerosol particles with coagulation affecting the lower size range. From Junge (1963).

v_t is given by the Stokes–Cunningham expression:

$$v_t = \frac{2}{9}\frac{r^2}{\eta} g(\rho_p - \rho_a)\left(1 + \frac{B\bar{l}}{r}\right) \tag{3-4}$$

where ρ_p and ρ_a are the densities of the particle and air, and B is a factor whose value lies between 1.25 for $\bar{l}/r \leq 0.1$ and 1.65 for $\bar{l}/r \geq 10$.

Junge et al. (1961), in studying the exchange of materials between the stratosphere and troposphere, used (3-4) in computing terminal velocities of small spheres in a model atmosphere similar to USSA-1962. These velocities are plotted in Figure 3.4. The effects of reduced air density at higher altitudes, with a consequent reduction in viscosity, are outstanding. For example, a 0.1-μm particle falls faster at 40 km than a 1.0-μm particle at sea level, although the mass of the latter is 1000 times greater. On a global scale, and as a long-term average, there is a net sedimentation flux of particles from the upper atmosphere toward the ground.

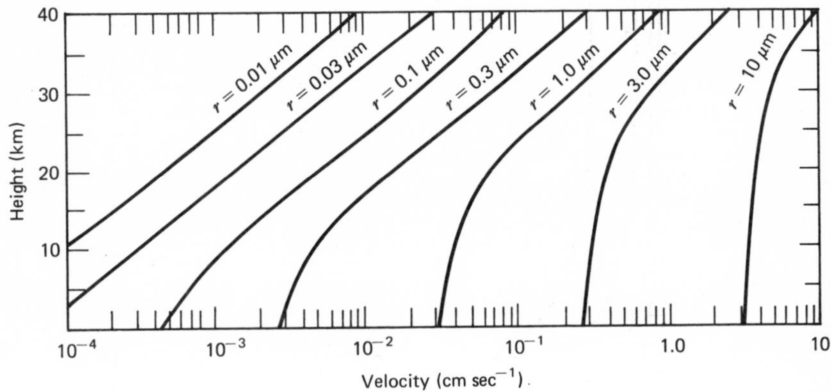

Figure 3.4 Terminal velocities of small particles as a function of altitude. Particle density is assumed to be 2 g cm^{-3}. From Junge et al. (1961).

Washout or scavenging is the removal of particles from portions of the atmosphere by rain and snow. The overall process often is effective in clearing the air, as can be observed most often after a heavy snowfall. Washout implies some kind of joining action between haze particles and raindrops and snowflakes. Actual physical contact between these objects of such dissimilar size, however, may not be necessary. In fact, a small particle likely is pushed aside by the airstream of a falling raindrop. It appears that washout is accomplished in several steps. First, some of the nuclei grow into cloud droplets by condensation of water vapor. Next, some of the particles that have not grown coagulate with these droplets in the manner just described. Many of these droplets, having masses far greater than those of the particles, then merge with falling raindrops by the complex processes responsible for raindrop growth. In this manner the aerosol particles in a cloud are carried to earth.

Greenfield (1957) investigated the several stages of washout by employing various coagulation equations. Figure 3.5 shows his results for the efficacy of cloud droplets in collecting aerosol particles. The droplets were assumed to have radii of 10 μm and a concentration of 100 cm^{-3}.

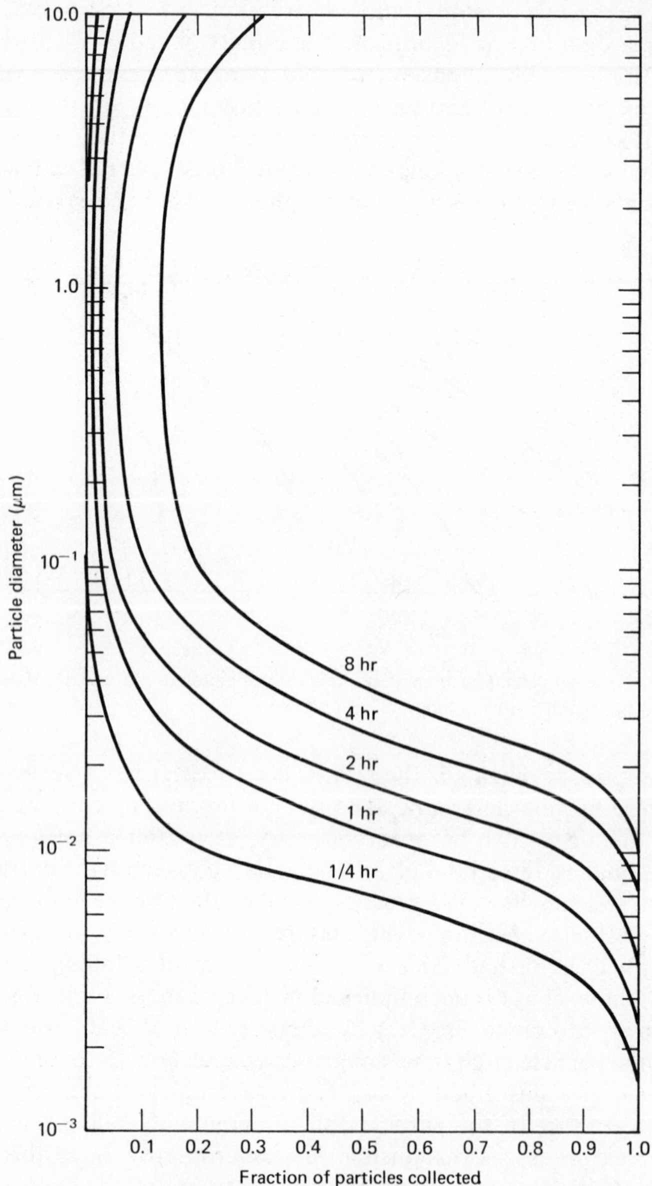

Figure 3.5 Coagulation of small aerosol particles and cloud droplets. From Greenfield (1957).

These assumptions do not represent an especially dense cloud, yet it is seen that nearly all the smallest particles are gathered up by the droplets within a few minutes. This bears out the earlier statement that coagulation between small and large particles proceeds rapidly. With the collection accomplished as in Figure 3.5, the actual removal of aerosol particles from the cloud by rainfall takes place according to Figure 3.6. The air beneath the cloud, however, is not necessarily cleared so completely by the rainfall. Snowflakes, having larger cross sections and smaller terminal velocities than raindrops, and also rougher surfaces, may be more effective in scavenging the air beneath the cloud. Thus coagulation, fallout, and washout continually carry both dust and hygroscopic particles to the ground, tending to prevent an unlimited increase in the atmospheric particle content. Further investigations of these processes are reported by Beard (1974), Hampl et al. (1971), Kerker and Hampl (1974), and Storebø and Dingle (1974).

There seems little doubt, however, that the particle content of the atmosphere has increased during recent decades, at least in the lower troposphere. The overall amount of the increase and any effect on the global climate have not been determined with satisfactory assurance. Only long-term observations are meaningful, but these are made at only a few locations. According to McCormick and Ludwig (1967), at Washington, D.C., the average particle content of a zenith path, as expressed by the average turbidity, increased 57% during the period 1903–1907 to 1962–1966. At Davos Observatory in the Alps, at an altitude near 1600 m, the increase was 88% during the period 1914–1926 to 1957–1959. Meas-

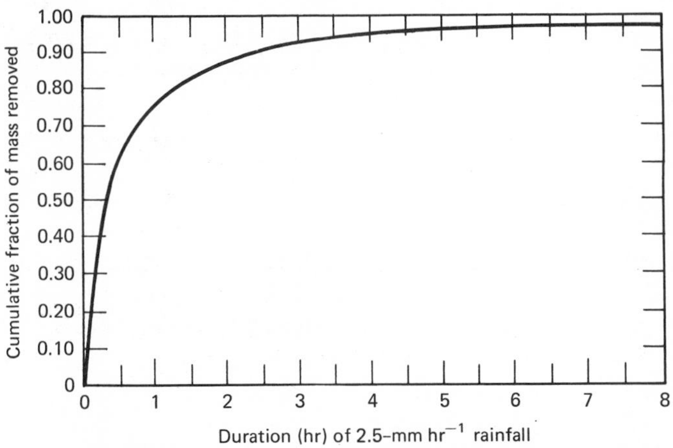

Figure 3.6 Removal of aerosol particles by rain washout in 2.5 mm hr^{-1} rainfall occurring after 1 hr of particle residence in cloud. From Greenfield (1957).

urements at Washington, D.C., have showed that a turbidity increase of 100%, at present values, produces a 5% decrease in the transmission of solar flux to the ground. In apparent support of such observations and inferences is the deterioration of visibility in the eastern United States since 1965, as reported in *Bull. Am. Meteorol. Soc.* (1972).

Contrary to the Davos measurements, however, are those made at other and higher stations. As reported by Machta (1972), solar observations over a 13-year period at a 4170-m altitude station near the summit of Mauna Loa in Hawaii show no long-term increase in turbidity above that altitude. Equally significant are the data from the Smithsonian Astrophysical Observatory gathered from 13 remote, high-altitude stations in North and South America and Africa. All the stations used essentially the same equipment and techniques in their solar observations. According to Roosen et al. (1973), no long-term increases in turbidity were found over a period of 50 years. In discussing the observations, these investigators state: "Our best estimate is that, except for sporadic perturbations due to volcanic activity, there has been no detectable change in the global atmospheric transmission measured from remote, high altitude sites in the last half century." This argues that the indicated increases in turbidity, such as those measured at Washington and Davos, may be occurring principally at the lower levels of the atmosphere.

3.2 SIZE DISTRIBUTIONS OF HAZE PARTICLES

Hazes are polydisperse aerosols in which particle sizes vary across two and three orders of magnitude. This dispersal of sizes is the complex result of the processes that create the particles, the interactions of the particles with other materials of the atmosphere, and the processes that remove the particles. The characteristic gray hue of haze in nonpolluted air, and nearly all the optical attenuation due to haze, are attributable to numerous "large" particles having radii between 0.1 and 1.0 μm. Smaller members of this group, along with the more numerous Aitken nuclei, impart a blue tinge to thin hazes by preferentially scattering the shorter wavelengths of sunlight. Particles having a radius near 0.3 μm are usually the most effective in determining visibility. Giant particles having a radius greater than 1.0 μm are much less numerous but strongly influence forward scattering. Particles larger than 10 μm tend to be so few in genuine haze, as distinguished from fog, that they can usually be disregarded.

The manner in which the particle population is spread over the range of sizes is defined by the *size distribution function*. From the scattering standpoint this is the most important characteristic of any aerosol, be it haze, fog, or cloud. From an astonishing number and diversity of haze investigations, many workers take the distribution of sizes to conform either to an exponential or to a power-law function. These two functions are widely used, because of reasonable fidelity to measurement data and suitability for analytic studies. Each function is explained in this section, following a review of several measures of particle size and concentration that are applicable to any type of distribution. The section concludes with a discussion of size distributions either as found by direct sampling and measurement of particles, or as deduced from measurement of attenuation.

3.2.1 Measures of Particle Size and Concentration

The techniques employed to count and measure particles usually provide data on the number of particles per specified interval of radius. The intervals are called *size classes*, and those for haze particles are typically 0.1 or 0.2 μm in width and may be several micrometers for fog and cloud droplets. Such data are presented as histograms which give the population n_i of each size class r_i, as illustrated in Figure 3.7. A unit volume of aerosol is implied, though not always stated. The concentration N, which is the total number of particles per unit volume, is equal to the sum of the class populations, or

$$N = \sum_{i=0}^{k} n_i(r_i) \tag{3-5}$$

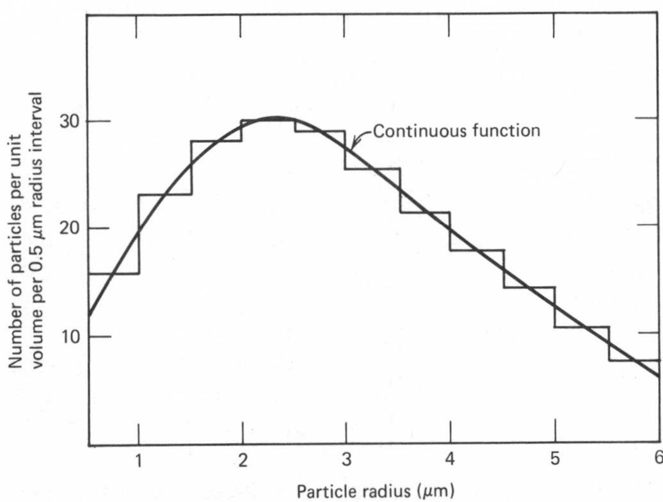

Figure 3.7 Histogram of particle size data.

which is proportional to the histogram area. If the size classes are narrow and the populations are large, a continuous function can often be devised to fit the histogram, as suggested by the curve in Figure 3.7.

When a distribution can thus be expressed by a continuous function, the number $n(r)$ of particles per unit interval of radius and per unit volume is given by

$$n(r) = \frac{dN}{dr} \tag{3-6}$$

The differential quantity dN expresses the number of particles having a radius between r and $r + dr$, per unit volume, according to the distribution function $n(r)$. The dimensions of $n(r)$ are L^{-4}. Equation (3-6) implies a definition of N:

$$N = \int_0^\infty n(r)\, dr \tag{3-7}$$

which is the analytic counterpart of (3-5). Necessarily with some types of functions, or sometimes for convenience, cutoff or limiting values of radii r_1 and r_2, with $r_1 < r_2$, are used as integration limits instead of 0 and ∞. The *cumulative concentration* N_c is defined as the number of particles having radii less than, or greater than, r according to

$$N_c = \int_{0 \text{ or } r}^{r \text{ or } \infty} n(r)\, dr \tag{3-8}$$

Particle technology employs several measures of size. The linear mean radius $\bar{r}$, sometimes denoted by r_1, is obtained by integrating over the sizes and dividing by the concentration:

$$\bar{r} = \frac{\int_0^\infty rn(r)\, dr}{\int_0^\infty n(r)\, dr} \approx \frac{1}{N} \sum_{i=0}^{k} r_i n_i \tag{3-9}$$

where the integral in the denominator represents the total number of particles. Equation (3-9) shows that $\bar{r}$ is a *weighted mean* value. The rms radius r_{rms}, also called the *surface mean radius* and sometimes denoted by r_2, is defined by

$$r_{\text{rms}} = (\bar{r^2})^{1/2} = \left[\frac{\int_0^\infty r^2 n(r)\, dr}{\int_0^\infty n(r)\, dr}\right] = \left(\frac{1}{N} \sum_{i=0}^{k} n_i r_i^2\right)^{1/2} \tag{3-10}$$

In normal probability distributions, r_{rms} is the square root of the *variance*, which is the mean of the squares of the deviations from the arithmetic average. Hence it is strongly affected by the larger deviations. In size distributions of atmospheric particles, r_{rms} is similarly affected by the presence of very small and very large particles.

The total cross-sectional area of the particles within a unit volume is evidently

$$A = \pi \int_0^\infty r^2 n(r)\, dr \qquad (3\text{-}11)$$

The scattering by an aerosol is proportional to this area. When $n(r) \propto r^{-2}$, which is often true with fog and cloud droplets over part of the size distribution, the cross-sectional area per increment in radius remains constant across the range of sizes. Since the term A in (3-11) represents area per unit volume, its dimension is L^{-1}. The total volume of the particles within a unit volume of space is given by

$$\phi = \tfrac{4}{3}\pi \int_0^\infty r^3 n(r)\, dr \qquad (3\text{-}12)$$

For haze aerosols, ϕ is measured experimentally by deposition and filtration techniques. For fog and cloud, ϕ represents the liquid water content and has great meteorological importance, but its accurate measurement is difficult. When $n(r) \propto r^{-3}$, which is true for many hazes, the particle volume per increment in radius remains constant across the range of sizes. Since ϕ represents volume per unit volume, it is dimensionless.

In dealing with the total scattering by a polydispersion, it is sometimes convenient to employ an *effective mean radius* r_e:

$$r_e = \frac{\int_0^\infty r^3 n(r)\, dr}{\int_0^\infty r^2 n(r)\, dr} = \frac{\sum_{i=0}^{k} r_i^3 n_i}{\sum_{i=0}^{k} r_i^2 n_i} \qquad (3\text{-}13)$$

Thus r_e is, except for a missing factor $\frac{1}{3}$, equal to the total volume of the particles divided by their total surface area, that is, to their volume/surface ratio. Values of this ratio for representative particle sizes may be appreciated from Table 3.1, where the surface/mass ratio for water droplets is the reciprocal of their volume/surface ratio. For a polydispersion having a relatively narrow size distribution, r_e is equivalent to the radius required for a monodispersion to exhibit the same total scattering characteristic. Thus the use of r_e enables one to handle, within limits best defined empirically, scattering problems of a polydispersion in terms of an equivalent monodispersion.

Closely related to r_e is the volume/surface mean diameter D_{32}, called the *Sauter mean diameter* by Mugele and Evans (1951) who applied it to droplets from spray nozzles. This diameter is defined as the ratio of the third moment to the second moment of the size distribution, or

$$D_{32} = \frac{\int_0^\infty ND^3 f(D)\, dD}{\int_0^\infty ND^2 f(D)\, dD} \tag{3-14}$$

where $f(D)$ is the distribution function normalized so that its integral over a given diameter interval represents the probability of occurrence of particles within that interval. The quantity $D_{32} = 2r_e$. It is easy to verify that D_{32} is equal to six times the volume/surface ratio. The use of D_{32} to find the total scattering characteristic of a polydispersion is subject to the same limitations as those obtaining for r_e.

The following list summarizes the foregoing concepts of radius and several others encountered in the literature. The concepts are equally applicable to diameter.

$\bar{r}$: Linear mean radius.

r_{rms}: Root-mean-square radius or surface mean radius.

r_e: Effective mean radius.

Δr: Width of the distribution, in radius measurement units, at the half-maximum values of $n(r)$.

r_m: Median radius. One-half of the particles have radii smaller than r_m.

r_c: Critical or mode radius. This is the radius of the size class containing the greatest population and corresponds to the maximum of $n(r)$. Some size distributions, especially those of fog and cloud, may be bimodal.

r_1, r_2: Lower and upper limits for integration of the size distribution.

3.2.2 Exponential Size Distribution Function

A function much used for model aerosols in studies of scattering is the one proposed by Deirmendjian (1963a, 1964, 1969). It has the form

$$n(r) = ar^\alpha \exp(-br^\gamma) \tag{3-15}$$

where a, b, α, and γ are positive constants. The function vanishes at $r = 0, \infty$ and is called by Deirmendjian a *modified gamma distribution* "in analogy with the gamma distribution to which it reduces when $\gamma = 1$." The concentration is given by the integral of (3-15) between the limits zero and infinity,

$$N = a\gamma^{-1} b^{-(\alpha+1)/\gamma}\, \Gamma\left(\frac{\alpha+1}{\gamma}\right) \tag{3-16}$$

where Γ represents the gamma function.

The mode radius is found by setting the derivative of (3-15) equal to zero, which yields

$$r_c^{\gamma} = \frac{\alpha}{b\gamma} \tag{3-17}$$

When this is substituted into (3-15), the latter becomes

$$n(r_c) = ar_c^{\alpha} \exp\left(-\frac{\alpha}{\gamma}\right) \tag{3-18}$$

where $n(r_c)$ is the number of particles per unit radius interval at the value r_c per unit volume.

Because it has four adjustable constants, (3-15) can be fitted to various models of haze, cloud, and even rain. General methods of determining these constants should be noted. Experimental data from particle sizing and counting yield values for N and r_c and allow a distribution curve to be plotted. This curve is a histogram in which the size classes, or radius intervals, are kept as narrow as may be consistent with the task of particle counting. It is seen from (3-16) that the constant a is essentially determined by the experimental value of N. If α and γ are estimated from the shape of the curve, or selected by insight, the constant b is determined by r_c according to (3-17).

Versatility of the function (3-15) can be appreciated from Table 3.2 which shows its application to several model dispersions. The concentrations listed are nominal values which may be shifted up or down by altering the value of the constant a. The distributions for haze models H, L, and M are shown in Figure 3.8, where the evident preponderance of small particles is characteristic of all haze aerosols, whatever the particle size distribution. Model H applies to either stratospheric dust particles or to hailstones with a suitable choice of units, as indicated in the table. Haze model L generally represents continental aerosols; with a change in units it becomes rain model L for light and moderate precipitation. Haze model M, one of the early models in the series, was first developed for maritime and coastal aerosols. The function (3-15) has also been applied to fog by Chu and Hogg (1968) and Rensch and Long (1970), as shown in Section 3.4.2. The cloud models based on (3-15) are discussed in Section 3.4.3.

3.2.3 Power-Law Size Distribution Function

The second distribution function, developed by Junge (1958, 1960b, 1963) and also described by Manson (1965), is

$$n(r) = \frac{dN}{d\log r} = cr^{-v} \tag{3-19}$$

TABLE 3.2 PARAMETERS OF EXPONENTIAL-TYPE SIZE DISTRIBUTIONS

Distribution type	N	a	r_c	α	γ	b	$n(r_c)$
Haze M	$100\ cm^{-3}$	5.3333×10^4	$0.05\ \mu m$	1	$\frac{1}{2}$	8.9443	$360.9\ cm^{-3}\ \mu m^{-1}$
Rain M	$1000\ m^{-3}$	5.3333×10^5	0.05 mm	1	$\frac{1}{2}$	8.9443	$3609\ m^{-3}\ mm^{-1}$
Haze L	$100\ cm^{-3}$	4.9757×10^6	$0.07\ \mu m$	2	$\frac{1}{2}$	15.1186	$446.6\ cm^{-3}\ \mu m^{-1}$
Rain L	$1000\ m^{-3}$	4.9757×10^7	0.07 mm	2	$\frac{1}{2}$	15.1186	$4466\ m^{-3}\ mm^{-1}$
Haze H	$100\ cm^{-3}$	4.0000×10^5	$0.10\ \mu m$	2	1	20.0000	$541.4\ cm^{-3}\ \mu m^{-1}$
Hail H	$10\ m^{-3}$	4.0000×10^4	0.10 cm	2	1	20.0000	$54.14\ m^{-3}\ cm^{-1}$
Cumulus cloud, C.1	$100\ cm^{-3}$	2.3730	$4.00\ \mu m$	6	1	$\frac{3}{2}$	$24.09\ cm^{-3}\ \mu m^{-1}$
Corona cloud, C.2	$100\ cm^{-3}$	1.0851×10^{-2}	$4.00\ \mu m$	8	3	$\frac{1}{24}$	$49.41\ cm^{-3}\ \mu m^{-1}$
MOP cloud, C.3	$100\ cm^{-3}$	5.5556	$2.00\ \mu m$	8	3	$\frac{1}{3}$	$98.82\ cm^{-3}\ \mu m^{-1}$

Source: Deirmendjian (1969).

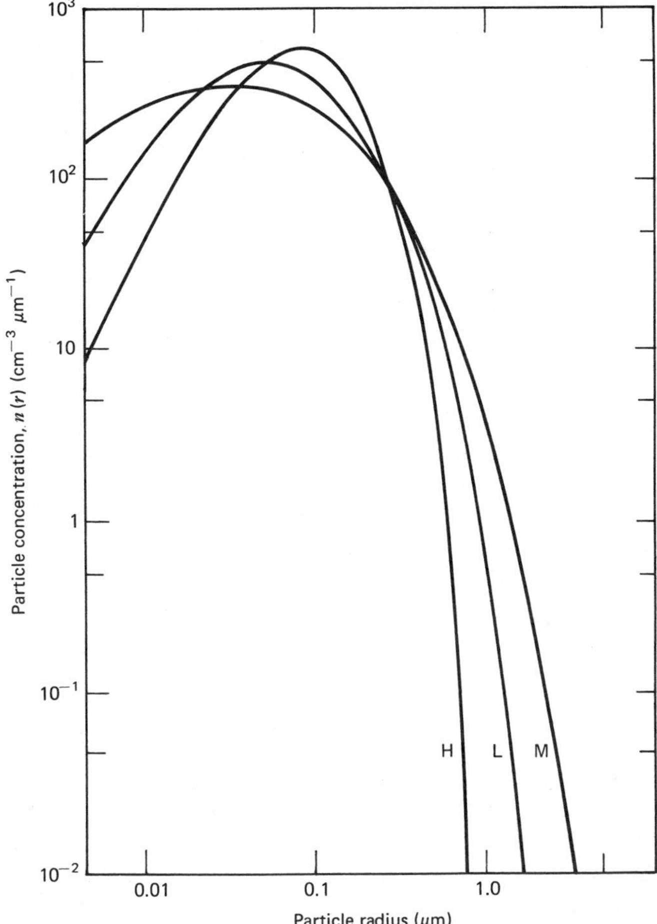

Figure 3.8 Particle size distributions for haze models H, L, and M of Table 3.2. From Deirmendjian (1969).

Here N is the number of particles per unit volume, from the smallest size limit up to size r, c is a constant whose value depends on the concentration, and the exponent v determines the slope of the distribution curve. Since the derivative is with respect to log r, dN is the number of particles per increment in log r. A plot of (3-19) for a typical continental aerosol is shown in Figure 3.9a in which the cutoff radii r_1 and r_2 define the valid range of the expression. A striking feature of this distribution is its linearity over a size range of about 1 to 100.

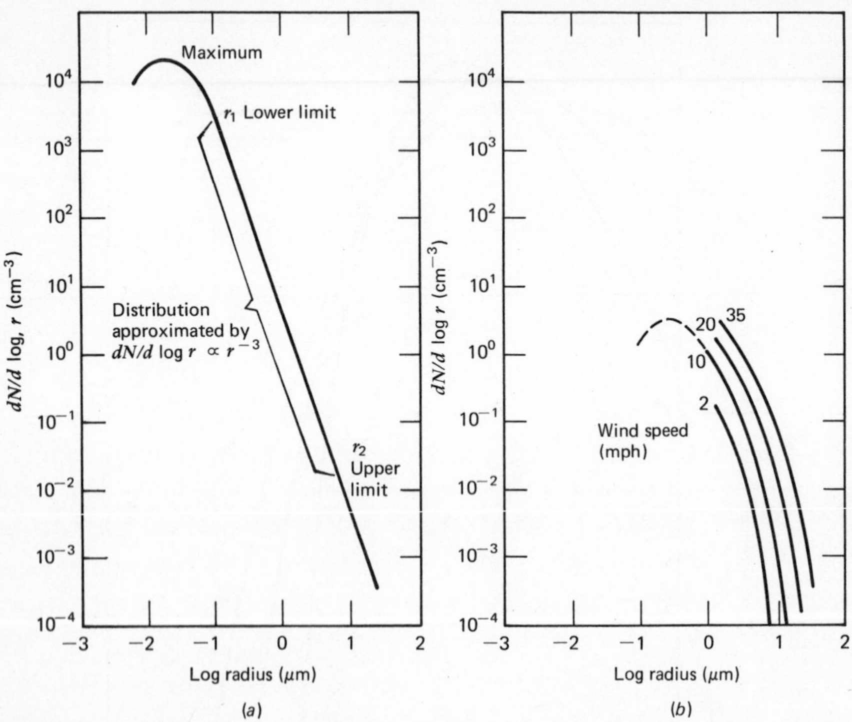

Figure 3.9 Power-law size distributions of particles in continental and maritime aerosols according to the power law, Eq. (3-19). (*a*) Continental type, (*b*) maritime type. From Junge (1960b).

This type of plot is explained by Junge (1960b):

> To obtain the number of particles per cm^3 found in a given size interval, one multiplies the average ordinate value over the interval by the positive change in the logarithm of the radius. For example, over the size interval from 0.1 to 0.2 μm, the average ordinate value [in Figure 3.9a] is approximately 10^3 particles per cm^3. The change in logarithm of the radius is $-0.699-(-1.0)$, which gives 0.301. The product is thus approximately 300 particles per cm^3. Similarly for the radius interval from 1 to 2 μm, the average ordinate value is about 1, thus 0.3 particle per cm is indicated.

Equation (3-19) can be put in a nonlogarithmic form. Since

$$d \log r = 0.434 d \ln r \qquad \text{and} \qquad d \ln r = \frac{dr}{r}$$

we can write

$$n(r) = \frac{dN}{dr} = 0.434 c r^{-(v+1)} \tag{3-20}$$

The power-law distribution is much used in aerosol investigations and deserves close consideration. It is clear that the number of larger particles, relative to the number of smaller particles, increases as the value of the exponent decreases. Typical hazes conform to $3 < v < 4$, while many fogs are characterized by $v \approx 2$. When r is expressed in centimeters, the factor c necessarily has a very small value, as may be seen by looking ahead at Table 6.2. The dimension of c depends on the value of v and is equal to L^{v-3}. The particle concentration is given by the integral of (3-20):

$$N = 0.434c \int_{r_1}^{r_2} r^{-(v+1)} \, dr \qquad (3\text{-}21)$$

When $n(r)$, v, and N are found experimentally by particle sizing and counting, c can be determined from (3-20) or (3-21).

Also shown in Figure 3.9 are the distribution curves for typical maritime aerosols. Comparing continental and maritime aerosols, the greater concentration and smaller size of particles of the first type are noteworthy. The curve in Figure 3.9*a* has a slope of about -3 which, from (3-19), is the value of v for this distribution. Then from (3-20), the number of particles per unit interval of radius, per unit volume, is proportional to r^{-4}. This is a strong dependence. When $v > 3$, there are relatively more small particles than are shown by Figure 3.9*a*. Maritime aerosols, in which sea salt nuclei may predominate, tend to have fewer but larger particles and are influenced by wind speed, as shown by Figure 3.9*b*. Usually there are no sharp distinctions between the two aerosol types in broad coastal regions, however, because continental and oceanic air masses are not kept apart by coastlines.

Actual particle size distributions on an individual basis may differ considerably from a strict power-law form. On the average, however, the power law seems to be a good representation of aerosols having a wide variety of origins and compositions. Although the concentrations may differ greatly, and the exponent v may vary between 2 and 4, the form of the distribution over the size range from 0.10 to 20 μm tends to remain the same. Presumably this indicates a quasi-equilibrium condition between the processes that create the particles and the processes of coagulation, fallout, and washout. Much current attention is centered on this matter, because of the growing concern with air pollution. Even in polluted air, however, the power-law distribution is often applicable. Apparently the particulate atmosphere has the capability of maintaining its status quo, within limits not yet understood, as long as sufficient mixing motions and meteorological activities occur. Friedlander (1961, 1965) has proposed a theory of *self-preserving size distributions*, based on interactions of coagulation and sedimentation, to explain the prevalence of the power-law relationship. Junge (1969) has commented on that theory.

3.2.4 Measured and Deduced Size Distributions

Aerosol investigations at a variety of locations have verified that the power law is a good representation of particle size distribution. First, there were the early measurements by Junge (1958, 1963) in Germany, which led to formulation (3-19). Pasceri and Friedlander (1965) sampled many hazes at Baltimore, finding that $n(r)$ varied as r^{-4} over an approximate size range 0.2 to 20 μm. Clark and Whitby (1967) obtained similar results at Minneapolis, as did Pueschel and Noll (1967) near Seattle. Blifford and Ringer (1969) sampled tropospheric aerosols over Scottsbluff and determined that $dN/d \log r$ varied approximately as r^{-3}. Whitby (1971) and Whitby et al. (1972a,b) (see also Hidy 1972) made about 300 measurements of Los Angeles aerosols, day and night, over a 4-week period. Their data, which cover the size range from about 0.02 to 2 μm show that $n(r)$ had an r^{-4} dependence in most cases. Many samplings of maritime aerosols in Atlantic and Pacific regions are effectively represented by the curves in Figure 3.9*b*.

One set of data from Clark and Whitby (1967) is discussed here to illustrate the power law and its application. Employing instrumentation to measure the volume of particles contained in a measured volume of aerosol, they found that

$$n(r) = 0.05\phi r^{-4} \tag{3-22}$$

with r in micrometers. This expression is equivalent to (3-20) with $v = 3$. The term ϕ is the volume concentration of particles, in cubic micrometers per cubic centimeter of aerosol; the product 0.05ϕ corresponds to c in (3-20). The relationship (3-22) was valid for particles in the size range 0.05 to 3 μm, which thus included larger Aitken nuclei. From averaged data, a representative value of ϕ was such that

$$n(r) = 2.49 r^{-4} \tag{3-23}$$

which is plotted in Figure 3.10.

Particle concentrations for given size ranges are readily calculated from the power law when the data are such that the factor c is known, as in (3-23). Integration of this expression between radii limits of 0.1 and 1.0 μm yields $N \approx 830\ \text{cm}^{-3}$. When the limits are 0.2 and 1.0 μm, $N \approx 104\ \text{cm}^{-3}$, or only one-eighth of the first value, thus indicating the strong influence of smaller particles on concentration. This influence is strikingly shown in Figure 3.10. Also plotted is the cumulative concentration N_c defined by (3-8) and obtained here by integrating (3-23) between the

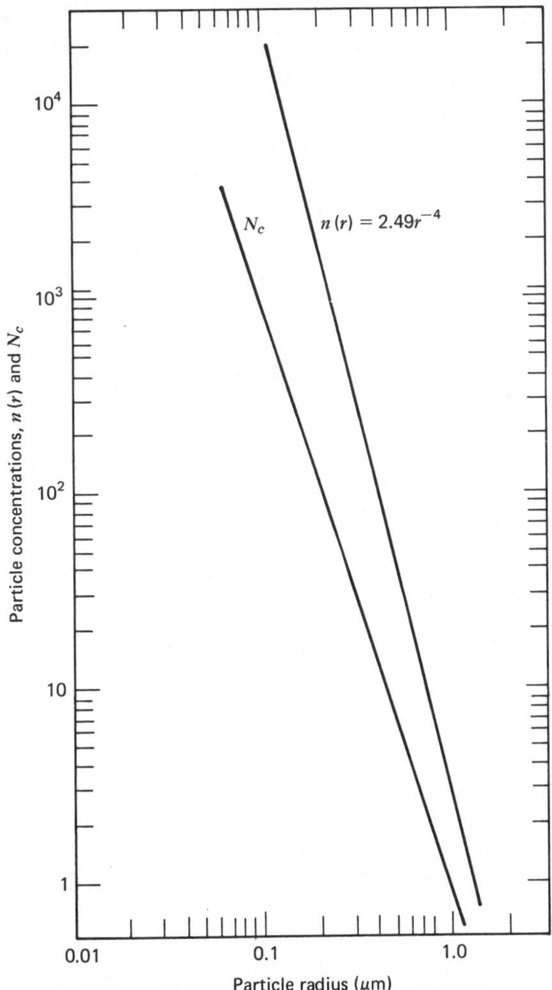

Figure 3.10 Particle size distribution and cumulative concentration of a Minneapolis aerosol. Data from Clark and Whitby (1967).

limits r and ∞. (We ignore any difficulties this upper limit might cause elsewhere.) With these limits, N_c is the number of particles having radii greater than r, so that the number within any radius interval is equal to the difference between the corresponding ordinates.

Many workers have studied the relationship between size distribution and scattering characteristics, since these two observables provide independent views of aerosol characteristics. Two general methods are employed in such studies: computation of scattering attenuation coefficients from measured size distributions, and computation of size distributions from measured attenuation coefficients. In either case, the computed values are then compared with values measured for that aerosol or a similar one. Considering the variability of the atmosphere, one would expect the strongest correlation when the size distribution and the coefficient are measured in the same sample. This was seldom done until recently, as described below.

Attenuation coefficients related to atmospheric spectral transmittance by (1-49) are determined from measurements of the latter in several narrow bands, using either the sun or a searchlight as a source. When the sun is the source, transmittance necessarily is measured over a slant path from the observer's altitude to the atmospheric limit. When a searchlight is used, transmittance usually is measured over a path of several kilometers. The coefficient also can be found more easily but less accurately from estimates of the visual range, according to (1-71). Whatever the method, the value of coefficient so obtained is an average for a relatively long path. In contrast, direct measurements of particle size and of volume and mass concentrations are applicable to a discrete site or a relatively small region where the aerosol exhibits temporal variations peculiar to that locality. Hence the two observables—particle content and attenuation coefficient—are not really measured in the same sample.

Despite this basic difficulty, which can be alleviated by averaging, fair correlations between the two observables have been disclosed. Pueschel and Noll (1967) measured the size distributions of Seattle aerosols, for which they generally found that $n(r) \propto r^{-3}$, and then computed corresponding values of the attenuation coefficient and meteorological range R_m. Comparisons of R_m with estimates of the visual range R_v for 16 test runs yielded fair agreements between these basically different measures. In analyzing their results these investigators discuss several sources of error that are more or less common to all such experiments. Averaging over the range of refractive indices of 1.33 (water) and 1.6 (salt) introduced errors not greater than $\pm 20\%$ in computations of attenuation coefficients. Errors due to using observer estimates of visual range rather than values (of transmittance, for example) measured by instrumentation were not greater than 20%. The effects of aerosol inhomogeneities along the paths used for estimating visual ranges were difficult to evaluate.

Using an inverse procedure, Yamamoto and Tanaka (1969) computed particle size distributions and then the volume concentrations of Chesapeake Bay aerosols for which Knestrick et al. (1962) had measured attenuation coefficients. The computed distributions follow the power law (3-20), with $2 < v < 3$, but in some cases slight curvatures of the plots indicated that v was not constant over the size range. Figure 3.11 shows

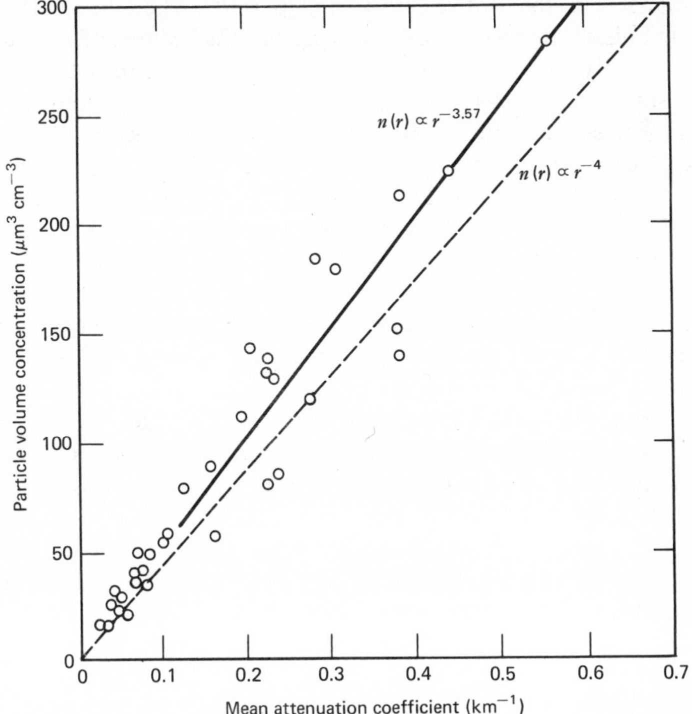

Figure 3.11 Volume concentration of aerosol particles versus attenuation coefficient of a Chesapeake Bay haze. From Yamamoto and Tanaka (1969).

some of their data as averaged over the size range 0.1 to 5 μm and the wavelength range 0.35 to 5 μm. A best fit to the data points is provided by the function

$$n(r) \propto r^{-3.57}$$

shown as a solid line, which is not greatly different from the function indicated for the dashed line. When the shape or slope of the size distribution curve is invariant with changes in concentration, a linear relationship exists between attenuation coefficient and concentration. The data points in the figure follow such a course to a good approximation.

Many cases occur in which the size distribution inferred from attenuation measurements departs notably from a single power-law distribution. Curcio (1961) and Curcio et al. (1961) found that their data were explained most easily by a two-component aerosol in which nearly all the particle sizes conformed to a continental-type distribution and the remaining few to a maritime type. Volz (1970c) made spectral measurements of solar and sky radiances in the Caribbean during periods when the trade winds carried in Sahara dust as evidenced by particle sampling at Barbados. For this presumably two-component aerosol, the optical data were such that no very simple size distribution could be inferred. It appeared, however, that a power-law function with $1.5 < v < 2.5$, enhanced by a few large particles, could explain the data. From similar types of measurements in southern Thailand hazes, Volz and Sheehan (1971) inferred a power-law distribution with $v = 3$ for $0.3\ \mu\text{m} < r < 2\ \mu\text{m}$ and $v = 4$ for $0.1\ \mu\text{m} < r < 0.3\ \mu\text{m}$. These are indeed steep distributions, as may be appreciated from (3-20).

In some cases attenuation data indicate that size distributions can be approximated best by normal and log-normal functions. Gebbie et al. (1951) investigated the spectral attenuation by haze along the east coast of Scotland and determined the characteristics of a size distribution that would explain their data. This turned out to be essentially a normal distribution centered at the particle radius occurring most frequently rather than at the mean radius. These investigators pointed out, however, that this was not necessarily a unique answer to the problem. Quenzel (1970) measured the spectral attenuation of solar flux in the tropical Atlantic, where westerly trade winds prevail, and found that it *increased* with wavelength in the visual and near infrared. Such behavior is often called *anomalous extinction.* The particle sizes inferred from the data showed log-normal distributions having maxima in the range 0.3 to 0.9 μm. These distributions may have been superposed on a power-law characteristic. Porch et al. (1973) analyzed other cases of anomalous extinction that in the extreme would produce a blue sun or moon such as that reported in Section 3.1.2. These investigators found this extinction could be explained by log-normal distributions having mean radii in the range 0.4 to 0.9 μm. Normal and log-normal distributions are treated by Green and Lane (1957), Fuchs (1964), Cadle (1965, 1975), Kerker (1969), and Butcher and Charlson (1972).

Development of the integrating nephelometer, described by Charlson et al. (1967) and Butcher and Charlson (1972), makes possible the measurement of attenuation coefficients without recourse to long paths. This is done by forcing a stream of aerosol through a small illuminated chamber in the instrument and measuring the resulting scattering. The volume and mass concentration of the particles, in micrograms per cubic meter, are measured by particle sampling techniques at the nephelometer site. Thus the optical and physical measurements are made on essentially the same aerosol sample at about the same time. One early result from such experiments is shown in Figure 3.12, where we see that the ratio of particle

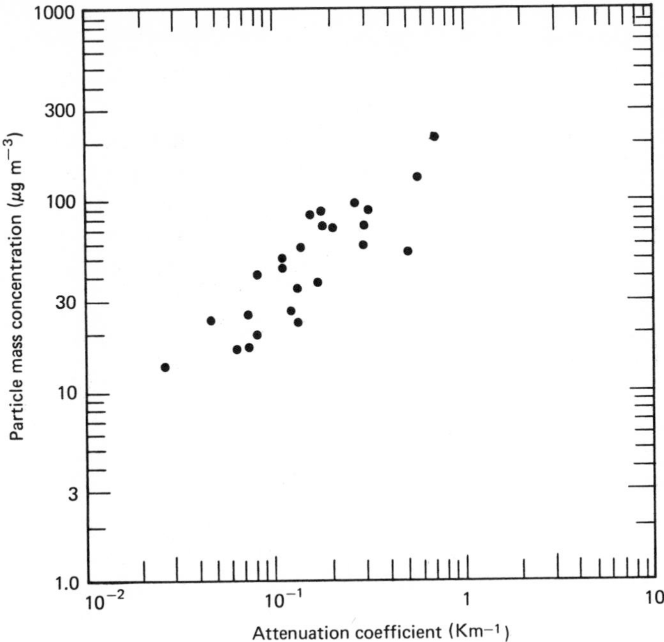

Figure 3.12 Mass concentration of aerosol particles versus attenuation coefficient of a Seattle haze. From Charlson et al. (1967).

mass concentration to attenuation coefficient tends to have a constant value. Investigations of Los Angeles smog with the integrating nephelometer are described by Charlson (1972), Ensor et al. (1972), Thielke et al. (1972), and Charlson et al. (1972).

3.3 VERTICAL DISTRIBUTIONS OF HAZE AEROSOLS

Just as the density of the permanent gases and water vapor of the atmosphere steadily decreases as altitude increases, so do the concentrations of haze particles. This decrease, primarily an exponential one, is an important factor in atmospheric scattering. In this section we review typical data obtained by physical sampling and optical probing techniques. Analytic functions that define the altitude dependence of particle concentration are given, and the concept of a homogeneous haze atmosphere is presented. Attention is then called to several published tabulations of model vertical profiles that have great utility in atmospheric optics.

3.3.1 Data from Sampling and Optical Probing

Although many aerosol investigations have been made at ground level and in the stratosphere, relatively few have been made in the troposphere where most of the particles reside. Blifford and Ringer (1969) measured particle sizes and concentrations by airborne sampling techniques at altitudes to 9 km near Scottsbluff. They found that size distributions in these rural aerosols, over the size range 0.13 to 5.5 μm, followed the power law (3-20), with the exponent v varying between 2 and 3. Average concentration for the complete size range was 11.0 cm^{-3} at 1.5 km, and 0.04 cm^{-3} at 9 km. The value of v generally decreased with altitude above 3 km. At a given altitude the concentration frequently changed by a factor as great as 4 in a 1-hr period, indicating both temporal and spatial variations. Seasonal effects were observed, the greatest concentrations occurring in midsummer, the time of greatest biochemical activity.

Blifford (1970) made similar measurements over the Pacific Ocean, Death Valley, and Barbados. The last-named location is marked by large amounts of dust carried from the Sahara by trade winds, as mentioned previously. He found greater particle concentrations at Barbados than at the other locations, and quite variable size distributions in the low altitudes at all the locations. Above a few kilometers, however, trends toward uniformity were apparent, and the distributions became similar to those of the midcontinent aerosols near Scottsbluff. By chemical analyses of the particle samples obtained during these and similar programs, Gillette and Blifford (1971) determined the element compositions of the aerosols. These investigators present such data for several altitudes from 15 m to 9.1 km at the following locations: Scottsbluff, Pacific offshore, mid-Pacific (Palmyra), Death Valley, Orinoco Delta, and Chicago.

The tenuous aerosols in the stratosphere have been studied extensively, partly in connection with radioactive fallout. Rosen (1964, 1967, 1968, 1971), Junge (1961, 1963), Junge and Manson (1961), and Junge et al. (1961) investigated size distribution, composition, and origin to altitudes of 30 km by airborne sampling methods. A representative size distribution of stratospheric particles is shown in Figure 3.13. Although the concentration of Aitken nuclei decreases rapidly with altitude, often becoming practically zero near 20 km, the size distribution of larger particles does not appear to change greatly with altitude. Particles having a radius in the range 0.1 to 1.0 μm usually reach the greatest concentration at about 20 km; many of these are the sulfate particles and sulfuric acid droplets mentioned previously. Particles having a radius greater than 1 μm and correspondingly greater terminal velocities according to Figure 3.4 are

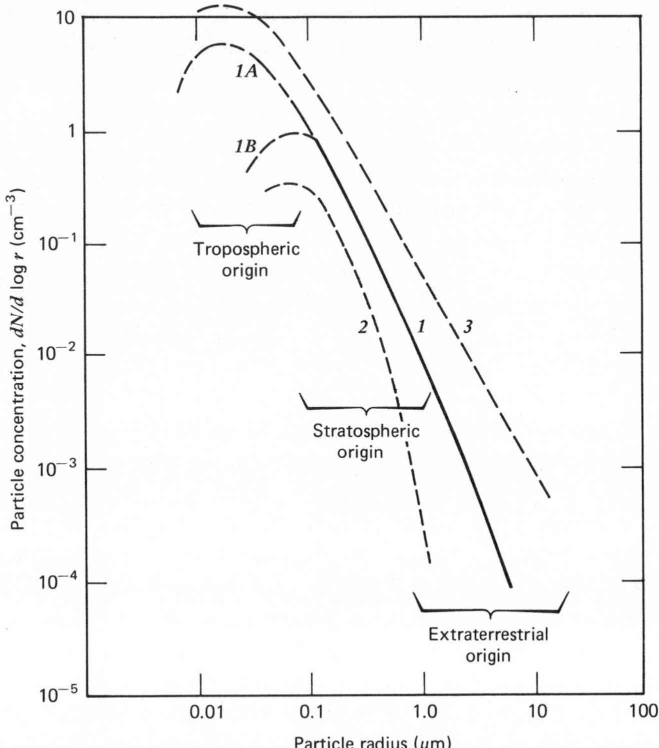

Figure 3.13 Representative size distribution of stratospheric particles. Curve *1* is overall average. Curve *1A* is for $r < 0.1$ μm in lower stratosphere. Curve *1B* is for $r < 0.1$ μm at altitudes above 20 km. Curves *2* and *3* are estimated confidence limits. From Junge et al. (1961).

thought to have an extraterrestrial origin. Recent and comprehensive investigations of stratospheric aerosols, on a global basis, are reported in a series of papers by Hoffman et al. (1975), Rosen et al. (1975), and Pinnick et al. (1976).

Optical probes operating from ground sites have been increasingly used to investigate the atmospheric particle content and its distribution with altitude. Compared to airborne sampling methods, they are able to monitor continuously the vertical aerosol structure over periods of several hours—a desirable feature, for example, where pollution buildup is to be studied. In addition, by avoiding requirements for balloons or aircraft, they tend to be more economical and convenient in use. Two general classes of optical probing techniques have been developed: (1) searchlight beams, first employed about 1936, and (2) lidar beams, first used in the early 1960s. Because scattering is the operational basis of these techniques, we call attention here to their essential features and the kinds of data they provide.

The early use of searchlights for atmospheric probing is reviewed by Middleton (1952). Basically the searchlight is pointed vertically, or at known zenith angles, and the radiance of a selected portion of the path is measured either by photographic or electrooptical means from a detection site sufficiently far from the searchlight to produce a favorable geometry. For a given beam candlepower, the radiance of the observed portion depends on the volume angular scattering coefficient at that altitude and on the vertical distribution of the total coefficient. The particle concentration is then deduced from the radiance data. Experiments of this type were conducted by Hulburt (1937, 1946), who found that the beam was visible to an altitude of 20 km on clear nights and could be photographed to 28 km with lengthy exposures. Analysis of the photographs indicated that at 5 km the scattering by particles was about seven times the Rayleigh value, while above 10 km most of the scattering was Rayleigh.

Further experiments with searchlights were performed by Johnson et al. (1939) and Stewart et al. (1949). Romantzov and Khvostikov (1946) enhanced the technique by using polarizing filters with the camera. Mironov et al. (1946) employed ultraviolet light which permitted daytime operation, in what may have been an early "solar blind" technique. Elterman (1966a,b) used an intensity-modulated searchlight and synchronous detection at a site 30 km distant. Scattering data were obtained to altitudes as great as 70 km, and operations over periods of months yielded reliable profile data to 35 km. These were employed in revisions of his atmospheric attenuation model described in Section 3.3.3. One of the profiles, in terms of the attenuation (total scattering) coefficient versus altitude, is shown in Figure 3.14. Quite evident is aerosol stratification at 18 km, identified with the layer of sulfate particles (or sulfuric acid droplets) noted in Section 3.1.2. Above this layer aerosol scattering decreases rapidly, becoming less than the Rayleigh value at about 22 km.

Lidar instruments provide a direct method of investigating the vertical structures of haze aerosols. As usually emplaced, the transmitter and receiver are at the same site, so that the backscattering coefficient of the particles and the two-way attenuation of the path determine the ratio of received signal to transmitted signal. Early investigations in this field were made by Fiocco and Smullin (1963) and Fiocco and Columbo (1964). Since then many workers have developed lidar into a practical instrument for probing the atmosphere. Among these investigators are Barrett and Ben-Dov (1967), Clemeshaw et al. (1967), Fiocco and De Wolf (1968), Collis (1969), and Schuster (1970). Survey articles with extensive bibliog-

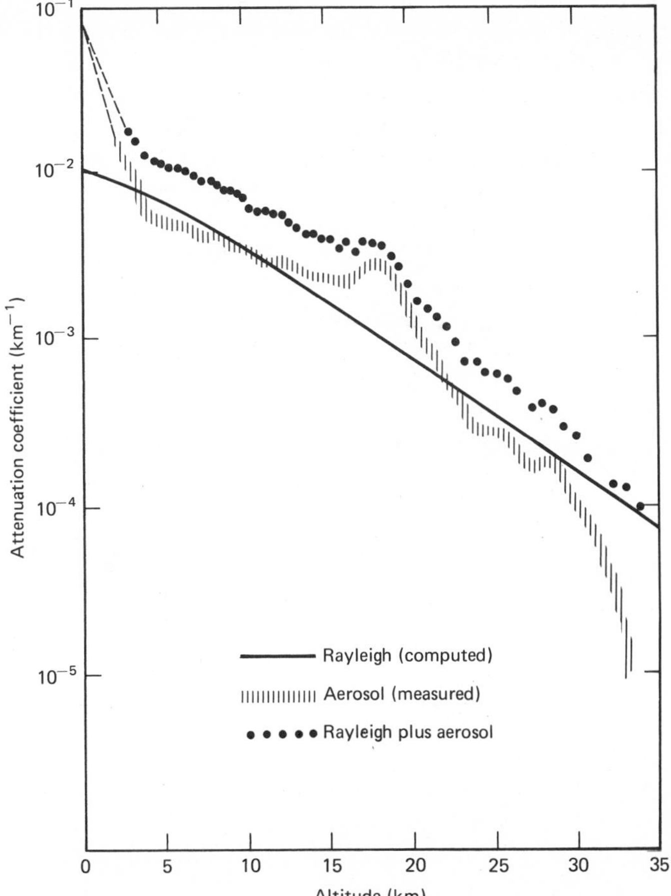

Figure 3.14 Vertical distribution of attenuation coefficient for New Mexico aerosol. From Elterman (1966b).

raphies have been written by Goyer and Watson (1968), Collis (1970), and Evans and Collis (1970). An interesting comparison of aerosol profiles measured concurrently by lidar and balloon-borne sampling over Laramie is given by Northam et al. (1974). Good agreement is shown between the profiles, all of which exhibit considerable structure. Recent measurements over Jerusalem are reported by Cohen and Graber (1975).

Other optical methods also provide information on the haze aerosol. Even in the clearest air the sun's disk appears surrounded by a halo of bright sky called the *aureole*, which is several degrees in diameter. This is caused by the forward-scattering characteristics of the particles along the solar ray path. From the patterns in Figure 1.5 we can appreciate that measurements of aureole radiance analyzed in terms of scattering theory will give information on the average particle size. Also, measurements at several altitudes should provide data from which a vertical distribution of concentration can be inferred. Investigations of this type have been made by Newkirk (1956), Newkirk and Eddy (1964), and Deirmendjian (1969), who also gives a broad survey of this field. Recent work, both experimental and theoretical, has been accomplished by Green et al. (1971, 1972), who provide an extensive bibliography, and by Ward et al. (1973).

Measurements of atmospheric transmittance and sky radiance, with particular reference to stratospheric dust content, have been made by Volz and Goody (1962), Volz (1969b, 1970a,b), Quenzel (1970), and Kondratyev et al. (1971). Further investigations dealing with the stratification of high-altitude aerosols are reported by Kondratyev et al. (1967), Volz (1969a, 1971), and Elterman et al. (1973).

3.3.2 Average Altitude Dependence of Concentration

Although vertical distributions of haze aerosols always exhibit irregularities, they tend to conform to average long-term patterns. Penndorf (1954) analyzed the data from airborne measurements of the attenuation of solar flux and found that particle concentration decreased exponentially as altitude increased in the first 5 km, or about halfway to the tropopause. Size distributions, although not invariant with altitude, did not appear to change greatly in this region, and this view was generally adopted by later workers. Hence on the average the particle concentration N_Z at altitude Z can be represented by

$$N_Z = N_0 \exp\left(-\frac{Z}{H_p}\right) \tag{3-24}$$

where N_0 is the concentration at ground level, and H_p is the scale height for the aerosol. Penndorf's study of the data available at that time showed that H_p varied between 1 and 1.4 km and that 1.2 km was a good average. This value has been used in many studies since that time.

The concept of equivalent path, developed in Section 2.5.3 for slant paths in the molecular atmosphere, is applicable to the average haze atmosphere. Writing (2-83) in terms of the scale height H_p, we have

$$\bar{R} = H_p \sec\zeta\left[\exp\left(-\frac{R_1\cos\zeta}{H_p}\right) - \exp\left(-\frac{R_2\cos\zeta}{H_p}\right)\right] \tag{3-25}$$

for the general expression. Taking the common situation in which the lower end of the path is at the ground so that R_1 is equal to zero, and using

1.2 km as the scale height value, (3-25) becomes

$$\bar{R} = 1.2 \sec \zeta [1 - \exp(-0.83 R_2 \cos \zeta)] \tag{3-26}$$

This is plotted in Figure 3.15 for several zenith angles and for values of range sufficient to produce asymptotic convergence. Although these expressions were developed for the flat earth approximation with an associated nominal restriction to zenith angles less than about 75 deg, we should realize that the transmission of haze slant paths at large zenith angles is usually poor. Thus the transmittances likely to be of interest, except when the sun is being observed, generally do not involve large values of the geometric range. This fact often justifies the (cautious) use of the flat earth approximation at large zenith angles.

Basic reasons for the exponential dependence of particle concentration on altitude are not far to seek. The lower troposphere is usually well stirred by winds and convection currents, so that the mixing ratio of particles to air tends to remain constant in a given weather state. This mixing region extends from the ground to an altitude of several kilometers. Blifford and Ringer (1969), for example, found a fairly constant mixing ratio to about 5 km for particles in the size range 0.5 to 2.0 μm. The mixing ratio may be expressed either as total mass of particles to unit mass of air, or as particle concentration to molecular number density. Air density, however, decreases exponentially as a given parcel of air rises

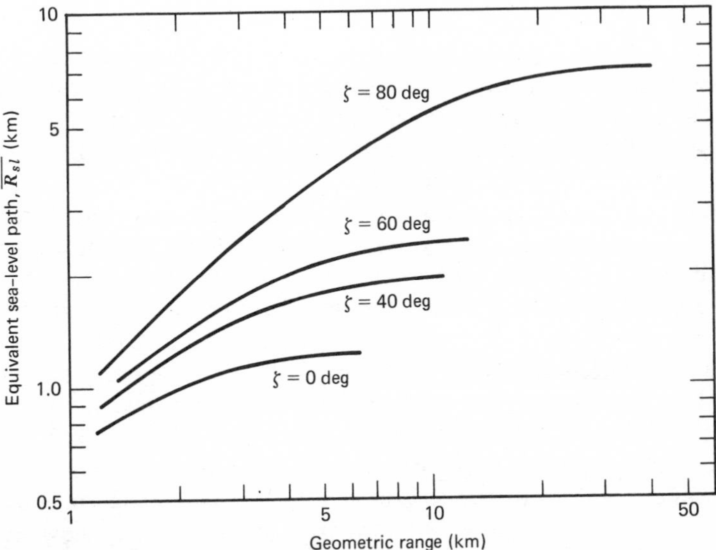

Figure 3.15 Equivalent path length versus geometric range in the haze atmosphere for the indicated zenith angles.

and expands, so that from this cause alone the particle concentration decreases with altitude. Additional factors in the vertical distribution are the particle removal processes described in Section 3.1.4 and the fact that many of the particle sources are at or near the earth's surface.

Related to the exponential form of the distribution is the concept of a homogeneous haze atmosphere or layer analogous to the constant-density model of the gas atmosphere discussed in Section 2.3.1. In the present concept the particle size distribution is assumed not to change with altitude, and all the particles aloft are considered to be uniformly distributed throughout the layer. The particle concentration in the layer is equal to the actual value N_0 at ground level, and the thickness of the layer is equal to the scale height H_p of the real haze atmosphere. This equality can be verified by integrating (3-24) between the limits 0 and ∞, which yields

$$N_{\text{total}} = N_0 H_p \tag{3-27}$$

This same result is obtained, of course, by multiplying N_0 by the thickness of the homogeneous layer. The homogeneous haze atmosphere is a useful model in which scattering by the entire vertical extent of the atmosphere is involved. Bullrich (1964) and de Bary et al. (1965) have based their computations of atmospheric scattering on a homogeneous haze layer having a vertical extent of 1.25 km.

3.3.3 Model Vertical Profiles of Haze

Based on the considerations in the two preceding sections, several model vertical profiles of haze have been established and are in extensive use. As part of his atmospheric attenuation model, Elterman (1964) set up such a profile from measurements by Dunkelman (1952), Baum and Dunkelman (1955), and Curcio et al. (1961) in the ultraviolet, visual, and infrared at sea-level locations. In the last-named set of measurements, when the meteorological range was 25 km, the haze aerosol was characterized by a power-law distribution having $v \approx 2.8$ and $N \approx 200\,\text{cm}^{-3}$. Employing these values, a scale height of 1.2 km from Penndorf (1954), and stratospheric aerosol data from Chagnon and Junge (1961), Elterman constructed the profile shown by the solid line in Figure 3.16. We may note the exponential decrease in concentration from $200\,\text{cm}^{-3}$ at sea level to about $0.02\,\text{cm}^{-3}$ at 10 km, and the layer near 20 km.

The dust content of the atmosphere exhibits long-term variations; stratospheric effects of the Krakatoa and Genung Agung volcanic eruptions are noted in Section 3.1.1. In recent years many new data on stratospheric aerosols have become available from the techniques described in Section 3.3.1. Accordingly, Elterman (1968) revised the

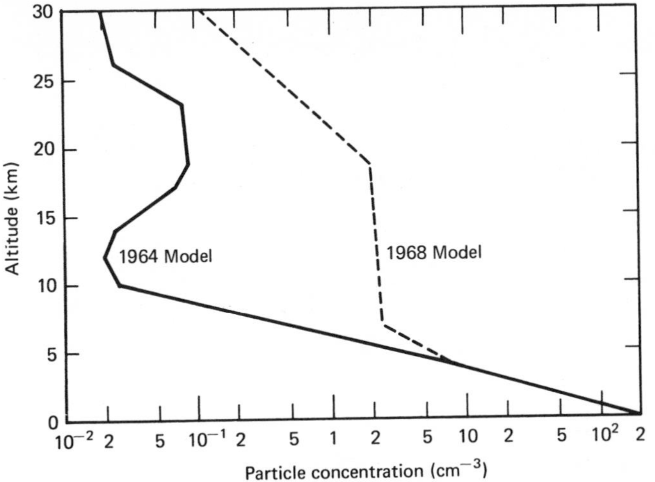

Figure 3.16 Vertical profile of particle concentration for an atmospheric attenuation model. From Elterman (1964, 1968).

aerosol attenuation profile of the 1964 model above 3 km to incorporate the new information. The corresponding concentration profile, as deduced from the attenuation profile, is shown by the dashed line in Figure 3.16. The revision is significant mostly for very long paths at large zenith angles.

The average thickness of the aerosol mixing region, as found from several types of investigations, is often taken as approximately 5 km. Within this region the aerosol profile tends to be dependent on the actual aerosol conditions at ground level. This means that the aerosol scale height itself should be a function of particle concentration at the ground, as manifested in the meteorological range, for example. A reasonable question then is: Can the aerosol profile in the first 5 km, where most of the attenuation occurs, be related in a meaningful way to the value of the surface meteorological range?

Elterman (1970a,b) found that such an average relationship tends to exist and constructed an additional model of attenuation for eight values of surface R_m between 2 and 13 km. Each value of R_m has an associated value of H_p, listed in Table 3.3, which also gives the corresponding value of the total scattering coefficient at the wavelength 0.55 μm. This 1970 model embodies the same parameters of molecular scattering and ozone absorption in the previous models, and the same parameters of aerosol scattering above 5 km in the 1968 model. McClatchey et al. (1971), in an atmospheric attenuation model that takes account of seasons and latitude

TABLE 3.3 ATTENUATION COEFFICIENTS AND AEROSOL SCALE HEIGHT VERSUS METEOROLOGICAL RANGE

R_m (km)	β_{sc} (km^{-1})	β_m (km^{-1})	β_p (km^{-1})	H_p (km)
2	1.955	0.0116	1.943	0.84
3	1.303	0.0116	1.291	0.90
4	0.978	0.0116	0.966	0.95
5	0.782	0.0116	0.770	0.99
6	0.652	0.0116	0.640	1.03
8	0.489	0.0116	0.476	1.10
10	0.391	0.0116	0.379	1.15
13	0.301	0.0116	0.289	1.23

Source: Elterman (1970a).

regions, also give tabulations of particle concentration versus altitude. Their data are based on a size distribution function devised by Deirmendjian (1963a) for a continental aerosol, but the altitude dependence is the same as that employed by Elterman (1968).

3.4 DROPLET CHARACTERISTICS OF FOG, CLOUD, AND RAIN

Fog droplets develop from condensation nuclei when the air layer near the ground becomes saturated, hence fog and haze have a common origin. Perceptible haze, however, extends to an altitude of several miles, whereas a typical fog bank is only a few hundred feet thick. In addition, there are great disparities between the size distributions of haze particles and fog droplets. Cloud droplets develop from condensation nuclei when lower-altitude air rises and becomes saturated; thus the droplets in lower-altitude clouds are similar to those in fog. But, by definition, fog rests on the ground and is often considered a misplaced cloud, a term that emphasizes the similarity of composition.

Although fog and cloud have been studied extensively, surprisingly little is known about the detailed processes of formation and dissipation from direct observation at the droplet level. This lack of knowledge reflects the difficulties in observing very minute droplets of water in a free atmosphere. The difficulties are especially great with respect to clouds, as can be imagined. Measurement methods have been based on sampling, replication, and determination of volume concentration. In all these methods, the droplets are snatched from their environment and then either immobilized or destroyed in the measurement, so that their dynamic behavior cannot be observed. Some of the sampling methods are known to discriminate against droplets having a radius of less than about 3 μm, thereby raising questions as to the statistical validity of a sample. The discrimination is due to largely unavoidable instrumental factors such as alterations in droplet trajectories near the collecting point, coalescence of droplets at the collecting point, and changes in relative humidity of the air sample. Various methods of sampling and measurement and their problems are discussed by Houghton (1951), Johnson (1954), Green and Lane (1957), Byers (1965), Speyers-Duran and Braham (1967), Averitt and Rushkin (1967), and Kumai (1973).

3.4.1 Origins and Classes of Fog

The environment factors associated with fog are described by Myers (1968) with many photographs and diagrams. Byers (1974), Tverskoi (1965), Petterssen (1969), and most meteorological textbooks describe fog in some technical detail, while George (1951) reviews the geographical factors. Fog evolves when the relative humidity of an air parcel is brought to an approximate saturation value. Some of the nuclei then grow by condensation into water droplets, as described in Section 3.1.3. If the nuclei are especially large or very hygroscopic, growth may take place at less than the saturation value. Because very hygroscopic nuclei are quite plentiful in a metropolitan region, fogs may form there more easily and persist longer than in the country. True fog should be distinguished from dense metropolitan haze which is usually loaded with pollutants.

The increase in relative humidity required to turn haze into fog can be brought about by any of several meteorological processes. Each processes, when dominant, produces a particular class of fog: advection, radiation, advection-radiation, evaporation, upslope, and frontal.

The term *advection* refers to the horizontal motion of a mass of air, and *advection fog* is formed when warm, moist air moves across water or land having a lower temperature. Such fog is common off the Grand Banks, where air with a long track over the Gulf Stream blows across the Labrador Current, making this one of the world's foggiest regions. Similar fog is an almost daily feature of the Pacific coastline of the United States. *Radiation fog* is formed when the ground loses heat at night by radiation through a clear atmosphere and chills the overlying moist air. Fog of this class, previously noted in Section 3.1.3, occurs frequently in inland areas and may be intensified by cold air drainage from sloping terrain. *Advection-radiation* fog is formed by a combination of the two processes and is common in the Great Lakes region and the Atlantic coastal plains.

Evaporation fog, often called warm water fog, is produced when the vapor from a water surface rises into colder, comparatively quiet air. Depending on the temperature difference, a shallow layer of dense fog then develops. Fog of this type often is observed over lakes and streams during autumn mornings, especially when an overnight cold snap has occurred. Evaporation fog is also produced by warm rain falling through a layer of cold air. *Frontal fog* associated with a weather front is formed when a mass of warm moist air slides over and mingles with a colder surface layer. Frontal fog thus is related to warm rain fog, because the temperature inversion required for the latter frequently is provided by a weather front.

Apart from meteorological interest, the practical distinction between haze and fog lies in the greatly reduced visibility imposed by the latter. The International Visibility Code recognizes different degrees of each condition, as listed in Table 1.8. It is evident that the visual range changes by about 10^{-3} in going from *exceptionally clear* to *dense fog*, but the table itself does not provide a sharp distinction between haze and fog. Very often either term is employed when the visual range is several kilometers, although in aviation usage fog is said to exist when the daylight meteorological range in a horizontal direction is 1 km or less. Physically, haze does not change abruptly into fog. Instead, the transition is manifested in an intermediate condition called *mist*, although this term is also frequently applied to a light drizzle. The distinctive, sometimes puzzling, optical properties of mist are discussed by Eldridge (1969).

3.4.2 Fog Droplet Size Distributions

The development of haze into a radiation fog is noted in Section 3.1.3 as a grossly observed phenomenon. We now look more closely at the droplet growth in a model fog to provide a theoretical basis for the observed size distributions. The theoretical studies of cloud development by Neiburger and Chien (1960) show the manner in which a distribution of condensation nuclei is transformed into a broader one which encompasses cloud droplets. Eldridge (1966, 1969) considers their stratus cloud study to be generally applicable to fog. The transformation of the distribution is shown by the sequence of curves in Figure 3.17. At time zero and at moderate relative humidity, the nuclei distribution is described by (3-20) and typically illustrated by the curve for $n(r)$ in Figure 3.10. The condition *haze* exists. Now let the temperature of the air parcel be reduced slowly so that the relative humidity increases and a small value of supersaturation, with respect to a plane surface of water, is reached at about 44 min. The larger droplets grow by condensation throughout this period, and the smaller droplets also grow when sufficient supersaturation is reached. The distribution function, as shown for 40 min, begins to

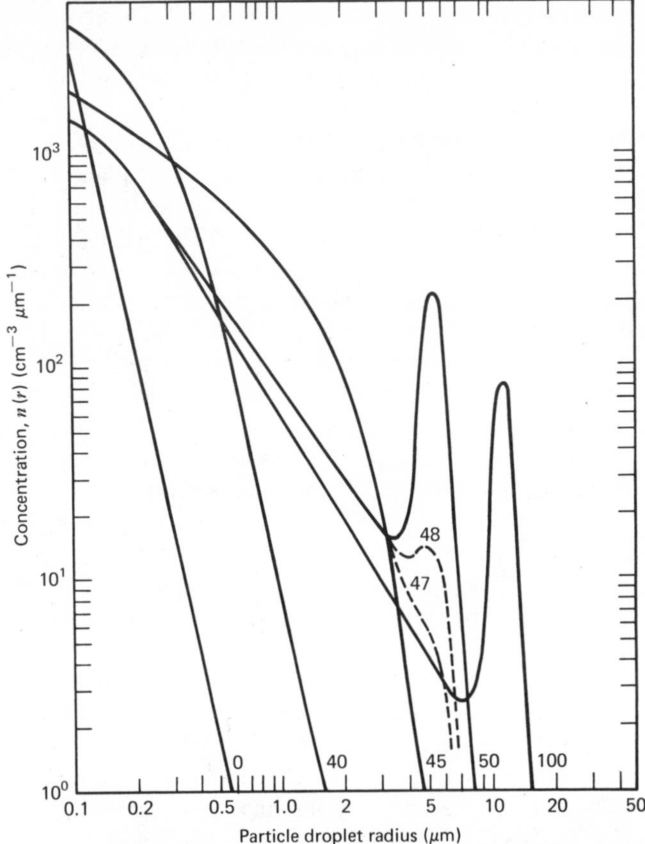

Figure 3.17 Theoretical size distribution of fog droplets as a function of elapsed time. The time in minutes after start is shown for each curve. After Neiburger and Chien (1960) and Eldridge (1969).

depart from a straight line, thus indicating that (3-20) is losing validity for the smaller sizes.

Droplet growth continues until a sufficient amount of the latent heat of vaporization is released by condensation to reverse the temperature trend. This reduces the supersaturation. As can be deduced from Figure 3.1, droplets smaller than some critical size now shrink, but droplets larger than this size continue to grow. As a result of these and other processes described in the cited reference, a secondary maximum begins to develop at about 47 min. At 50 min the distribution becomes fully bimodal, with the second mode occurring near $r = 5\ \mu m$. This rapid change may correspond to the transition condition called *mist* in the preceding section. At 100 min the second mode shifts to about 12 μm, and the fog (stratus cloud) is now fully evolved. Barring outside influences, the size distribution now changes only slowly with time as a result of coagulation and fallout.

It is interesting that bimodal distributions found from physical theory, as in the above discussion, must be invoked to explain some of the measured spectral attenuation by fogs. Among such measurements, those made by Arnulf et al. (1957) in four types of fog near Paris, over the wavelength range 0.35 to 13 μm, are outstanding with respect to thoroughness and number of observations. They employed the spider web technique of Dessens (1949), described in Section 3.1.3, for capturing droplets; the associated difficulties in measurement and sources of error are described in their paper. From a study of their data, Eldridge (1966) synthesized for each type of fog the size distributions that would produce spectral attenuations resembling the measured values as closely as possible. The resulting synthetic distributions were bimodal and generally similar to the one in Figure 3.17. Eldridge (1969) discusses other spectral attenuation measurements which are explained most easily by bimodal distributions of this type.

Ferrara et al. (1970) measured the angular scattering of laser light by artificial fogs grown from smoke nuclei in a humidity chamber. They assumed that the size distribution, although changing with the relative humidity, was always defined by (3-15). When the relative humidity was increased from 85% to saturation, the entire distribution shifted to larger values of radii, with the $n(r)$ value 1 cm^{-3} μm^{-1} moving from radius 1 to 5 μm. Concurrently the maximum value of $n(r)$ decreased, as in Figure 3.17. These investigators suggest that the preselected function (3-15) represented only the main feature of the distribution and that a bimodal characteristic might have been present. Harris (1969) calculated the polarization characteristics of angular scattering at several laser wavelengths using the evolving Nieburger–Chien distributions of Figure 3.17. The results indicate that these polarization characteristics, changing with time, can provide a sensitive method for following the evolution of clouds (fogs). Recent experimental investigations of the formation and dissipation of artificial fogs, by means of laser scattering, are reported by Tonna (1975).

Chu and Hogg (1968) applied function (3-15) to fogs, modifying it to the form

$$n\left(\frac{r}{r_c}\right) = a\left(\frac{r}{r_c}\right)^{\alpha} \exp -\left[b\left(\frac{r}{r_c}\right)^{\gamma}\right] \tag{3-28}$$

where the ratio r/r_c replaces r in the original, and b has the value α/γ instead of that given by (3-17). Figure 3.18 shows four plots of (3-28) for

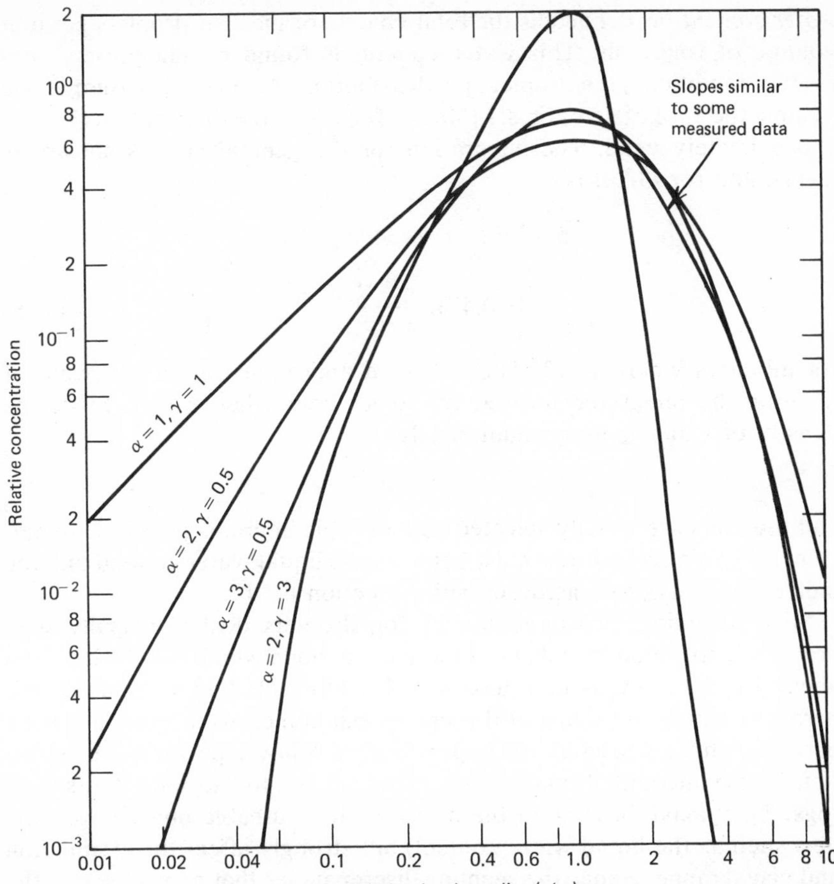

Figure 3.18 Fog droplet size distributions according to Eq. (3-28). In the figure r_c is the radius at which the maximum concentration occurs. From Chu and Hogg (1968). (Reprinted with permission from *The Bell System Technical Journal*, Copyright 1968, The American Telephone and Telegraph Company.)

the indicated combinations of α and γ. The slopes of three of the curves on the right side of their maxima are similar to those of the size distributions measured by Eldridge (1961). The remaining curve, which has a sharp maximum, corresponds to an asymptotic distribution of coagulating particles as found by Erkovich (1965). The distributions in Figure 3.18 have been used by Chu and Hogg (1968) and Rensch and Long (1970) in computations of scattering at laser wavelengths of 0.63, 3.5, and 10.6 μm.

A frequently used index of fog droplet size and number is the *liquid water content lw*, defined as the total volume or mass of droplets per unit volume of foggy air. This water content is found by integrating, with respect to volume, the droplet size distribution. As a simple example, we assume the case of an optically thin fog for which the distribution (3-20) is approximately valid. The integration for the general case is shown by (3-12), and for (3-20) is

$$\phi = \tfrac{4}{3}\pi \int_{r_1}^{r_2} r^3 0.434 c r^{-(v+1)}\, dr$$

$$\approx \tfrac{4}{3}\pi 0.434 c \frac{r_2^{(3-v)}}{3-v} \tag{3-29}$$

for all cases where $v < 3$. The indicated approximation is the result of ignoring the integrated term at the lower limit when $r_1 \ll r_2$. Since the density of water is unity, numerically

$$lw = \phi$$

and the units are usually selected so that lw is in grams per cubic meter. Generally fair correlations exist between the liquid water content and the meteorological range, as described in Section 6.3.1.

Most sampling measurements of fog droplets deal principally with droplets distributed in and about the second mode which may lie between about 4 and 10 μm, as in Figure 3.17. Usually sampling methods do not permit counting and sizing of the smaller but numerous droplets and haze particles whose sizes may still approximately follow a power-law distribution. These uncounted small members are present to varying extents in all fogs. They make little contribution to the measureable physical properties, such as the liquid water content, but strongly affect the attenuation and visual range. Hence the seeming discrepancies that may appear in the data obtained by different measurement techniques have understandable causes. An example of such a discrepancy and its resolution are given in Section 6.3.1.

Houghton (1934) and Houghton and Radford (1938) sampled many mountain slope and coastal fogs, finding droplet concentrations of 1 to 10 cm^{-3} and a liquid water content of 0.5 to 0.25 g m^{-3} over the approximate size range 3 to 70 μm. Silverman et al. (1964) and Thompson et al. (1967) describe a laser holographic instrument for counting and sizing fog droplets without recourse to physical sampling. Figure 3.19 shows a typical histogram of their data averaged over an 11-min period for a morning inland fog at Otis Air Force Base, Massachusetts. From the populations of the size classes shown, we deduce a concentration of about 8 cm^{-3} and a liquid water content of 0.07 g m^{-3}. Extremely thorough investigations of valley fog, through the cycle of formation, maturity, and

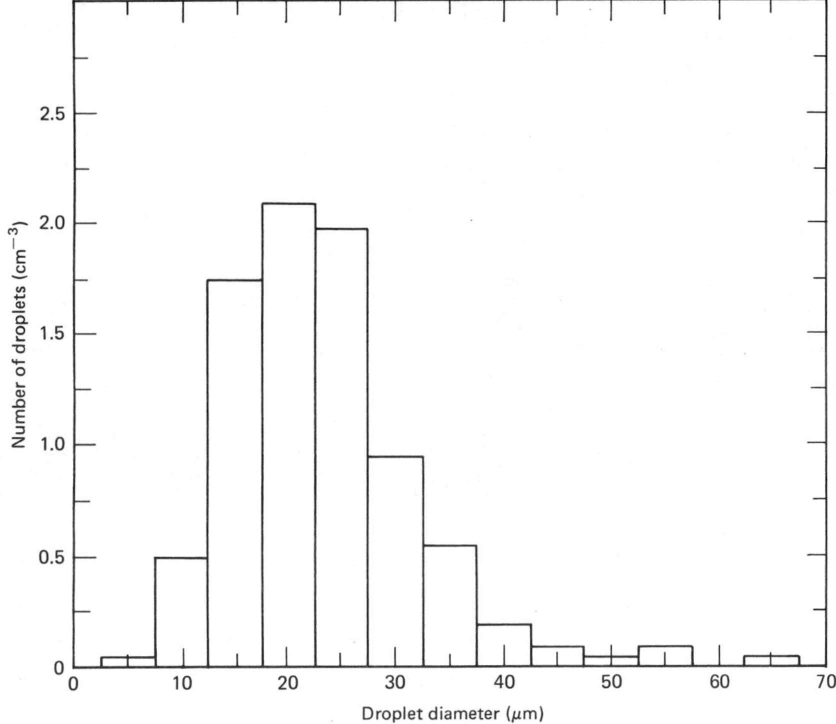

Figure 3.19 Laser holographic data on fog droplets. From Thompson et al. (1967).

dissipation, have been made by Pilie et al. (1975a,b). Their data cover measurements of meteorological factors, droplet size distribution, and liquid water content. Kumai (1973) investigated arctic advection fogs at Barrow, Alaska, measuring droplet size distribution, liquid water content, and visual range. The mean size distributions for two of these fogs are shown in Figure 3.20, where the ordinate scale is percent of droplets per 1.3-μm interval of radius. The relationships found by several investigators to exist between liquid water content and visual range are discussed in Section 6.3.1.

3.4.3 Cloud Classes and Droplet Size Distributions

The principal classes of clouds and their descriptions are listed in Table 3.4, which encompasses a wide range of physical conditions. From the standpoint of optics as well as meteorology, most of the important classes of clouds exist within the troposphere. Only passing notice can be given here to two classes found at much higher altitudes: *nacreous* and

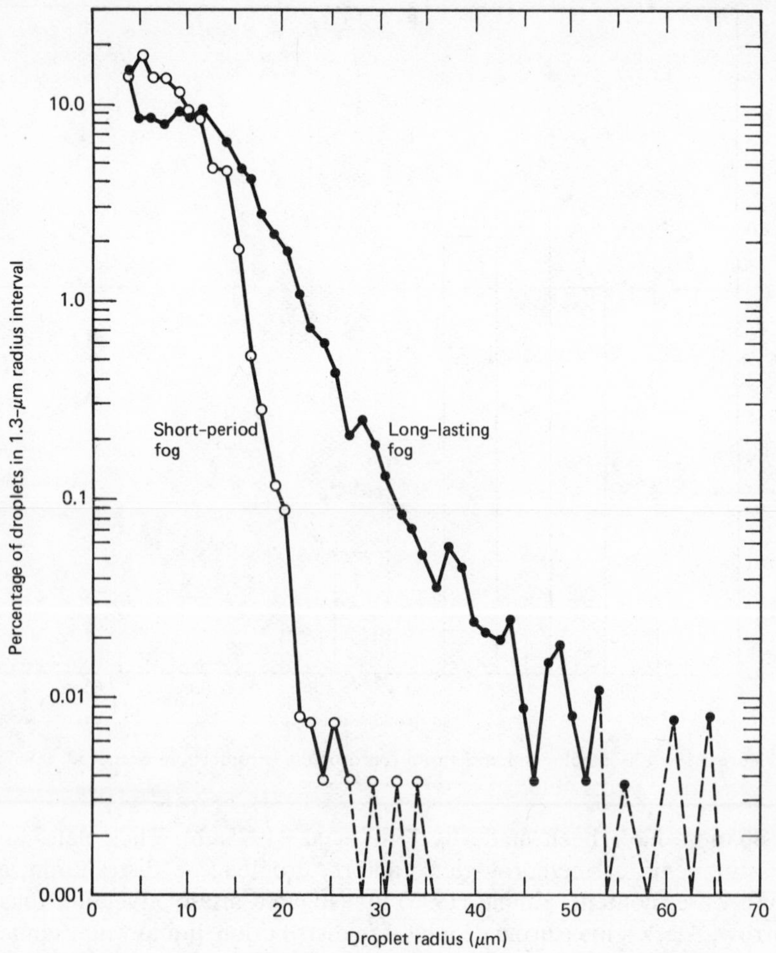

Figure 3.20 Droplet size distributions for arctic fogs. From Kumai (1973).

noctilucent clouds. The former is a special case of the *lenticular* class listed in Table 3.4 and is observed infrequently at altitudes near 25 km over mountainous terrain. Composed of long, prismatic ice crystals, nacreous clouds are brightly iridescent, from which they are often called *mother of pearl.* Noctilucent clouds, apparently composed of ice-coated dust grains, occasionally appear during summer dusk at high north latitudes. Floating near the mesopause at 90 km, they are considered indicators of high-altitude moisture regimes and circulation patterns. Christie (1969) discusses such aspects, while Soberman (1963) and

TABLE 3.4 PRINCIPAL CLASSES AND DESCRIPTIONS OF CLOUDS

Cloud type	Abbreviation	Description	In temperate regions: Height range	In temperate regions: Temperature range	Type of vertical air motion
		I. Stratiform or layer clouds			
High-level clouds					
Cirrus	Ci	Detached clouds composed of delicate white fibers and appearing in tufts, streaks, trails, feather plumes, or bands			
Cirrocumulus	Cc	A dappled layer or patch of cloud forming among cirrus. Composed of small white flakes or very small globules arranged more or less regularly in groups or lines, or more often as ripples resembling those of sand on the seashore	Above 20,000 ft	Below −25 °C	
Cirrostratus	Cs	A fused sheet of cirrus cloud which does not obscure the sun or moon, but gives rise to halos. Sometimes it appears as a diffuse white veil across the sky			Widespread, prolonged and regular ascent with vertical velocities of typically 5–10 cm sec^{-1}
Medium-level clouds					
Altostratus	As	A grey striated or fibrous veil, like thick Cs but without halo phenomena, through which the sun is seen only as a diffuse bright patch or not at all	7000–20,000 ft	0 to −25 °C	
Altocumulus	Ac	A dappled layer or patch of cloud composed of flattened globules which may be arranged in groups, lines, or waves collectively known as billows			

TABLE 3.4 (*Continued*)

Cloud type	Abbreviation	Description	In temperate regions: Height range	In temperate regions: Temperature range	Type of vertical air motion
Low-level clouds					
Stratocumulus	Sc	A layer of patches composed of laminae or globular masses arranged in groups, lines, or waves and having a soft, grey appearance. Very often the rolls are so close together that their edges join and give the undersurface a wavy character. Stratocumulus (cumulogenesis) is formed by the spreading out of the tops of cumulus clouds, the latter having disappeared	Below 7000 ft	Usually warmer than −5 °C	Widespread irregular stirring with vertical velocities usually less than 10 cm sec^{-1}
Stratus	St	A uniform, featureless layer of cloud resembling fog but not resting on the ground. When this very low layer is broken up into irregular shreds it is designated fractostratus	Usually within 1000 or 2000 ft of the ground	—	Widespread irregular stirring and lifting of a shallow layer of cool, damp air formed near the ground
Nimbostratus	Ns	An amorphous, dark gray, rainy cloud layer reaching almost to the ground		As for St	Widespread regular ascent with vertical velocities of 20 cm sec^{-1}
		II. Cumuliform or heap clouds			
Clouds with marked vertical development					
Cumulus	Cu	Detached, dense, clouds with marked vertical development; the upper surface is dome-shaped with sharp-edged rounded protuberances, while the base is nearly horizontal	Extend from 2000 to 20,000 ft or more	—	Convective motion in which large bubbles of warm air rise with vertical speeds of 1–5 m sec^{-1}

Cumulonimbus	Cb	Heavy masses of dense cloud, with great vertical development, whose cumuliform summits rise in the forms of towers, the upper parts having a fibrous texture and often spreading out into the shape of an anvil. These clouds generally produce showers of rain and sometimes of snow, hail or soft hail, and often develop into thunderstorms	May extend up to 40,000 ft	Summits may be as cold as −50 °C	Strong convective motions with vertical upcurrents of 3 to more than 30 m sec^{-1}
		III. Special types of cloud			
Fracto clouds: fracto-cumulus, fractostratus, fractonimbus		Fragments of low cloud associated with cumulus, stratus, or nimbostratus, as the case may be	—	—	Indeterminate
Castellanus	Ac-cas	Miniature turreted heap clouds forming at medium levels usually in lines. In summer they are symptomatic of the approach of thundery weather		As for Ac	Convective motions released at middle levels by the slow lifting of unstable air often ahead of cold fronts
Orographic clouds: Lenticular and wave clouds		When air is forced to ascend a hill or mountain barrier, a smooth, lens-shaped cloud with well-defined edges may form over the summit. This is a lenticular cloud. If the air flow is set into oscillation by the hill, a succession of such clouds may form in the crests of the stationary waves produced in the lee of the mountain. These are designated wave clouds	—	—	The upcurrents in these clouds are usually quite strong—of order 1–10 m sec^{-1}

Source: Mason (1975)

Hemenway et al. (1965) describe the appearance of these clouds and the techniques employed to capture and analyze their particles. Data by Farlow et al. (1970) indicate that most of the particles are in the size range 0.1 to 0.2 μm. Additional studies of the composition and physics of these clouds have been made by Donahue et al. (1972) and Reid (1975). A bibliography of reports on stratospheric clouds is provided by Stanford and Davis (1974).

Accounts of the complex processes occurring in clouds are given by Mason (1975) in sufficient detail for many optical purposes. Houghton (1951) reviews the physics of clouds and precipitation, emphasizing those problem areas where knowledge was scarce and difficult to obtain at the time. With respect to droplet growth behavior, the situation has not changed greatly since then. Johnson (1954), aufm Kampe and Weickman (1957), Byers (1965), and Mason (1971) treat the physics of clouds in considerable detail. Weickman (1960), as editor, presents a veritable compendium on the subject. Valley (1965) provides much data on the properties of clouds and the occurrence of precipitation.

With respect to scattering and absorption, the important cloud properties are droplet (or ice crystal) size distribution and the amount of precipitable water along a given path in the cloud. Generally, clouds such as cirrus in its several forms are composed of ice crystals, because of the prevailing low temperatures at their altitudes. Intermediate-altitude clouds such as altostratus and altocumulus consist of ice crystals and undercooled droplets, with water alternating between these phases. Lower-altitude clouds, such as cumulus, stratus, and their derivatives, mostly consist of droplets which become undercooled when carried upward by convective currents. Droplets in these clouds have a radius from about 1 to 30 μm, and the liquid water content varies between 0.5 and 1.5 g m^{-3}. Thus clouds are usually wetter than fog, whose liquid water content is between 0.05 and 0.25 g m^{-3}.

Because the droplets evaporate and a cloud disappears unless the air is saturated, the relative humidity in a cloud may be taken as 100%. The resulting water vapor density as a function of temperature may be read from the saturation values in Table 2.4. It is evident that the amount of water stored as vapor greatly exceeds the liquid water content, except possibly at very low temperatures. The total water content is the sum of the amounts in the gas and liquid (or solid) phases. According to Valley (1965), the total water must be greater than 1.7 g m^{-3} before precipitation can occur, whereas the maximum amount that can exist without precipitation is between 6 and 10 g m^{-3}. A nonprecipitating cloud can maintain itself as long as the parcel of air remains saturated. The consequences of saturation within a cloud and nonsaturation beneath are strikingly shown when rain falls from a cloud base but evaporates before reaching the ground.

A widely used analytic function for cloud droplet size is (3-15), which can represent several cloud models by suitable choices for the constant terms, as noted in its discussion and listed in Table 3.2. When the values for cumulus model C.1 are employed, (3-15) becomes

$$n(r) = 2.373 r^6 \exp(-1.5r) \tag{3-30}$$

which is shown in Figure 3.21. For the assumed concentration $N = 100\ \text{cm}^{-3}$, a reference value only, the liquid water content is $0.063\ \text{g m}^{-3}$, which is less than that of many fogs. The mode radius r_c is 4.0 μm, and $n(r_c)$ is equal to $24\ \text{cm}^{-3}\ \mu\text{m}^{-1}$. Also shown in the figure are the narrow distributions for models C.2 and C.3, developed from studies of high-altitude scattering. Deirmendjian (1969) employed these three models in tabulations of scattering at several wavelengths, as described in Sections 6.4.1 and 6.4.2. In a review of infrared scattering, Bauer (1964) used model C.1 with alternative values of the constants.

Considering droplet size measurements next, we should bear in mind that physical sampling methods tend to discriminate against sizes smaller than about 3 μm. Further, the composition of the same generic cloud varies considerably from time to time and place to place. Thorough investigations of droplet size and concentration have been made by Diem (1942, 1948), aufm Kampe (1950), Weickman and aufm Kampe (1953), and

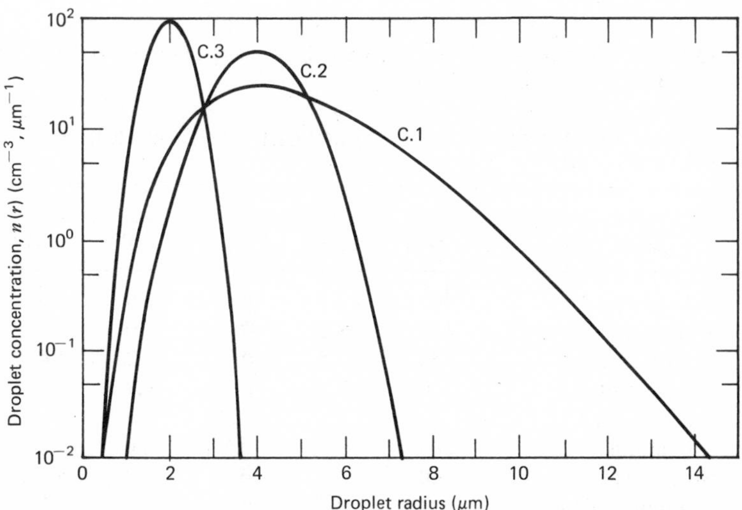

Figure 3.21 Droplet size distributions for the three cloud models of Table 3.2. From Deirmendjian (1969).

Warner (1969). After an extensive literature survey, Carrier et al. (1967) established models of the major cloud classes for the computation of scattering at laser wavelengths. The size distributions of these models are shown in Figure 3.22, while the parameters are listed in Table 3.5. In

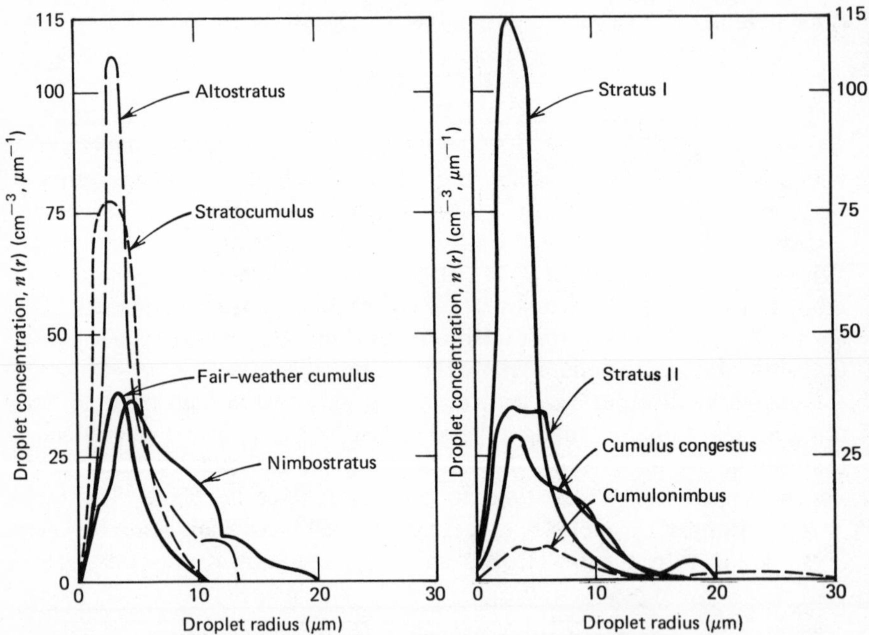

Figure 3.22 Droplet size distributions for major cloud models. From Carrier et al. (1967).

TABLE 3.5 PARAMETERS OF DROPLET SIZE DISTRIBUTIONS SHOWN IN FIGURE 3.22

Cloud type	N (cm^{-3})	r_c (μm)	r_{min} (μm)	r_{max} (μm)	Δr (μm)
Stratus I	464	3.5	0	16.0	3.0
Altostratus	450	4.5	0	13.0	4.5
Stratocumulus	350	3.5	0	11.2	4.4
Nimbostratus	330	3.5	0	19.8	9.5
Fair-weather cumulus	300	3.5	0.5	10.0	3.0
Stratus II	260	4.5	0	20.0	5.7
Cumulus congestus	207	3.5	0	16.2	6.7
Cumulonimbus	72	5.0	0	30.0	7.0

Source: Carrier et al. (1967).

Figure 3.23 are shown the distributions for the three types of clouds employed by Yamamoto et al. (1971) in their extensive computations of angular and total scattering functions for the wavelength range 5.0 to 40.0 μm. Histogram-type data on droplet size classes and populations for

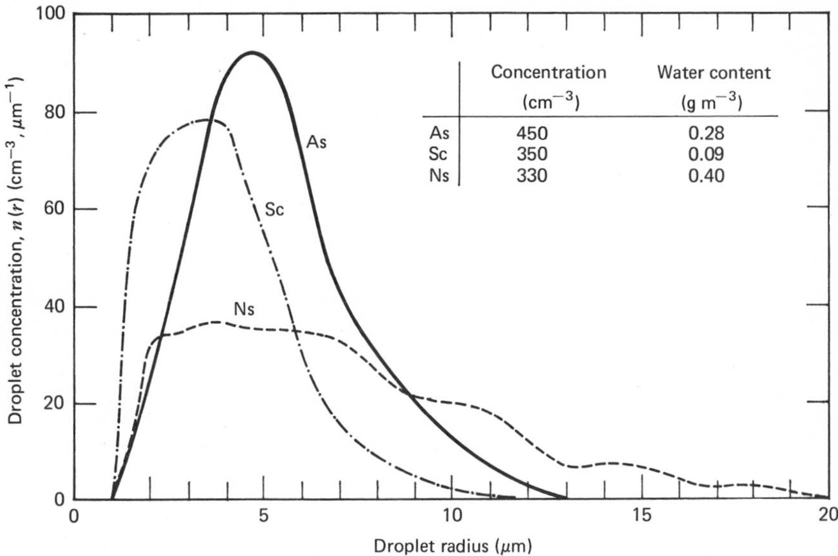

Figure 3.23 Droplet size distributions for altostratus, stratocumulus, and nimbostratus clouds. From Diem (1948) and Yamamoto et al. (1971).

major cloud types are listed in Table 3.6. Data on arctic clouds are provided by Witte (1968). Information on ice crystals in clouds can be found in Ono (1969, 1970), Auer and Veal (1970), and Mossop (1970).

3.4.4 Size Distributions of Raindrops

The processes whereby cloud droplets become raindrops are treated in standard texts on meteorology, such as those cited at the beginning of Section 2.2. In addition, more detailed treatments are given by Weickman (1957), Mason (1971, 1975), and Byers (1965). According to the Bergeron–Findeisen theory of precipitation, in many cases the droplets become drops by way of an intermediate stage of ice crystal formation. The physical basis for this theory is twofold. (1) Both ice crystals and droplets of undercooled water, to temperatures of about −40 °C, exist together in many types of clouds, and (2) the saturation vapor pressure over an ice surface is less than that over a water surface, as may be seen from Table 2.3. When the relative humidity of cloud air is greater than 100%,

TABLE 3.6 POPULATIONS OF SIZE CLASSES OF CLOUD DROPLETS AS PERCENTAGE OF TOTAL POPULATION

	Pacific coast region			Other regions of the United States		
Mean effective diameter of droplet (μm)	Ac, Ac-As, 112 obs.	St, Sc, 60 obs.	Cu, Cb, 220 obs.	Ac, Ac-As, 128 obs.	St, Sc, 267 obs.	Cu, Cb, 110 obs.
0–9	8	5	5	20	32	6
10–14	22	36	19	32	43	31
15–19	28	25	25	30	16	35
20–24	22	17	28	12	6	20
25–29	7	7	15	5	2	5
>29	13	10	8	1	1	3
	(μm)	(μm)	(μm)	(μm)	(μm)	(μm)
Lower quartile	13.5	12.5	14.5	10	9	13
Median	18	16	19.5	14	11	16
Upper quartile	23	22	24	18	14.5	29
	(cm^{-3})	(cm^{-3})	(cm^{-3})	(cm^{-3})	(cm^{-3})	(cm^{-3})
Representative concentration	35	100	90	75	320	160

Source: Lewis (1951).

supersaturation for the ice crystals is greater than that for the water droplets, so that the crystals grow in size at the expense of the droplets. The crystals continue to grow until they become heavy enough to fall into a warmer region of the cloud, where they melt into raindrops. Any tendency of the crystals or drops to fall is governed of course by the updrafts and downdrafts within the cloud. As a drop falls it grows rapidly by accretion, that is, by collisionally collecting the droplets in its fall path.

When the drops finally leave the cloud and enter the air beneath, some evaporation occurs unless this air is saturated. When the relative humidity is low and the drops are initially small, they may evaporate completely before reaching the ground, producing an effect of wispy veils or streamers hanging from the base of the cloud. Known as *virga*, such streamers can be seen frequently in the prairie regions of the West, where long-distance views of the horizon sky are possible. In all cases the amount of water a drop loses by evaporation is a function of the relative humidity, the size and terminal velocity of the drop, and the altitude of the cloud base. Large drops having terminal velocities that may lie between 5 and 10 m sec^{-1} lose their spherical shapes because of air resistance and break up into smaller drops. The process of coalescence, however, always works to make fewer but larger drops whenever they come within grazing distance of each other. Thus any observed size distribution of raindrops, whether they are sampled at ground level or above, is the result of several complex processes. By their nature and habitat, these processes are difficult to observe.

Raindrops are counted and sized by a variety of ingenious techniques, several of which are described by Blanchard (1967). Patience and finesse are required to obtain accurate data, and care must be taken to ensure that the samples are statistically valid. Measurements reported in the literature include those by Laws and Parsons (1943), Marshall and Palmer (1948), Blanchard (1953), Hudson (1963), Mueller and Sims (1966), and Sekhon and Srivastava (1971). Figure 3.24 shows measured size distributions for several precipitation rates as quoted by Blanchard and Spencer (1970) in their investigations of raindrop breakup during fall. The number beside each curve indicates the rainfall rate in mm hr^{-1}. The dashed curves represent smoothed data from Hudson (1963), while the solid curves for 300 and 500 mm hr^{-1} (a torrential downpour) are averaged values from Mueller and Sims (1966). The curve for 1500 mm hr^{-1} is from the experiment described in the concluding paragraph.

Marshall and Palmer (1948) found that raindrop concentration per unit interval of drop diameter generally conforms to

$$N(D) = N_0 \exp(-\Lambda D) \tag{3-31}$$

in which

$$N_0 = 8 \times 10^3 \qquad m^{-3}\,mm^{-1}$$
$$\Lambda = 4.1R^{-0.21} \qquad mm^{-1}$$

where N_0 is the value of N_D where the curve crosses the axis $D = 0$, and R is the rainfall rate in millimeters per hour. The function (3-31) is also compatible with radar measurements of rainfall according to Sekhon and Srivastava (1970), who discuss several parameter relationships for exponential size distributions of this type. Gunn and Marshall (1958) found that (3-31) also represents fairly well the size distribution of snowflakes, but with

$$N_0 = 3.8 \times 10^3 R^{-0.87} \qquad m^{-3}\,mm^{-1}$$
$$\Lambda = 2.55R^{-0.48} \qquad mm^{-1}$$

Additional investigations of snowflake size spectra are reported by Ohtake (1970).

The function (3-31), with the associated values of the constants, has been subjected to many tests since it was proposed. Because the function

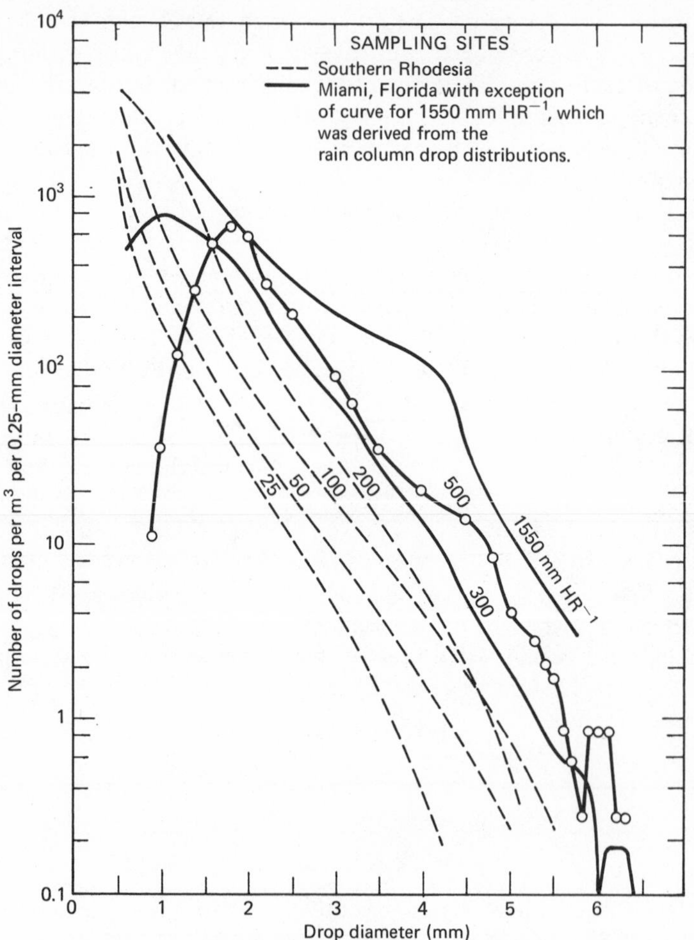

Figure 3.24 Raindrop size distribution data. From Blanchard and Spencer (1970).

is exponential, it usually overestimates the number of very small drops, because it has no lower cutoff limit. Also, since the diameters of raindrops are physically limited to 5 or 6 mm by drop breakup, the function overestimates the number of large drops. Cole et al. (1969) state that this function has the greatest validity for drops having diameters between 0.75 and 2.25 mm for rainfall rates of about 1 mm hr^{-1}, between 1.25 and 3 mm for rates near 5 mm hr^{-1}, and between 1.5 and 4.5 mm for rates greater than 25 mm hr^{-1}. For these size ranges and rates, (3-31) provides values that are reasonable averages for rainfalls having diverse origins and locations. Actually a log-normal distribution is often thought to be more widely applicable to raindrop size distributions. Also, function (3-15), which is an approximation to a log-normal distribution, has been proposed by Deirmendjian (1969) for light to moderate rainfall. Exponential distributions such as (3-31), however, are the most widely used because of their simplicity.

Several investigators have called attention to the similarities among averaged drop size distributions of different rainfalls for a given rainfall rate. The similarities hold for rainfalls from clouds have different physical characteristics in widely separated geographical regions, and they suggest that the processes of drop breakup and coalescence may balance out to create a near-stationary size distribution. Blanchard and Spencer (1970) simulated a rain column by allowing a stream of water to fall 62 m in an aircraft hangar to avoid wind disturbances. Size distribution data were obtained from drop sampling, and drop breakup was observed by high-speed photography. One of their distributions is shown as the curve 1550 mm hr^{-1} in Figure 3.24. Their data on drop sizes, when extrapolated to greater heights of fall and translated to equivalent rainfall rate, are similar to those from measurements in actual heavy rainfall.

4

Rayleigh Scattering by Molecules

The blue of the clear daytime sky and its changing brightness from dawn to dusk long stood as a challenge to scientific thought. Lord Rayleigh (1871a,b, 1899) showed that scattering by air molecules alone was sufficient to produce the observed effects, and this type of scattering usually bears his name. Rayleigh scattering is marked by several characteristics:

- The amount of scattered light varies nearly as the inverse fourth power of wavelength, hence a clear sky is predominantly blue and not the color of sunlight.
- Spatial distribution of the scattered light bears a simple relationship to the direction of observation.
- Equal amounts of light are scattered into the forward and back hemispheres.
- The light scattered at 90 deg is almost completely polarized.

The objective of this chapter is to encourage an understanding of Rayleigh scattering by atmospheric gases. This is an important subject in its own right and a helpful introduction to scattering by particles, which is discussed in the following two chapters. We start by calling attention to the origin of the molecular scattering theory, which has endured with but little change to the present time. Next, we look at the essence of the theory by considering a model elemental scatterer, its behavior under a passing light wave, and the resulting creation of a secondary wave. The intensity and polarization characteristics of the secondary or scattered wave are then derived and translated into scattering coefficients. Numerical values of these coefficients for a broad range of wavelengths are presented, along with guidance for their use in everyday problems. Finally, the chapter deals with the large-scale molecular scattering in the atmosphere and concludes with a description of studies and tabulated functions applicable to this scattering.

4.1 LORD RAYLEIGH AND SCATTERING THEORY

The manifold visual phenomena in the sky have always produced wonder and speculation. Rainbows, halos, mock suns, red skies, and other spectacular sights for centuries were frequently taken as portents of things to come, giving rise to a mass of weather lore among farmers and sailors. The correlation between the prediction and the event often is slight, as may be appreciated by testing any of the several weather proverbs that survive to this day. Nevertheless, the proverbs grew out of studying the changing appearance of the sky, and to that extent they mark a beginning of the study of atmospheric optics. In this introductory section we trace the historical growth of the subject of molecular scattering; readers will find interesting the review by Middleton (1960). The development of atmospheric scattering theories is discussed by Rozenberg (1960) in considerable technical detail.

The blue of the sky has always stimulated the imagination, and most of the speculations have invoked the scattering of sunlight by some type of particle. Scattering by dust particles, for example, was a familiar phenomenon, having been observed and identified centuries ago. Thus the erratic movements of dust motes, seen as wandering points of light when the sun's ray are admitted to a darkened room, were cited by Lucretius about 70 B.C. as a hint that the small "primal seeds" of matter were in constant motion. His insight, which strangely anticipated the idea of brownian motion, is right in line with modern kinetic theory.

In thus trying to account for skylight, any explanation had to deal with a basic question: What kind of particle was responsible? Several early workers, including Bouguer and de Saussure, suggested that the blue of the sky was caused by the clear air itself, but their prescient guesses did not produce theories that could be tested. In contrast, Newton and many of his followers thought that water droplets dispersed throughout the atmosphere and reflecting as "thin plates" were responsible. Clausius, a leading physicist of the nineteenth century and a founder of the science of thermodynamics, attributed the light of the sky to reflections from small water bubbles. Tyndall, whose original experimental work on scattering by hydrosols remains a classic, likewise held the view that water droplets were responsible, but he did not specify their form.

It remained for Lord Rayleigh (J. W. Strutt, third Baron of Rayleigh) to provide the correct explanation in 1871. He assumed only that the particles were spherical, isotropic, far smaller than the wavelength of light, and denser than the surrounding medium. Employing the elastic-solid ether theory, and straightforward dimensional reasoning, he arrived at the central features of what is universally called Rayleigh scattering. The scattering varies directly as the square of the particle volume, and inversely as the fourth power of the wavelength. His dimensional reasoning is set forth with such clarity that it can be followed by a schoolboy. Rayleigh then cast his theory into a mathematical form quite similar to the one we know. He seemed indifferent at the time to the actual nature of the atmospheric particles responsible for the scattering, but he showed that they could not be water bubbles. He suggested that they might be bits of salt, but indicated no preferences. His approach to the problem by way of dimensional analysis is summarized by Humphreys (1940) and is presented in detail by Tricker (1970).

In 1899 Rayleigh returned to the subject of skylight. As an alternative to the old elastic-solid theory, he employed electromagnetic theory which by that time was firmly established through the work of Maxwell and Hertz. Replacing particle density, appropriate to the old theory, with refractive index as the analogous parameter in the new theory, he derived the scattering law in its present form. By comparing calculations from his law with the best available data on atmospheric transmittance and sky radiance, he concluded that "the light scattered from the molecules would suffice to give us a blue sky, not so very greatly darker than that actually enjoyed." In addition to his monumental work in scattering and other branches of optics, Rayleigh enriched all fields of physical science to which he turned his attention. He reported his highly original theories and investigations, often made with the simplest of apparatus, in more than 400 technical papers published over a period of 50 years. Readers who wish to become better acquainted with this remarkable man will enjoy the special issue AO (1964) and the book by Lindsay (1970).

We note briefly the course of his scattering theory during the 100 years since its appearance. Only one emendation has been found necessary, and that possibility was foreseen by Rayleigh himself. Careful laboratory measurements of scattering made by Rayleigh's son (R. J. Strutt, fourth Baron) and others about 1920 disclosed that gas molecules are not quite isotropic. Cabannes and others developed a correction factor to take account of the resulting slight effect on scattering. Rayleigh theory has been used and confirmed in innumerable investigations of skylight intensity and polarization, notably those by Hulburt and his associates, recounted by Sanderson (1967), and those by Sekera and co-workers. The above discussion has implicitly assumed single scattering. In 1950 Chandrasekhar greatly extended Rayleigh's theory by developing general solutions to multiple scattering by a molecular atmosphere. This has led to several tabulations of the intensity and polarization values of skylight that take multiple scattering into account.

4.2 RADIATION BY A MOLECULAR DIPOLE

The individual molecule is the elemental scatterer of electromagnetic waves. Because of its electric charges or dipole nature, the molecule interacts with the passing primary wave, abstracting energy therefrom and sending it into space. The molecule is a point source of this scattered energy which proceeds outward as though the primary wave were not present. This section describes the pertinent characteristics of the gas molecule, the nature of the secondary wave, and the parameter role of the refractive index of the gas.

4.2.1 Model of an Elemental Scatterer

The mechanism of Rayleigh scattering is conveniently examined by means of the model pictured in Figure 4.1. This is an idealized gas molecule which could pass for the "billiard ball" model used in elementary kinetic theory. The molecular mass resides almost entirely in the atomic nuclei at the center, considered here as a single mass carrying a net positive charge. The orbital electrons shared by the constituent atoms and represented by a concentric shell provide a balancing negative charge. Although the electron mass is relatively small, it is significant in scattering theory. An elastic binding force due to attraction between these charges holds these electrons captive. This force, everywhere directed toward the central mass, is characterized by a spring constant analogous to that of a mechanical system. Such an arrangement of unequal masses and an elastic constraint forms a mechanical oscillator, with the added feature that the central and peripheral masses carry opposite charges.

The model is assumed to be nonionized, nonpolar, isotropic, linear, and lightly damped. Such assumed properties are not greatly at variance with the actual properties of nitrogen and oxygen molecules in the troposphere. The first assumption means that there is no overall charge on the molecule; hence it does not experience a net force in an electric field. The second means that the negative charge is uniformly distributed over the shell and may be treated as though it were at the center. The third assumption, equivalent to the second, states that the spring constant of the binding force is the same in all directions. Thus there is no preferred orientation of oscillation. The linearity assumption ensures that the spring constant is invariant for all extensions of the bond, that is, Hooke's law is obeyed. Finally, the assumption of small damping means that the amplitude does not become unduly large at frequencies near resonance, and that the resonant frequency itself is little changed by the damping.

As noted above, the negative charge is considered to be at the center, along with the equal positive charge. The electric dipole moment, which is

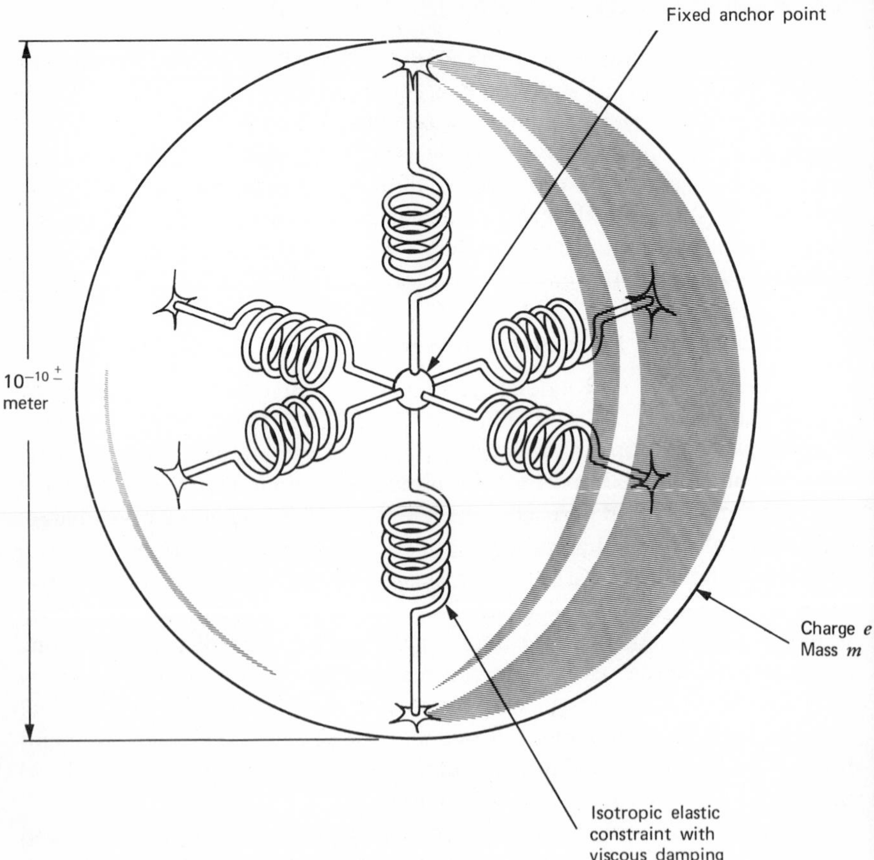

Figure 4.1 Model of an elemental scatterer. Adapted from J. Strong, *Concepts of Classical Optics*, W. H. Freeman, San Francisco. Copyright © 1958.

equal to the product of charge and separation distance, thus is zero. When the molecule is subjected to an electric field, the charges are forced apart, as shown in Figure 4.2, and an induced dipole moment is created. The nonpolar molecule now is *polarized* under the influence of the field; this molecular polarizability is the basis of all scattering and refraction. The force that separates the charges is opposed of course by the elastic force that binds the charges together. When the field strength varies periodically, as in an electromagnetic wave, the value of the dipole moment oscillates synchronously with the field. Because the molecule is isotropic, the axis of the dipole moment always aligns itself with the electric vector of the passing wave. And because the molecule is very small compared to typical

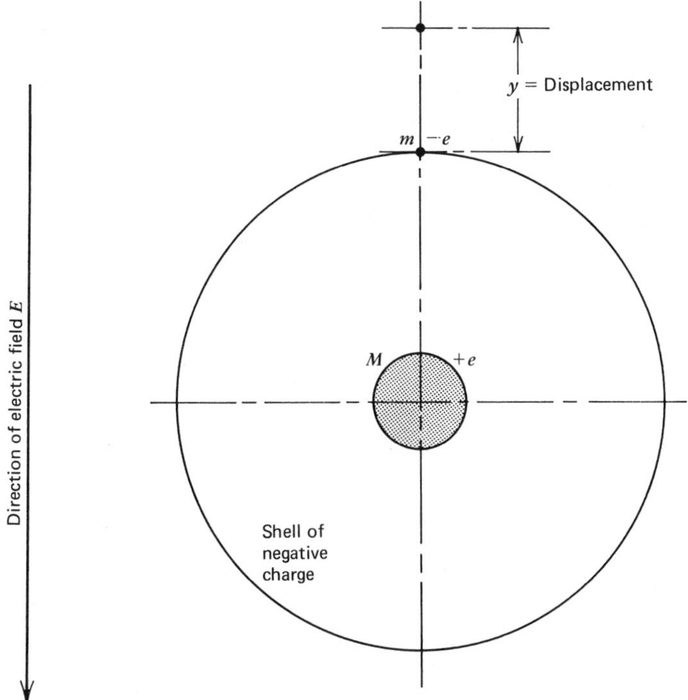

Figure 4.2 Creation of an induced dipole moment by an electric field, which displaces the plus and minus charges of a molecule.

wavelengths, the instantaneous phase of the wave is uniform over the molecule.

It is relevant to inquire into the natural or resonant frequency of this model scatterer. Considering that the central mass remains spatially fixed because of its large value, while the lighter electronic mass vibrates with respect to it, the resonant angular frequency is given by

$$\omega_0 = \left(\frac{k}{m}\right)^{1/2} \tag{4-1}$$

where k is the spring constant, or restoring force per unit displacement, and m is the mass of the electron. For simplicity it is assumed that only one electron, the outermost shell for example, is involved. The restoring force F for a displacement y of the electronic shell from equilibrium is found from electrostatic theory to be

$$F = -\left(\frac{e^2}{4\pi\epsilon_0 r^3}\right) y \tag{4-2}$$

where e is the charge of an electron, ϵ_0 is the permittivity or absolute dielectric constant of free space, and r is the radius of the electronic shell.

Equation (4-2) is in the form

$$F = -ky \tag{4-3}$$

so that the resonant frequency according to (4-1) is

$$\omega_0 = \left(\frac{e^2}{4\pi\epsilon_0 r^3 m}\right)^{1/2} \tag{4-4}$$

Substituting values from Appendix B, along with $r \approx 1\ \text{Å} \approx 10^{-10}$ m into (4-4), we find that

$$\omega_0 = 1.58 \times 10^{16} \text{ rad sec}^{-1}$$

The vibrational frequency and wavelength then are

$$\nu_0 = 2.53 \times 10^{15} \text{ Hz}$$

$$\lambda_0 = 0.118\ \mu\text{m}$$

Thus the resonant frequency of the simple model, in which only one electron was assumed active, lies in the middle ultraviolet.

Two important points are noted. In the real case, more than one electron may be driven by the passing wave. Multiple resonances and frequency interactions are then produced, but in general the resonant frequencies are increased by the square root of the number of driven electrons. Such frequencies are higher than that just computed and are much higher than those of interest here. Therefore the resonant frequency of the dipole oscillator is not dealt with numerically in this treatment of molecular scattering. The second point is concerned with damping. A completely undamped or frictionless oscillator has an infinite amplitude when driven at its resonant frequency. For a dipole oscillator, however, damping is provided by the radiation resistance of free space. This holds the vibration amplitude within bounds, while broadening the resonant frequency only slightly.

4.2.2 Dipole Moment and the Secondary Wave

Now let the model scatterer described in the preceding section be engulfed by a passing electromagnetic wave. The wave is assumed to be monochromatic and linearly polarized, with its electric field vector in the XY plane, as indicated in Figure 4.3. The force acting to separate the nuclear $+e$ and electronic $-e$ charges is parallel to the Y axis and is given by the product of charge and field strength:

$$F = eE = eE_0 \sin \omega t \tag{4-5}$$

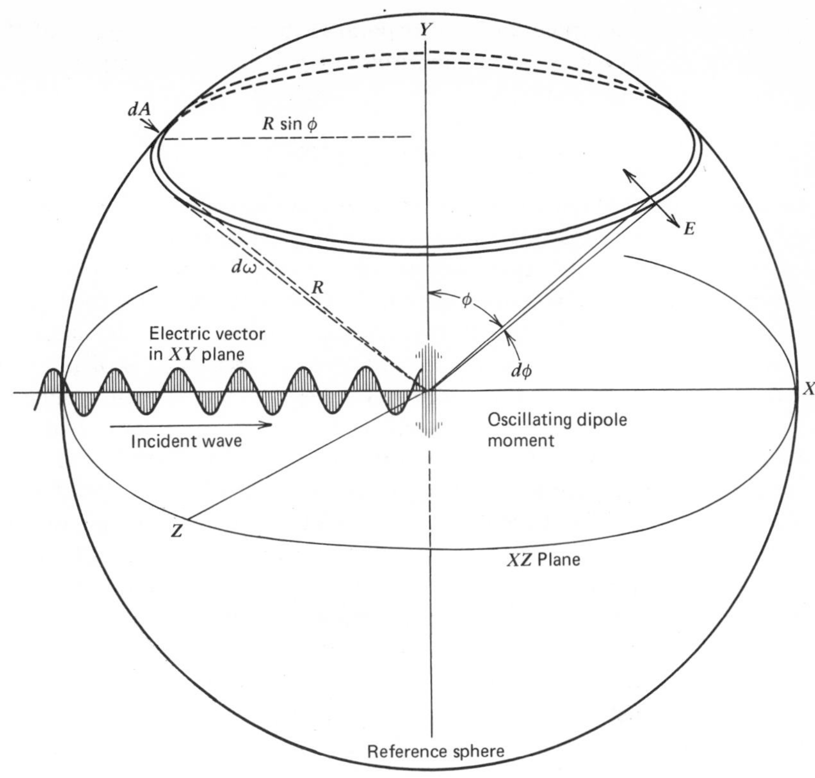

Figure 4.3 Geometry of molecular dipole radiation.

where E is the instantaneous value of the field, E_0 is the maximum value, and ω is the electromagnetic angular frequency. If it is assumed that the required damping is small enough to be ignored, the equation of motion for the electronic charge is

$$m\frac{d^2y}{dt^2} + ky = eE_0 \sin \omega t \tag{4-6}$$

where m is the mass of the charge, taken as one electron, and ky is the restoring force. This is a differential equation of second order and first degree and has the steady-state solution

$$y = \frac{e}{m(\omega_0^2 - \omega^2)} E_0 \sin \omega t \tag{4-7}$$

for the displacement of the charge at any instant.

The oscillating negative charge and the relatively fixed positive charge form an oscillating dipole moment p parallel to the electric vector of the exciting or primary wave. The amplitude of p is equal to the product of charge and displacement:

$$p = ey = \frac{e^2}{m(\omega_0^2 - \omega^2)} E_0 \sin \omega t \tag{4-8}$$

The amplitude of p synchronously follows the sinusoidal variation of the field, with a phase angle that depends on the value of ω relative to ω_0. When $\omega < \omega_0$, the phase angle lies between 0 and $\pi/2$ rad, becoming equal to the latter at $\omega = \omega_0$. This phase angle is of no consequence in scattering but is a basic factor in refractivity and dispersion. When $\omega = \omega_0$, the damping force, although small enough to disregard at our frequencies of interest, prevents the amplitude of p from becoming infinite, which would be required by (4-8). When ω is reduced far below ω_0, as in going from the ultraviolet to the infrared, the amplitude of p becomes very small.

The oscillating dipole moment driven by the primary wave generates a secondary wave. The to-and-fro motion of electronic charge on the molecule is a small-scale analog of electronic current flow in a hertzian dipole antenna of macroscopic dimensions. Although the radiation pattern in the immediate vicinity of a dipole is complex, an outgoing spherical wave emerges at a distance of a few wavelengths. In considering the power carried by this secondary wave, it is convenient to deal with the maximum rather than the instantaneous value of the dipole moment. From (4-8) this maximum value is

$$p_0 = \frac{e^2}{m(\omega_0^2 - \omega^2)} E_0 \tag{4-9}$$

where, as a reminder, E_0 refers to the primary wave. The electric vector E in the secondary wave has, according to Frank (1950), an instantaneous amplitude

$$E = \frac{\omega^2 p_0 \sin \phi}{4\pi\epsilon_0 c^2 R} \sin \omega \left(t - \frac{R}{c}\right) \tag{4-10}$$

where ϕ is the angle between the dipole axis and any direction of interest, as in Figure 4.3, and R is any distance along this direction. As a consequence of the term $\sin \phi$, the dipole cannot radiate in the direction of its axis.

The average power or time rate of energy flow in an electromagnetic wave per unit area normal to the direction of propagation is defined by the Poynting vector $\bar{S}$,

$$\bar{S} = \tfrac{1}{2} c\epsilon_0 E_0^2 \tag{4-11}$$

where E_0 is the maximum value of the electric field strength. The Poynting vector for the scattered flux at angle ϕ through an infinitesimal area on the reference sphere is then found by substituting the value of E from (4-10) into (4-11), where the factor $\frac{1}{2}$ in the latter takes care of the average value of the term $\sin^2 \omega(t - R/c)$ introduced by the substitution. This gives

$$\bar{S} = \frac{\omega^4 p_0^2 \sin^2 \phi}{32\pi^2 \epsilon_0 c^3 R^2} \tag{4-12}$$

This average power in the secondary wave is related to the power in the primary wave by the maximum value of dipole moment p_0 induced by the primary wave. Substituting the value of p_0 from (4-9) into (4-12) and using the relationship $\omega = 2\pi c/\lambda$, we find that

$$\bar{S} = \frac{\pi^2 c \sin^2 \phi}{2\epsilon_0 \lambda^4 R^2} \left[\frac{e^2}{m(\omega_0^2 - \omega^2)}\right]^2 E_0^2 \tag{4-13}$$

in which appears the dependence of molecular scattering on the inverse fourth power of wavelength.

Equation (4-13) defines the flux per unit wavelength interval and per unit area, that is, the *irradiance*, produced at distance R by the secondary wave. Because of this distance factor, however, the irradiance does not of itself uniquely express the angular scattering characteristic. Usually this is expressed more handily for atmospheric situations by the *intensity*, or flux per unit solid angle. The conversion between these two radiometric quantities, discussed in Section 1.2.3, is straightforward. The Poynting vector $\bar{S}$ is equivalent to the irradiance E defined by (1-8), while intensity I is defined by (1-9). From Figure 4.3 it is seen that

$$dA = R^2\, d\omega$$

where $d\omega$ is the infinitesimal solid angle lying between ϕ and $\phi + d\phi$. Substituting this into (1-8) gives

$$E = \frac{d\Phi}{R^2\, d\omega} = \frac{I}{R^2} \tag{4-14}$$

Substituting this for $\bar{S}$ in (4-13) we obtain

$$I(\phi) = \frac{\pi^2 c \sin^2 \phi}{2\epsilon_0 \lambda^4} \left[\frac{e^2}{m(\omega_0^2 - \omega^2)}\right]^2 E_0^2 \tag{4-15}$$

for the intensity of the scattered flux, called *scattered intensity* in this book.

4.2.3 Dipole Moment and Refractive Index

It is convenient at this point to relate the molecular dipole factors e, m, and ω to the refractive index of a gas in macroscopic or bulk form. This relationship involves the *polarizability* of gas, a subject treated by von Hippel (1954), Feynman et al. (1963), Sommerfeld (1964a), and Chu (1967).

The polarizability is related to the Lorenz–Lorentz expression, discovered independently by the two similarly named men at about the same time. One form of the expression, in mks units, is

$$\frac{n^2-1}{n^2+2}\frac{3\epsilon_0}{N}=\frac{e^2}{m(\omega_0^2-\omega^2)} \tag{4-16}$$

where n is the refractive index, and N is the number of dipole oscillators per unit volume. Strictly speaking, the electronic charge e and the mass m refer to a single electron, as in (4-6). However, the effective oscillator formed by a typical gas molecule may include several of its outer electrons. It is convenient in a first approximation, whose principal consequence is an increase in the resonant frequency, to interpret the quantities e and m as spread over the mass points in question. With this interpretation of (4-16), N is considered the number of molecules per unit volume, or number density.

The term on the extreme right in (4-16) is identical to the bracketed term in (4-15). After making this substitution, (4-15) becomes

$$I(\phi)=\frac{9\pi^2\epsilon_0 c\,\sin^2\phi}{2N_s^2\lambda^4}\left(\frac{n_s^2-1}{n_s^2+2}\right)^2 E_0^2 \tag{4-17}$$

where the subscript s means that the value refers to standard air. A word of explanation is warranted. The refractive index term n^2-1 itself is proportional to number density N or mass density ρ. Therefore in all scattering expressions the values of these quantities must be mutually consistent. Henceforth the development is based on standard temperature and pressure conditions; with this understanding, the subscript s is not used further. Corrections for nonstandard conditions are given in Section 4.4.3. Also, the refractive index increases somewhat with frequency, as is evident from (4-16), so that the wavelength dependence of scattering is slightly greater than λ^{-4}. According to Middleton (1952), the dependence is about $\lambda^{-4.08}$ in the visual region.

In some derivations of scattering theory, for example in the rigorous treatment by Stratton (1941), the dipole moment is expressed in rationalized mks units as

$$p_0=4\pi\epsilon_0 r^3\left(\frac{n^2-1}{n^2+2}\right)E_0 \tag{4-18}$$

where r is the radius of the scatterer, and E_0 refers to the primary wave, as before. In the context of (4-18), this scatterer need not be a single molecule but may be a cluster of a few molecules or a condensation nucleus. Equation (4-18) indicates that a small sphere can behave as a dipole, and it is an alternative form of (4-9) as far as scattering is concerned. When (4-18) and $\omega = 2\pi c/\lambda$ are substituted into (4-12) and relationship (4-14) is applied, we have

$$I(\phi) = \frac{8\pi^4 \epsilon_0 c r^6 \sin^2 \phi}{\lambda^4} \left(\frac{n^2 - 1}{n^2 + 2}\right)^2 E_0^2 \tag{4-19}$$

Thus the scattered intensity from a single very small particle varies as the sixth power of its radius. According to Pendorf (1962), the error in thus applying Rayleigh, rather than Mie, theory to small particles is 1% or less when $r \leq 0.03\lambda$. This is a useful rule to remember.

4.3 RAYLEIGH SCATTERING CROSS SECTIONS

The concept of a scattering cross section is widely employed in dealing with the interactions of matter and radiant energy in any of its several forms. For example, the deflection (scattering) of high-velocity alpha particles by an atomic nucleus is described in terms of a scattering cross section. Such scattering, brought about by repulsion in a central force field is a traditional problem in classical mechanics. In the domain of light scattering, the concepts of angular and total scattering cross sections are basic concepts which lead to several coefficients and expressions having great practical utility. These cross sections are now considered in some detail for cases in which the incident light is linearly polarized.

4.3.1 Angular Scattering Cross Section

The angular scattering cross section of a molecule is defined as that cross section of an incident wave, acted on by the molecule, having an area such that the power flowing across it is equal to the power scattered by the molecule per steradian at an angle ϕ. In symbols,

$$\sigma_m(\phi)E_0 = I(\phi) \qquad \text{or} \qquad \sigma_m(\phi)\bar{S} = I(\phi) \tag{4-20}$$

where E_0 is the irradiance of the incident wave, and $\bar{S}$ is, equivalently, the Poynting vector of that wave. Thus the angular scattering cross section represents the ratio of the scattered intensity to the incident irradiance.

According to (4-20), the angular cross section is found by dividing (4-11) into (4-17), giving

$$\sigma_m(\phi) = \frac{9\pi^2}{N^2\lambda^4} \left(\frac{n^2 - 1}{n^2 + 2}\right)^2 \sin^2 \phi \tag{4-21}$$

in which the mks electromagnetic constants have canceled out. Any convenient measure of length now can be used for λ and for establishing the unit volume associated with N. Because $n^2 - 1$ is closely proportional to N, the form of (4-21) is consistent with the necessity for the value of $\sigma_m(\phi)$, which refers to a single molecule, to be independent of gas temperature, pressure, and density. Following the convention that a solid angle is represented by a pure number, the dimensions of $\sigma_m(\phi)$ are L^2, corresponding to an area, and convenient units are $cm^2\ sr^{-1}$.

Two simplifications of (4-21) can be made based on the fact that the refractive index of air is only slightly greater than unity. For example, $n = 1.000293$ for $\lambda = 0.55\ \mu m$ and standard air. Hence $n^2 + 2 \approx 3$, and (4-21) reduces to

$$\sigma_m(\phi) = \frac{\pi^2(n^2-1)^2}{N^2\lambda^4} \sin^2 \phi \tag{4-22}$$

Also, another approximation states that

$$(n^2-1)^2 \approx 4(n-1)^2$$

which allows (4-22) to be written as

$$\sigma_m(\phi) = \frac{4\pi^2(n-1)^2}{N^2\lambda^4} \sin^2 \phi \tag{4-23}$$

To limit the number of expressions carried in parallel, only the first approximation is employed in further development of the theory. Thus (4-22) is taken as the basic equation for angular scattering.

When the scatterer is a very small particle to which (4-19) applies, its angular cross section also is found according to (4-20). Thus division of (4-19) by (4-11) yields

$$\sigma_p(\phi) = \frac{16\pi^4 r^6}{\lambda^4} \left(\frac{n^2-1}{n^2+2}\right)^2 \sin^2 \phi \tag{4-24}$$

which is the counterpart of the molecular expression (4-21). An identical expression is obtained from Mie theory when the particle meets the small-size requirement that $r < 0.03\lambda$. In the atmosphere, such a particle probably is a condensation nucleus having a refractive index $1.33 < n < 1.60$. Therefore approximations for the refractive index terms are not possible here. In terms of the volume V of the small scatterer, (4-24) becomes

$$\sigma_p(\phi) = \frac{9\pi^2 V^2}{\lambda^4} \left(\frac{n^2-1}{n^2+2}\right)^2 \sin^2 \phi \tag{4-25}$$

Several comments are in order at this point. The scattering expressions apply to molecular species having dipole resonant frequencies much higher than the frequencies that correspond to the near ultraviolet, visual, and infrared regions. Hence the expressions are valid in these spectral regions for nitrogen and oxygen molecules, which are responsible for about 99% of atmospheric molecular scattering. The irradiance of the incident flux may be expressed in any radiometric or photometric quantity, but the associated unit area must be the same as that in which $\sigma_m(\phi)$ is expressed. The scattered flux defined by $\sigma_m(\phi)$ is normalized by the operation indicated in (4-20) so that, when the incident flux has unit irradiance, the scattered intensity $I(\phi)$ is equal to the numerical value of $\sigma_m(\phi)$. This value, even at $\phi = 90$ deg, is quite small—less than $10^{-26}\ \text{cm}^2\ \text{sr}^{-1}$ at $\lambda = 0.55\ \mu\text{m}$—and is not measurable by any direct means.

The scattered intensity is always governed by $\sin^2\phi$, where ϕ is the angle between the dipole axis and the direction of interest, as shown in Figure 4.3. This direction may have any orientation about the dipole axis; hence the spatial pattern of scattered intensity has rotational symmetry about this axis. Figure 4.4 shows the patterns of scattered intensity in a plane containing the dipole axis and in a plane perpendicular to the dipole axis. The intensity is zero at the poles, which are on the Y axis, and a maximum

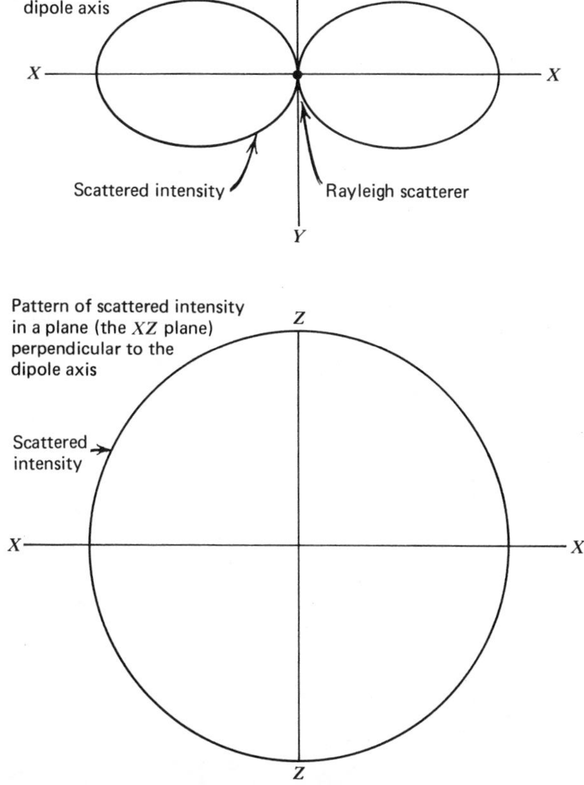

Figure 4.4 Patterns of scattered intensity for polarized incident light.

in any direction in the equatorial or XZ plane. It is evident that the pattern is shaped somewhat like a doughnut with a vanishingly small hole. Because the scatterer is assumed isotropic and the incident light linearly polarized, all the scattered light is linearly polarized. Its electric vector lies in the plane formed by the dipole axis and the direction of observation.

4.3.2 Total Scattering Cross Section

The total scattering cross section σ_m of a gas molecule is defined as that cross section of an incident wave, acted on by the molecule, having an area such that the power flowing across it is equal to the total power scattered in all directions. A comparison of this definition with the one given for $\sigma_m(\phi)$ in the preceding section indicates that σ_m is equal to the integral of $\sigma_m(\phi)$ over 4π sr. Thus from (4-22) and Figure 4.3 we have

$$\sigma_m = \int_0^{4\pi} \sigma_m(\phi)\, d\omega = \int_0^{4\pi} \frac{\pi^2(n^2-1)^2}{N^2\lambda^4} \sin^2\phi\, d\omega \tag{4-26}$$

Applying the geometric relationships (1-15) through (1-17), used with Figure 1.2, to Figure 4.3, we see that

$$d\omega = 2\pi \sin\phi\, d\phi \tag{4-27}$$

Substituting (4-27) into (4-26) and changing the limits of integration results in

$$\sigma_m = \frac{2\pi^3(n^2-1)^2}{N^2\lambda^4} \int_0^{\pi} \sin^3\phi\, d\phi \tag{4-28}$$

The definite integral is a standard form whose value is $\frac{4}{3}$. Substitution into (4-28) gives

$$\sigma_m = \frac{8\pi^3(n^2-1)^2}{3N^2\lambda^4} \tag{4-29}$$

for the total scattering cross section of an isotropic molecule illuminated by polarized light. The dimensions of σ_m are L^2, and the customary units are cm^2. When the incident flux has unit irradiance, the total amount of flux scattered in all directions is equal to the numerical value of σ_m, which is about 4.6×10^{-27} cm^2 at the wavelength 0.55 μm. Tabulated values of σ_m over the wavelength range 0.20 to 20.0 μm are provided by Penndorf (1957a).

The total scattering cross section of a small particle is obtained from a similar integration of (4-24). Applying the procedure used on (4-26), we find

$$\sigma_p = \frac{128\pi^5 r^6}{3\lambda^4} \left(\frac{n^2-1}{n^2+2}\right)^2 \tag{4-30}$$

For a wavelength of 1.0 μm and a condensation nucleus having a radius of 0.032 μm and a refractive index of 1.60, substitutions in (4-30) give $\sigma_p = 1.17 \times 10^{-14}$ cm^2.

The concept of a scattering efficiency factor Q_{sc}, widely used in Mie theory and discussed in Section 5.5.1, is applicable to such small particles. The factor is defined as the ratio of the total scattering cross section (area of the wave front acted on by the particle) to the geometric cross section of the particle. For a sphere this latter cross section is just πr^2. Dividing this into (4-30), we have

$$Q_{sc} = \frac{128\pi^4 r^4}{3\lambda^4}\left(\frac{n^2-1}{n^2+2}\right)^2 \tag{4-31}$$

which is dimensionless. Thus the total scattering efficiency of a small sphere, *in the size range considered here*, can be expressed as

$$Q_{sc} \propto \left(\frac{r}{\lambda}\right)^4 \tag{4-32}$$

which emphasizes that an important criterion of scattering is the ratio of particle size to wavelength, rather than the absolute value of either. Substituting into (4-31) the values of λ, r, and n used with (4-30) to obtain $\sigma_p = 1.17 \times 10^{-14}$ cm^2, we find that $Q_{sc} = 3.63 \times 10^{-4}$. This is the fraction of the cross-sectional area of the particle that acts on the incident wave front. Thus particles that are very small relative to the wavelength are inefficient scatterers, but the efficiency rises rapidly with particle size, as indicated by the exponent in (4-32). However, this simple relationship is not maintained much beyond $r = 0.03\lambda$, which is about the upper size limit for the validity of Rayleigh theory.

4.4 VOLUME SCATTERING COEFFICIENTS

For Rayleigh scattering to be discernible we must deal with the additive scattered fluxes from many molecules. The necessity for this may be seen in the very small numerical values of molecular scattering cross sections. In this section we develop the functions, or coefficients, that define the angular and total scattering by a volume of gas for the cases of polarized and unpolarized incident light. Correction factors for the effects of molecular anisotropy are reviewed, and the concept of phase function is explained.

4.4.1 Angular Scattering Coefficient for Polarized Light

The volume angular coefficient expresses the angular scattering characteristic of a unit volume of gas and is found directly from the angular scattering cross section of a molecule. Referring again to Figure 4.3, let the dipole moment at the center of the coordinates be replaced by a unit volume of gas containing N isotropic molecules of the same species. As before, let a linearly polarized wave of unit irradiance with its electric vector in the XY plane be incident on the unit volume from the left.

The random spacings and thermal motions of the molecules are such that scattering is independent and incoherent, as these characteristics are defined in Section 1.3.3. Thus there are no discernible phase relationships between the separately scattered fluxes except in the exactly forward direction. Here, in an infinitesimal cone centered about this direction, the scattered fluxes are coherently related to the primary wave. The resulting interferences cause a progressive retardation of the phase of this wave as it advances through the scatterers, and this phase retardation is the physical basis of refractivity.

In all other directions the scattering is incoherent, so that the individual intensities are additive. Hence the angular coefficient for the unit volume is just N times the angular cross section defined by (4-22):

$$\beta_m(\phi) = N\sigma_m(\phi) = \frac{\pi^2(n^2-1)^2}{N\lambda^4} \tag{4-33}$$

Because the refractive index term is closely proportional to N, the coefficient $\beta_m(\phi)$ varies directly with gas density. By similar reasoning, the angular coefficient for a unit volume of identical small particles, whose angular cross section is defined by (4-24), is

$$\beta_p(\phi) = N\sigma_p(\phi) = \frac{16\pi^4 r^6 N}{\lambda^4}\left(\frac{n^2-1}{n^2+2}\right)^2 \sin^2\phi \tag{4-34}$$

where N is the particle concentration.

When the incident flux has unit irradiance, the numerical value of $\beta_m(\phi)$ or $\beta_p(\phi)$ represents the scattered intensity in an infinitesimal solid angle $d\omega$ at the angle ϕ, as in Figure 4.3, from a unit volume having a unit cross section and a unit length. Thus the volume angular coefficient, like the angular cross section, is the ratio of the scattered intensity to the incident irradiance and is called the *Rayleigh ratio*. Because the dimensions of the angular scattering cross section are L^2, and the dimensions L^{-3} are implicit in the meaning of N, the dimension of $\beta_m(\phi)$ or $\beta_p(\phi)$ is L^{-1}.

For air at standard conditions and a wavelength of 0.55 μm, the maximum value of $\beta_m(\phi)$ is about 1.4×10^{-6} m^{-1} sr^{-1}. The assumptions are that the molecules or particles are isotropic and that the incident light is linearly polarized. The scattered light then is linearly polarized, having its electric vector in the plane formed by the incident electric vector and the direction of observation, as in Figure 4.3. The spatial pattern of scattered intensity is governed by $\sin^2 \phi$, so that the intensity is a maximum in all directions in the XZ plane and zero along the Y axis.

4.4.2 Angular Scattering Coefficient for Unpolarized Light

The volume angular coefficient for unpolarized light is found by extending the concept of the coefficient for polarized light. We recall that unpolarized or natural light can be treated as the sum of two orthogonal, linearly polarized waves having no coherent relationship, as stated in connection with Figure 1.7. The polarizations of these waves may have any orientation about the propagation axis, so long as they remain orthogonal. The basic relationship is

$$E = \tfrac{1}{2}(E_\perp + E_\parallel) \tag{4-35}$$

where

E = unit irradiance of the unpolarized light,
$E_\perp$ = unit irradiance of the polarized wave whose electric vector is perpendicular to a reference plane of arbitrary orientation about the propagation axis, and
$E_\parallel$ = unit irradiance of the polarized wave whose electric vector is parallel to this reference plane.

Referring to Figure 4.5, we note that two colinear waves having these characteristics and propagating along the X axis are incident on a unit volume of scatterers at point O. The electric vector of the first wave $E_\perp$ lies in the XY plane, and that of the second wave $E_\parallel$ lies in the XZ plane. The plane of observation is defined by the propagation axis of the incident waves and the observation direction OD. The angle between $E_\perp$ and OD is ϕ_1, and between $E_\parallel$ and OD is ϕ_2. Then if the incident irradiances are each unity, the resulting scattered intensity at point D is, from (4-33) and (4-35),

$$I(\phi_1, \phi_2) = \beta_m(\phi_1, \phi_2)E = \frac{1}{2} E \frac{\pi^2(n^2-1)^2}{N\lambda^4} (\sin^2 \phi_1 + \sin^2 \phi_2) \tag{4-36}$$

where the reason for the factor $\frac{1}{2}$ should be kept in mind.

The angular dependence in (4-36) is readily simplified to a dependence on the observation angle θ. In Figure 4.5 a joint solution of the two right spherical triangles YND and DNZ yields

$$\cos^2 \phi_1 + \cos^2 \phi_2 = \sin^2 \theta$$

By substitution of identities, this becomes

$$\sin^2 \phi_1 + \sin^2 \phi_2 = 1 + \cos^2 \theta$$

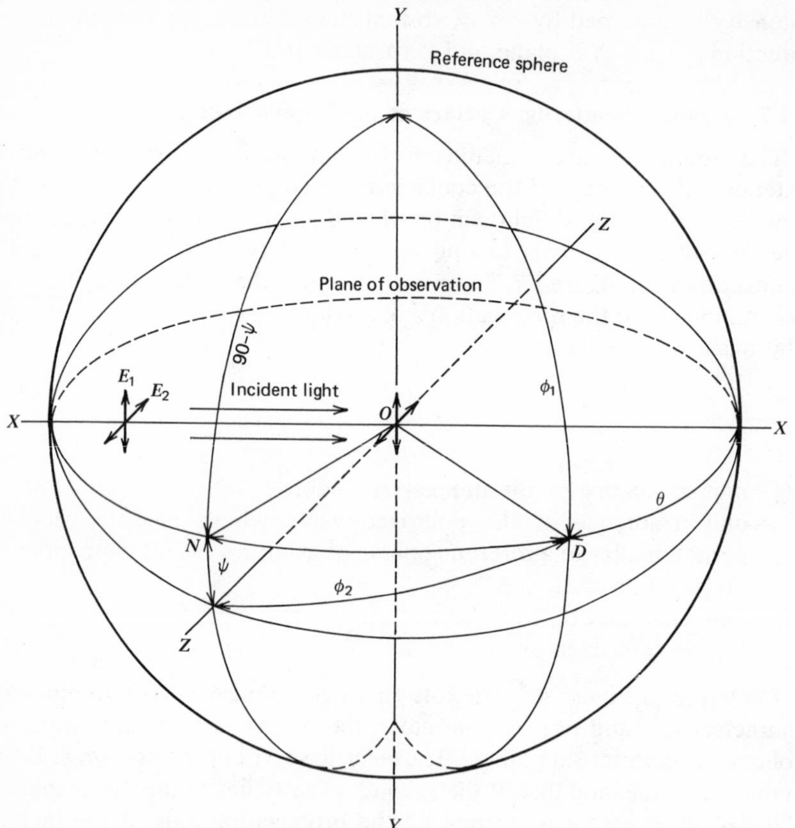

Figure 4.5 Geometry for the scattering of unpolarized incident light.

Thus (4-36) can be written

$$\beta_m(\theta) = \frac{\pi^2(n^2-1)^2}{2N\lambda^4}(1+\cos^2\theta) \tag{4-37}$$

as the volume angular scattering coefficient for unpolarized incident light. When this has unit irradiance, the numerical value of $\beta_m(\theta)$ is equal to the scattered intensity, in an infinitesimal solid angle $d\omega$ at observation angle θ, from unit volume of gas. The dimension of $\beta_m(\theta)$ is L^{-1}, as for $\beta_m(\phi)$. The foregoing treatment can also be applied to (4-34) to obtain the angular coefficient for a suspension of small particles illuminated by unpolarized light. This gives

$$\beta_p(\theta) = \frac{8\pi^4 r^6 N}{\lambda^4}\left(\frac{n^2-1}{n^2+2}\right)^2(1+\cos^2\theta) \tag{4-38}$$

The light scattered according to (4-37) or (4-38) consists of two linearly polarized components $I_\perp$ and $I_\parallel$, whose electric vectors are perpendicular and parallel to the plane of observation. The patterns of scattered intensity are shown in Figure 4.6. Because the $I_\perp$ component is governed by the factor unity in the second parentheses in (4-37), it remains constant for all values of θ. Similarly, the $I_\parallel$ component is governed by $\cos^2\theta$. Thus at $\theta = 90$ deg only $I_\perp$ is present, and the scattered light is completely polarized. When the scatterers are isotropic, the $I_\perp$ component is produced only by the $E_\perp$ incident wave, and the $I_\parallel$ component only by the incident $E_\parallel$ wave. That is, *depolarization* does not occur with isotropic scatterers. Because the plane of observation can have any orientation about the propagation axis of the incident wave, the patterns of Figure 4.6 have rotational symmetry about this axis. Equal amounts of light are sent into the forward and rear hemispheres; forward scattering and backscattering are equal. The spatial pattern of scattered intensity is shaped like a fattened dumbbell.

At all values of θ other than 0 and 180 deg, varying mixtures of polarized and unpolarized light are found. The degree of polarization P, or ratio of polarized to unpolarized light, is defined by

$$\mathrm{P} = \frac{I_\perp(\theta) - I_\parallel(\theta)}{I_\perp(\theta) + I_\parallel(\theta)} \tag{4-39}$$

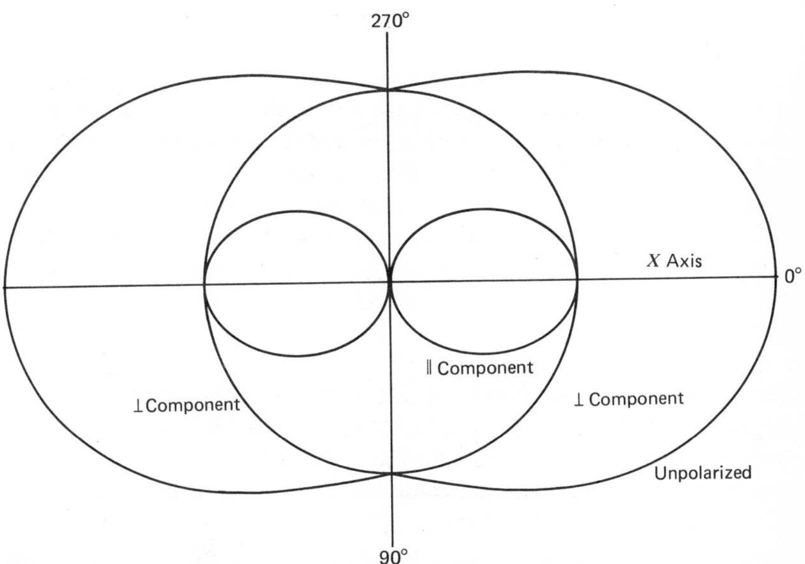

Figure 4.6 Patterns of scattered intensity, in the plane of observation, for unpolarized incident light.

For scattering by isotropic molecules and small particles, the dependences of $I_\perp$ and $I_\parallel$ on θ are noted above as 1 and $\cos^2\theta$, respectively. Substituting these factors into (4-39) gives

$$P = \frac{\sin^2\theta}{1+\cos^2\theta} \tag{4-40}$$

If scattering of only this type occurred in the atmosphere, (4-40) shows that the skylight from a great circle arc across the sky everywhere at 90 deg from the earth-sun line should be completely polarized. Actually this degree of polarization is never obtained even in the clearest air. This *polarization defect* is caused by light reflected diffusely from the earth's surface and scattered by air molecules, by Mie scattering from aerosol particles, and to a slight extent by anisotropy of air molecules.

4.4.3 Total Scattering Coefficient

The volume total scattering coefficient β_m is the ratio of the flux totally scattered in all directions, by a unit volume of a gas, to the irradiance of the incident flux. It describes the scattering efficacy of all the molecules and is the real-world, measurable counterpart of the total scattering cross section of a single molecule. In considering first the case of polarized light, this total cross section is defined by (4-29). Then the total coefficient for the unit volume containing N molecules is

$$\beta_m = \frac{8\pi^3(n^2-1)^2}{3N\lambda^4} \tag{4-41}$$

As with the angular coefficient, the total coefficient varies directly with gas density and can be regarded as the total area of the incident wave front acted on by N molecules. An expression identical to (4-41) can be obtained by integrating $\beta_m(\phi)$, defined by (4-33), over 4π sr. When the scatterers are small particles of the same size, each having a total cross section defined by (4-30), the total coefficient for a unit volume containing N particles is

$$\beta_p = \frac{128\pi^5 r^6 N}{3\lambda^4}\left(\frac{n^2-1}{n^2+2}\right)^2 \tag{4-42}$$

for polarized incident light.

The volume total coefficient β_m for a unit volume illuminated by unpolarized light is found by integrating the angular coefficient $\beta_m(\theta)$ defined by (4-37). The geometry for the integration is shown in Figure 4.7. From the meaning of $\beta_m(\theta)$, its value is independent of the orientation of the observation plane about the propagation axis of the incident light. Hence θ may be regarded as a polar angle, which means that

$$d\omega = 2\pi \sin\theta \, d\theta$$

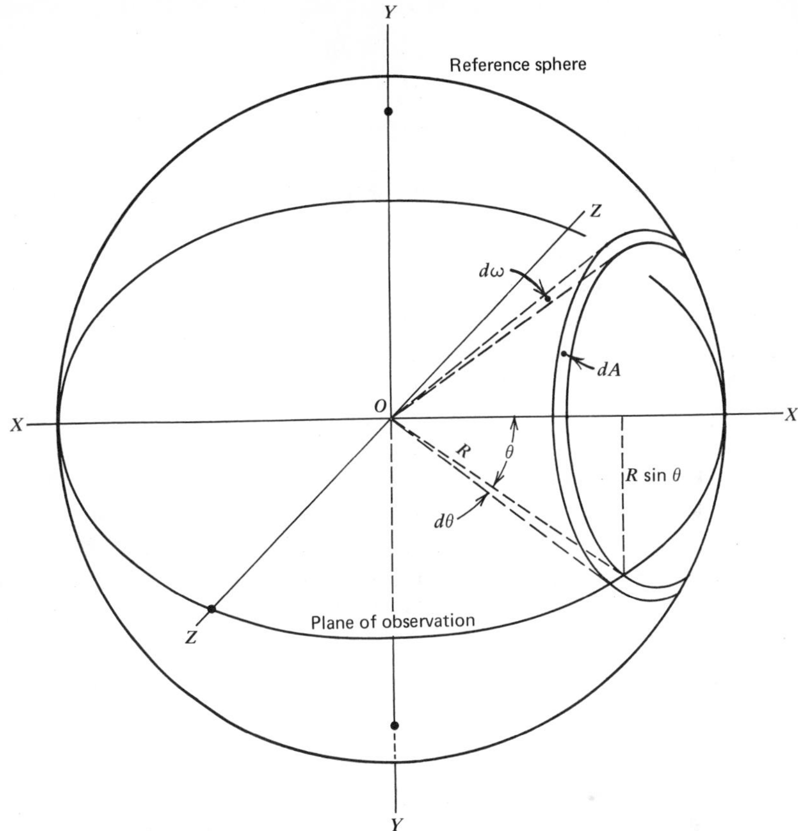

Figure 4.7 Geometry for integration of the angular scattering coefficient for unpolarized incident light.

Making this change in variable in (4-37) and integrating, we have

$$\beta_m = \int_0^{4\pi} \beta_m(\theta)\, d\omega = \frac{\pi^3(n^2-1)^2}{N\lambda^4} \int_0^{\pi} (1+\cos^2\theta) \sin\theta\, d\theta \qquad (4\text{-}43)$$

The value of the definite integral is $\frac{8}{3}$, so that

$$\beta_m = \frac{8\pi^3(n^2-1)^2}{3N\lambda^4} \qquad (4\text{-}44)$$

which is identical to the volume total coefficient (4-41) for polarized light. When the scatterers are small particles illuminated by unpolarized light, a similar integration of (4-38) gives

$$\beta_p = \frac{128\pi^5 r^6 N}{3N\lambda^4}\left(\frac{n^2-1}{n^2+2}\right)^2 \tag{4-45}$$

which is the same as (4-42) for polarized light. The dimension of either β_m or β_p is L^{-1}.

The fact that the total coefficients for polarized and unpolarized light are identical is not surprising, in view of the assumptions and concepts employed. The assumption of molecular isotropy means that the dipole moments are always aligned with the electric vector of the primary wave. Hence when the incident light is linearly polarized, the orientation of its electric vector about the propagation axis is immaterial. Unpolarized light is regarded as the resultant of two equal, orthogonal, linear polarizations having no coherent relationship. The resultant electric vector assumes all possible orientations about the propagation axis within a very short period of time. The instantaneous orientation of this resultant vector is likewise immaterial, because of the isotropic assumption. The irradiance of the unpolarized light is equal to one-half the sum of the irradiances of its two polarized components. Thus isotropic molecules scatter polarized and unpolarized light equally well. Actual air molecules are slightly anisotropic, and a correction factor is described in the following section.

4.4.4 Molecular Anisotropy and Phase Function

If gas molecules were isotropic, the light scattered at $\theta = 90$ deg would be completely polarized. That is, all the scattered light would consist of the component $I_\perp$, while the component $I_\parallel$ would be absent, and from (4-39) the degree of polarization would be unity. Actually, this is never quite true for any of the atmospheric gases; some *depolarization* always occurs because of molecular anisotropy. Qualitatively, we may suppose that anisotropy prevents the dipole moment from aligning itself exactly with the electric vector of the primary wave. The result is that a small component of $I_\parallel$ is found at $\theta = 90$ deg, where theory and the isotropic assumption predict that it should be zero. The subjects of anisotropy and depolarization are treated in considerable detail by Chandrasekhar (1950) and Kerker (1969), and more briefly by Goody (1964). Here we review only the effects on the scattering coefficients derived in the preceding sections.

The depolarization factor p_n is defined by

$$p_n = \frac{I_\parallel(\pi/2)}{I_\perp(\pi/2)} = \frac{2\gamma}{1+\gamma} \tag{4-46}$$

where the terms in γ are used below in connection with the phase function. The factor p_n is introduced into scattering theory by the expression

$(6+3p_n)/(6-7p_n)$ applied as a multipler to any of the standard coefficients. For example, the volume angular coefficient becomes

$$\beta_m(\theta) = \frac{\pi^2(n^2-1)^2}{2N\lambda^4}\left(\frac{6+3p_n}{6-7p_n}\right)(1+\cos^2\theta) \tag{4-47}$$

which can be refined further through the concept of phase function. Likewise, the volume total coefficient becomes

$$\beta_m = \frac{8\pi^3(n^2-1)^2}{3N\lambda^4}\left(\frac{6+3p_n}{6-7p_n}\right) \tag{4-48}$$

Values of p_n have been investigated by several workers, as summarized in Table 4.1, where a value of 0.035 is considered by Penndorf (1957a) to be representative for air. Thus the correction for molecular anisotropy is

$$\frac{6+3p_n}{6-7p_n} = 1.06$$

TABLE 4.1 DEPOLARIZATION FACTORS FOR ATMOSPHERIC GASES

	Values found by several workers					
Gas	1	2	3	4	5	6
Air	0.042	—	0.0415	0.031	0.0365	0.033
Oxygen	0.060	0.067	0.0642	—	0.060	0.054
Nitrogen	0.030	0.030	0.0357	—	0.031	0.0305
Carbon dioxide	0.080	0.073	0.097	—	0.0922	0.0805

Source: Penndorf (1957a).

Values of β_m computed from (4-48) by Penndorf (1957a) are noted in Section 4.5.2 and listed in Appendix I.

The concept of *phase function* expresses in a formal manner the angular dependence of scattering, such as that indicated by the term $(1+\cos^2\theta)$ in (4-47). This use of the word *phase* has no relevance to the phase of an electromagnetic wave but is akin to the concept of lunar phase (angular aspect) as employed in astronomy. The phase function, denoted here by $P(\theta)$ for unpolarized light, is defined by van de Hulst (1957) as "the ratio of the energy scattered per unit solid angle in this [a given] direction to the *average* energy scattered per unit solid angle in all directions." This definition requires that the integral of the phase function be normalized to unity, which is to say that

$$\frac{1}{4\pi}\int_0^{4\pi} P(\theta)\, d\omega = 1 \tag{4-49}$$

This requirement is satisfied by making

$$P(\theta) = \tfrac{3}{4}(1+\cos^2\theta) \tag{4-50}$$

as shown in the following.

The reference level needed for the formal phase function $P(\theta)$ is the average flux scattered per unit solid angle, and this is evidently equal to $\beta_m/4\pi$. Performing this operation on (4-48), we have

$$\frac{\beta_m}{4\pi} = \frac{2\pi^2(n^2-1)^2}{3N\lambda^4}\left(\frac{6+3p_n}{6-7p_n}\right) \tag{4-51}$$

for the reference level. Multiplication by the selected phase function (4-50) then results in an identity with the angular coefficient (4-47), as may be seen from

$$\frac{2\pi^2(n^2-1)^2}{3N\lambda^4}\left(\frac{6+3p_n}{6-7p_n}\right)\times\frac{3}{4}(1+\cos^2\theta) \equiv \frac{\pi^2(n^2-1)^2}{2N\lambda^4}\left(\frac{6+3p_n}{6-7p_n}\right)(1+\cos^2\theta) \tag{4-52}$$

Values of the reference level function $\beta_m/4\pi$ are noted in Section 4.5.2 and listed in Appendix I.

Molecular anisotropy affects the phase function slightly. Chandrasekhar (1950) gives the formula

$$P(\theta) = \frac{3}{4(1+2\gamma)}[(1+3\gamma)+(1-\gamma)\cos^2\theta] \tag{4-53}$$

where the relationship of γ to the depolarization factor p_n is defined by (4-46). A representative value for p_n is 0.035, as noted previously, so that $\gamma = 0.0178$. Substitution of this value into (4-53) yields

$$P(\theta) = 0.7629(1+0.9324\cos^2\theta) \tag{4-54}$$

for the normalized phase function corrected for molecular anisotropy. Equation (4-54) satisfies the normalization requirement of (4-49), as integration will show. Substitution of (4-54) for the uncorrected phase function $\frac{3}{4}(1+\cos^2\theta)$ in (4-47) produces an optimized angular coefficient:

$$\beta_m(\theta) = \frac{2\pi^2(n^2-1)^2}{3N\lambda^4}\left(\frac{6+3p_n}{6-7p_n}\right)\times 0.7629(1+0.9324\cos^2\theta) \tag{4-55}$$

Values of $P(\theta)$ according to (4-54), and of $\beta_m(\theta)$ according to (4-55), are noted in Section 4.4.1 and are listed in Appendix H.

4.5 TABULATIONS OF RAYLEIGH SCATTERING FUNCTIONS

In solving problems that involve Rayleigh scattering, much labor can be avoided by employing numerical values of the functions, corrected as need be for given conditions. Undoubtedly the most comprehensive tabulations of these functions available in the literature are those by Penndorf (1957a) for the spectral range 0.20 to 20.0 μm. From his work this section presents values of the phase function, angular coefficient, total coefficient, and refractivity term. Also given are the data and method for obtaining the angular coefficient at any of the listed wavelengths. Correction factors for temperature, pressure, density, and altitude are described, so that the listed values can be adapted to any atmospheric regime.

4.5.1 Phase Function and Angular Coefficient

Values of the normalized phase function (4-54) and the volume angular coefficient (4-55) are tabulated in Appendix H. Values of the coefficient refer to *unpolarized* incident light at the wavelength 0.55 μm, and air at $T = 288.15$ K, and $P = 1013.25$ mb. These are the sea-level conditions for USSA-1962. Values are given for 1-deg increments of the observation angle θ from 0 to 90 deg; the front-to-back symmetry of the scattering pattern, shown in Figure 4.6, makes the values applicable from 0 to 180 deg. Because of the left-to-right symmetry, the values apply to an observation direction on either side of the incident light direction.

It is easy to apply the tabulated values to cases in which the incident light is polarized either perpendicular or parallel to the observation plane. When the first condition exists, the scattered intensity is governed by the factor unity in the parenthetical terms of the phase function (4-54). This means that for all values of θ the phase function has the value 0.7629, which is its value at 90 deg, and that $\beta_m(\theta) = 0.7053 \times 10^{-3}\ \text{km}^{-1}$ for all values of θ. When the second condition exists, the scattered intensity is governed by the factor $\cos^2 \theta$ in the phase function, and the factor that is normally unity becomes zero. The tabulated values of the phase function can be corrected for this condition by subtracting 0.7629 from these values. Likewise, the angular coefficient can be corrected by subtracting 0.7053×10^{-3} from the tabulated values.

The wavelength 0.55 μm used in computing the listed values of $\beta_m(\theta)$ marks the maximum photopic response of the eye, as shown in Figure 1.3. Occasionally this wavelength is used to represent the entire visual region of the spectrum when approximate evaluations of scattering are wanted. For more accurate work, it is convenient to have available the values of the angular coefficient for many wavelengths, in the ultraviolet and infrared regions as well as in the visual. Such values may be obtained easily by the following procedure. It was shown previously that the angular coefficient is equal to the product of the normalized phase function and a "reference level." This level expresses the average flux scattered per steradian and is defined by (4-51) as $\beta_m/4\pi$. Values of this quantity are tabulated in Appendix I for wavelengths from 0.30 to 20.0 μm. Therefore, to find $\beta_m(\theta)$ for any of these wavelengths at any angle θ, it is necessary only to select the value of $\beta_m/4\pi$ for the desired wavelength and multiply it by the appropriate value of $P(\theta)$ from Appendix H.

4.5.2 Total Coefficient and Refractive Index Term

Values of the volume total coefficient (4-48) are tabulated in Appendix I for 89 wavelengths between 0.30 and 20.0 μm. The values refer to either unpolarized or polarized incident light, air at $T = 288.15$ K and $P = 1013.25$ mb. The increasing wavelength interval employed beyond 0.80 μm reflects the rapidly decreasing magnitude of Rayleigh scattering at longer wavelengths according to the λ^{-4} dependence. This means that molecular scattering at longer wavelengths can often be evaluated to sufficient accuracy from an approximate knowledge of wavelength. Values of the reference level defined by $\beta_m/4\pi$ are given for use in finding the angular coefficient at any of these wavelengths, as described in the preceding section. Also listed are values of the term $(n^2-1)^2$.

As a part of his atmospheric attenuation models, Elterman (1968, 1970a) has tabulated the values of the volume total coefficient for many altitudes from sea-level to 50 km. His models are based on the air density data of USSA-1962, and the tabulations cover 20 wavelengths within the range 0.27 to 2.17 μm. McClatchey et al. (1971) provide tabulations of the volume total coefficient from sea-level to 100 km in their attenuation models. These models are based on the air density data of USSA-1962 and USSAS-1966, and they cover the tropical, midlatitude, and subarctic latitude zones and the summer and winter seasons. Tabulations are given at 12 laser wavelengths from 0.3371 to 337 μm.

4.5.3 Corrections for Temperature, Pressure, and Altitude

The expressions for the scattering coefficients thus far developed are general, but the value of the refractive index n must correspond to the value of the mass density ρ or number density N employed. The tabulated values listed in Appendices H and I refer to air at the sea-level temperature and pressure conditions of USSA-1962. From the measurement standpoint, the use of temperature and pressure parameters is really only a convenient way of specifying air density, which is a basic quantity in molecular scattering. For example, it follows from the value of the gas constant (2-21) and the gas law (2-23) that air at 288.15 K and 1013.25 mb has a mass density of 1.225 kg m^{-3} or 1.225×10^{-3} g cm^{-3}. This corresponds to a number density of 2.547×10^{25} m^{-3} or 2.547×10^{19} cm^{-3}. It is desirable to provide correction factors so that the tabulated values of the scattering coefficients can be adjusted to the real-world values of temperature, pressure, and altitude. These factors are given in several forms in the following paragraphs, but the basic factor in every case is molecular number density.

For any nonstandard values of T, P, and ρ it follows from (2-23) that

$$\rho = \rho_0 \frac{P}{P_0} \times \frac{T_0}{T} \tag{4-56}$$

where the subscripts indicate any set of self-consistent reference values, and T must be expressed on the Kelvin scale. When a worker has a preference for temperature t on the Celsius scale, (4-56) can be put into the form

$$\rho = \rho_0 \frac{P}{P_0} \left(\frac{1}{1 + 0.00366t} \right) \tag{4-57}$$

where the multiplier of t is equal to the expansion coefficient of a perfect gas. In each of the above equations, pressure can be expressed in any consistent unit.

The Rayleigh coefficients are proportional to air density, so that

$$\beta_m(\theta) \text{ or } \beta_m = [\beta_m(\theta) \text{ or } \beta_m]_0 \frac{\rho}{\rho_0} \tag{4-58}$$

where the subscript 0 means that the value is with respect to a specified condition taken as a reference. Thus the corrections can be made directly from (4-58) if the air density for the given situation is known. If the molecular number density N is known, which seldom happens, the ratio N/N_0 may be substituted for ρ/ρ_0 in (4-58). More often, temperature and pressure are known, in which case

$$\beta_m(\theta) \text{ or } \beta_m = [\beta_m(\theta) \text{ or } \beta_m]_0 \frac{P}{P_0} \times \frac{T_0}{T} \tag{4-59}$$

A similar expression can be written from (4-57) for temperature on the Celsius scale. Finally, only altitude Z may be given, and tabulations of density, or temperature and pressure, versus altitude, such as are contained in USSA-1962 and USSAS-1966 may not be available. In such a case, the corrections can be made from the sense of (2-55) according to

$$\beta_m(\theta) \text{ or } \beta_m = [\beta_m(\theta) \text{ or } \beta_m]_0 \exp\left(-\frac{Z}{H_\rho}\right) \tag{4-60}$$

where H_ρ is the density scale height defined by (2-54). A correction of type (4-60) is easily made, since tables of the negative exponential are usually at hand.

4.6 RAYLEIGH SCATTERING IN THE ATMOSPHERE

Rayleigh scattering by molecules represents the irreducible minimum of scattering along an atmospheric path. At lower altitudes, Mie scattering by particles always predominates, but on the average decreases more rapidly with altitude than does Rayleigh scattering. This is to be expected, because representative scale heights of the haze aerosol are near 1 km, while scale heights of the permanent gas envelope are in the range 6 to 9 km and greater. Thus the haze atmosphere has a limited vertical extent, except for several tenuous layers of particles such as the one near 20 km. In contrast, the Rayleigh or molecular atmosphere continues upward to great altitudes, but its transparency is sufficient that lines of sight have lengths limited by geometric, not optical, factors. The Rayleigh atmosphere is a useful starting point for studying the optical complexities of the real atmosphere. Many aspects of Rayleigh scattering are revealed by considering the radiance and polarization of the daytime sky.

4.6.1 Optical Thickness and Atmospheric Turbidity

The outstanding optical features of a clear day sky are the intensity, color, and polarization of the scattered light arriving from a hemisphere of directions. These features are governed by the optical thickness of the atmospheric paths, the solar altitude, and the angle between the solar direction and the direction of observation. The molecular optical thickness T_m of an optical path expresses the combined total scattering effects of all the molecules along that path. Horizontal paths tend to be homogeneous in a molecular sense, so that β_m can be taken as constant, and T_m is defined by (1-51). When the path is vertical or slant, β_m is a function of distance x along the path. Hence in principle T_m must be found by integration:

$$\mathrm{T}_{m,R} = \int_0^R \beta_m(x)\, dx \tag{4-61}$$

where R is the path length. However, when only single scattering is concerned, it is the total number of molecules, but not their spatial distribution, that is significant. Therefore the reduced height Z' or the equivalent path $\bar{R}$, discussed in Sections 2.4.3 and 2.5.3, can be used with (1-51) to find the optical thickness of a vertical or a slant path.

The molecular optical thickness of a vertical path from a given altitude to the top of the atmosphere thus is given by

$$\mathrm{T}'_m = Z'\beta_m \tag{4-62}$$

where Z' is the reduced height for that altitude, and β_m is appropriate to the density of the reduced atmosphere. Figure 4.8 shows the vertical optical thickness of the molecular atmosphere computed from values of Z' in Table 2.8 and β_m in Appendix I. The optical thickness between any two altitudes is equal to the difference between the corresponding ordinates. Elterman (1968, 1970a) provides tabulated values of the molecular optical thickness of a vertical path, for many initial and terminal altitudes from sea-level to 50 km. His tabulations cover 20 wavelengths from 0.27 to 2.17 m.

The optical thickness of a horizontal path (assumed homogeneous) is equal to the product of the path length and the value of β_m for that altitude. For a slant path the optical thickness is found by multiplying the thickness of a vertical path by sec ζ, when ζ is not greater than about 75 deg. At larger angles the optical air mass, values of which are listed in Table 2.11, should be used instead of sec ζ. At high altitudes and zenith angles greater than 75 deg, the Chapman function tabulated in Table 2.12 should be used.

Several features of atmospheric scattering are illustrated by Figure 4.8. The plots are straight lines, except for altitudes in the troposphere, where

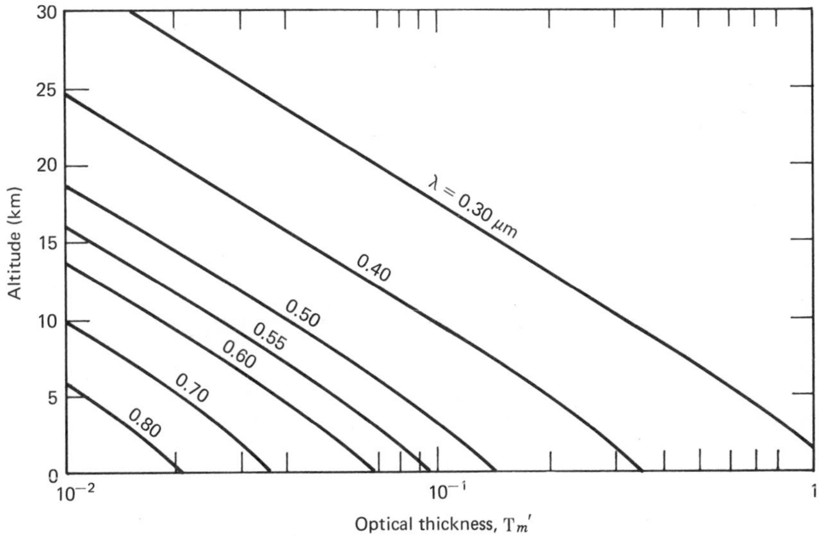

Figure 4.8 Optical thickness of the molecular atmosphere as a function of altitude and wavelength for a vertical path, above any given altitude.

slight curvatures may be seen. Such linearity on semilog scales is due of course to the near-exponential dependence of air density on altitude. The variation in T_m with λ is the same as the λ^{-4} variation in Rayleigh scattering itself. For example, in going from 0.8 to 0.4 μm the value of T_m increases by a factor of 16. Such increases explain why the overhead sun appears slightly yellow instead of white. As the sun approaches the horizon, the optical thickness of the path increases with the air mass values in Table 2.11. At a zenith angle of 89 deg, which puts the sun above the horizon by only twice its diameter, the value of air mass is about 27. The resulting optical thickness is 0.99 for a wavelength of 0.7 μm, but is 9.8 for a wavelength of 0.4 μm. From (1-52) the resulting transmittance at 0.7 μm is 0.37, but at 0.4 μm it is only about 6×10^{-5}, and the sun appears almost ruby red. Thus most of the energy at shorter wavelengths is scattered from the solar beam, producing the profound change in color, but reappears in all other directions as blue sky light.

A concept used frequently in studies of visibility and transmittance is that of *atmospheric turbidity*. Turbidity expresses the total scattering by the entire vertical extent of the atmosphere (molecules plus particles) in terms of molecular scattering. This scattering, being the irreducible amount, is a natural reference level for the aggregate amount. The turbidity T is defined by

$$T = \frac{T'_m + T'_p}{T_m} \tag{4-63}$$

where T'_p is the optical thickness of the haze atmosphere. For example, the value $T = 5$ means that the haze scattering (alone) is four times the molecular scattering. This value of turbidity corresponds to the condition "exceptionally clear" in Table 1.8.

4.6.2 Radiance of the Molecular Sky

Scattering by the molecular atmosphere was investigated early in this century as a part of determining the solar constant, that is, the value of solar irradiance at the top of the atmosphere. For example, Fowle (1914) determined the vertical transmittance of "dust-free" air above the observatory at Mt. Wilson, California, from measurements of solar irradiance at that location. An interesting consequence of his work was a determination of Loschmidt's (or Avogadro's) number. Since this value agreed with values obtained by other methods, the agreement argued for the essential correctness of Rayleigh scattering theory. High-altitude investigations, still continuing, of skylight intensity and polarization began after World War II. At that time the availability of large aircraft that could fly routinely in the lower stratosphere allowed extensive measurements to be made in relatively haze-free atmospheres. Outstanding among these programs were the ones carried out by Hulburt, Tousey, and their associates at the Naval Research Laboratory.

As an illustration we summarize one example of the investigations of skylight by Tousey and Hulburt (1947). The geometry of Figure 2.11 is generally applicable, but for simplicity here the earth and its atmospheric envelope are assumed to be plane-parallel, as shown in Figure 4.9. According to these investigators, this assumption does not cause errors greater than 2% for all zenith angles less than 80 deg. In considering the diagram, the value of sky radiance in the direction *OP* is to be determined. The point *P*, at altitude *Z*, lies within the atmosphere, and the observer is at point *O*, whose altitude is taken as zero. However, this need not be sea level, but may be any reference altitude. The reduced height of the atmosphere above *P* is denoted by Z', and the total reduced height above *O* by Z'_0. All the solar rays incident on the atmosphere are considered parallel, and the relative bearing of the line *OP* from the sun line *OS* is denoted by δ. The zenith angles of the sun and point *P* are, respectively, ζ_s and ζ_P. Only a narrow spectral interval of radiant flux is considered, and the wavelength is taken to be in a spectral region free of absorption.

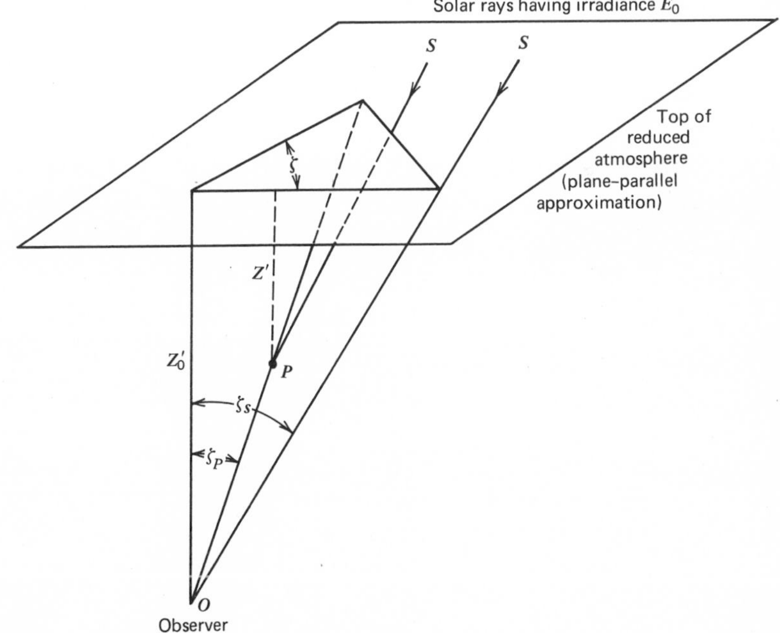

Figure 4.9 Geometry of atmospheric scattering of sunlight. After Tousey and Hulburt (1947).

The solar flux, having the irradiance E_0 at the top of the atmosphere, is attenuated by scattering as it travels the slant path through reduced height Z' to reach point P. The irradiance of a unit volume at P is

$$E = E_0 \exp(-\beta_m Z' \sec \zeta_s) \tag{4-64}$$

The scattered intensity in the direction PO from a differential element of path at P having a unit cross section and a length

$$ds = \sec \zeta_P \, dZ$$

and characterized by the angular scattering coefficient $\beta_m(\theta)$ is given by

$$dI(\theta) = E_0 \exp(-\beta_m Z' \sec \zeta_s)\beta_m(\theta) \sec \zeta_P \, dZ \tag{4-65}$$

The attenuation along the path PO is defined by

$$\exp[-\beta_m(Z'_0 - Z') \sec \zeta_P]$$

so that the differential intensity perceived at 0 is

$$dI(\theta) = E_0 \exp(-\beta_m Z' \sec \zeta_s)\beta_m(\theta) \exp[-\beta_m(Z'_0 - Z') \sec \zeta_P] \sec \zeta_P \, dZ \tag{4-66}$$

The intensity at O due to all elements along PO is given by

$$I(\theta) = E_0\beta_m(\theta) \int_0^{Z'_0} \exp(-\beta_m Z' \sec \zeta_s) \exp \times [-\beta_m(Z'_0 - Z') \sec \zeta_P] \sec \zeta_P \, dZ \tag{4-67}$$

The indicated integration was performed by Tousey and Hulburt, who found that

$$I(\theta) = \frac{E_0\beta_m(\theta)}{\beta_m} \left[\frac{\exp(-\beta_m Z'_0 \sec \zeta_s) - \exp(-\beta_m Z'_0 \sec \zeta_P)}{1 - \sec \zeta_s \cos \zeta_P} \right] \tag{4-68}$$

for the intensity of scattered solar flux at the bottom of a plane-parallel atmosphere in a specified direction such as OP. Because the cross section of the path is taken as unity, the value of $I(\theta)$ is equal to the sky radiance in that direction. Usually it is helpful to interpret the radiance of the sky, which has no measurable area or distance, according to (1-13) and the associated discussion. The observation angle θ, which must be known in order to select the value of $\beta_m(\theta)$, is defined by (2-75).

In a series of airborne measurements at altitudes to 18,000 ft, Tousey and Hulburt (1947) tested expressions of the type (4-68), including refinements that take account of ground-reflected light and multiple scattering. Measured values of skylight intensity and polarization were compared with the theoretical values. Even above 10,000 ft and in directions far from the sun, it was necessary to take $\beta_{sc} = 0.017$ instead of $\beta_m = 0.0126$, which they had calculated for the visual spectrum and pure air, in order to reconcile the measured intensities with the predictions. This indicated of course that aerosol particles were present in addition to air molecules.

In an extensive series of flights to 38,000 ft, Packer and Lock (1951) made further comparisons between measured and theoretical values. One set of their results is shown in Figure 4.10, where the observed brightness agrees well with the predictions when $\beta_{sc} = 0.017$ at scattering angles from the sun greater than 30 deg. The rapid increase in brightness as the line of sight is moved toward the direction of the sun is a phenomenon called the *solar aureole*, always a sign that aerosol particles are present. The scattering patterns for such particles are peaked in the forward direction, in strong contrast to the $(1 + \cos^2 \theta)$ pattern for molecular scattering, as shown in Figure 4.6, which does not produce an aureole. Thus measurement of sky radiance, in directions near that of the sun, is a sensitive method for detecting aerosol particles, whose concentrations may be so low that the tenuous haze itself is invisible to ordinary observation from the ground. Deirmendjian and Sekera (1953), in their study of the multiple scattering of

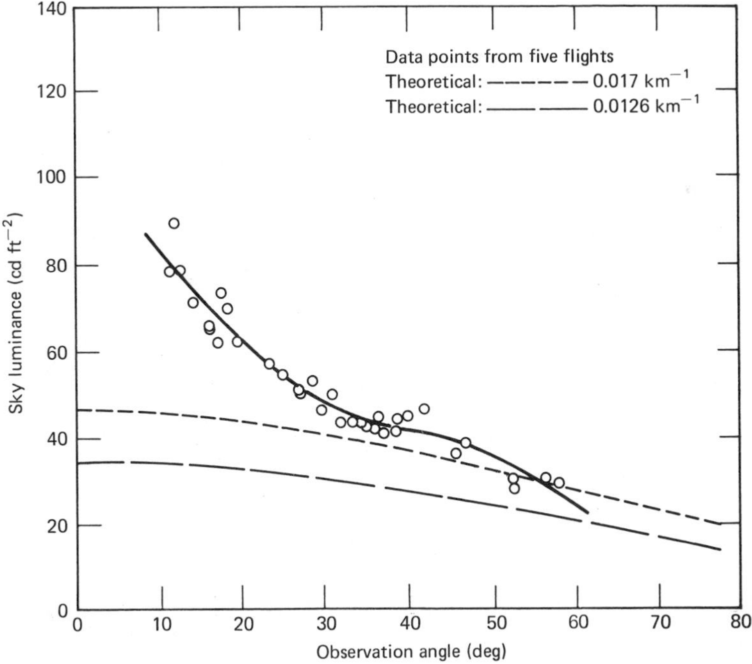

Figure 4.10 Sky luminance at 38,000 ft at various observation angles from the direction of the sun. Theoretical values are shown for comparison. From Packer and Lock (1951).

solar flux by a Rayleigh atmosphere, reached a similar conclusion regarding the presence of particles at high altitudes and their effects on sky radiance.

The light scattered upward or outward by the atmosphere has received much attention in recent years. This light consists of two components: (1) that which has been backscattered from extraterrestrial light incident on the atmosphere, and (2) that which has been forward-scattered from ground-reflected light. The spectral intensity of this outgoing light, which determines the luminance of the atmosphere, thus depends on the incident spectral irradiance, the characteristics of the ground reflection, the direction of irradiance and observation, and the optical thickness. Experimentally it may be difficult to distinguish between the outgoing light from molecular scattering and that from aerosol scattering and ground reflection. A spectral clue is furnished, however, by the fact that molecularly scattered light is predominantly blue, as expected from the greater optical thickness at shorter wavelengths. Color photographs of the earth taken from high altitudes and space usually show a pronounced blue tinge, and the horizon often appears as a bright-blue band.

4.6.3 Polarization of Sky Light

Sky light polarization, discovered by Arago about 1809, is as much a result of the scattering process as is the light itself. Comprehensive reviews of this subject are provided by Sekera (1951, 1956, 1957a,b). The degree of polarization is governed by (4-40), the plane of observation in which θ is measured having freedom of orientation about the observer-sun line. This proviso, mentioned in the discussion following (4-38), allows any point in the sky to be observed by employing suitable values of the observation and orientation angles. Such a capability can be visualized from Figure 2.11. According to (4-40), the sky light from a plane containing the observer, normal to the observer-sun line, is completely polarized. In addition, two neutral points exist where the scattered light is unpolarized. One point lies at $\theta = 0$ deg, which is in the direction of the sun; this scattered light is not observable. The second point lies at $\theta = 180$ deg and is called the *antisolar point*.

Actually, the degree of polarization never quite conforms to (4-40). This failure, called the *polarization defect*, is caused by reflected light from the earth's surface, multiple scattering, molecular anisotropy, and scattering by aerosol particles. As part of the defect, several points of neutral polarization exist in directions different from those predicted, although they usually lie in the vertical plane containing the observer and the direction of the sun. This plane, often called the *sun's vertical*, is designated ZSH_0O in Figure 2.11, where only one quadrant of this plane appears.

As shown in Figure 4.11, the three most common neutral points are: the Babinet, 15 to 25 deg above the sun; the Brewster, 15 to 20 deg below the sun; and the Arago, 15 to 25 deg above the antisolar point. Van de Hulst (1952) gives an approximate derivation for these angular positions, which vary with the solar altitude, the wavelength of observation, and the haze aerosol composition. For example, Neuberger (1950) found that the angle between the Arago and antisolar points is greater when the sky is observed through a blue filter than when observed through a red filter or no filter. Also, the angle changes rapidly during the early stages of cloud development, as when condensation nuclei are growing into cloud droplets. Large changes in the degree of polarization and in its distribution have been observed for periods of several years after large volcanic eruptions.

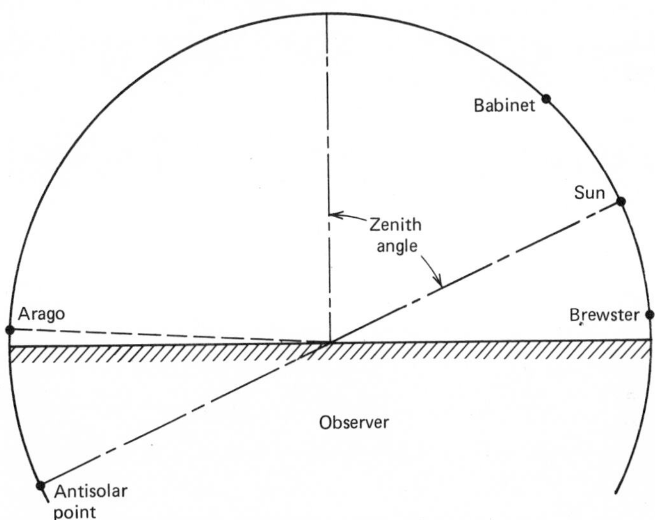

Figure 4.11 Relative positions of the neutral points of polarization along the sun's vertical. From Johnson (1954).

The following discussion of sky-light polarization is concerned only with the characteristics due to molecular scattering of sunlight. The modifications imposed by particle scattering and other factors are treated by the authors referenced in Section 4.6.4. Evaluations of sky-light polarization usually start with the distribution of the degree of polarization in the sun's vertical. This plane thereby becomes the plane of observation; therefore it is the reference plane for identifying the components $I_{\perp}$ and $I_{\parallel}$ and for applying (4-39) and (4-40). Such distributions of the degree of polarization for three optical thicknesses, a solar zenith distance of

53.1 deg, and a ground reflectance of zero are shown in Figure 4.12 from the detailed study by Sekera (1956). The two points of zero polarization are the Babinet and Brewster neutral points shown in Figure 4.11. The small amount of negative polarization in the region near the solar position means that the component whose electric vector is parallel to the sun's vertical slightly exceeds the perpendicular component. This meaning may be verified from (4-39). The principal effect of a greater optical thickness, which brings about greater multiple scattering, is a reduction in the degree of polarization. Lesser effects are the increase in negative polarization and the shifts of the neutral points. Not shown here are the effects of ground reflections which return light to the sky. Coulson (1968) examined in detail the effects of large reflectances such as are exhibited by desert sands.

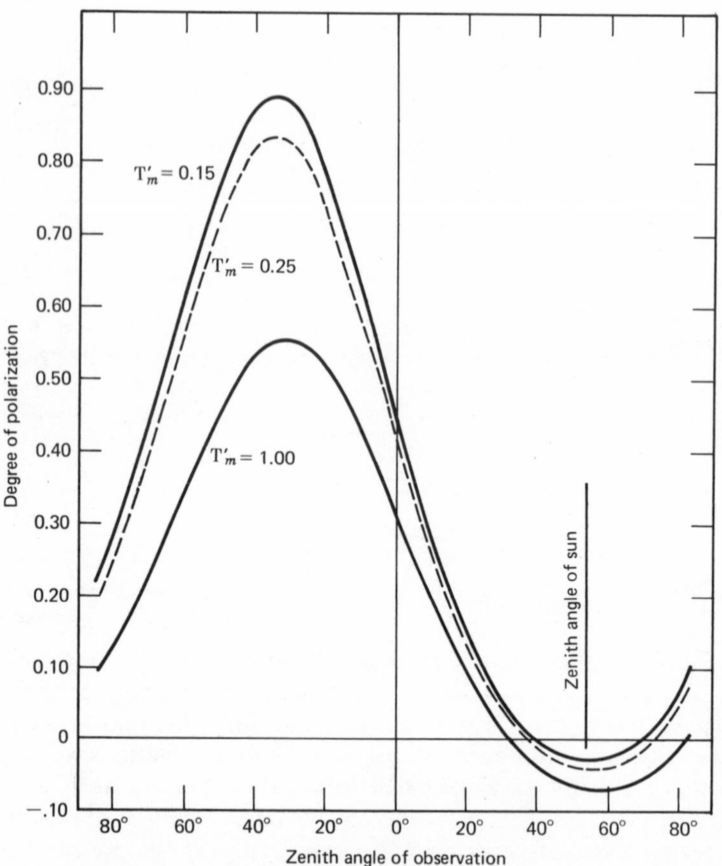

Figure 4.12 Distribution of the degree of polarization along the sun's vertical for three optical thicknesses of the molecular atmosphere. From Sekera (1956).

In any direction away from the sun's vertical, the polarization state of sky light is defined by the degree of polarization and by the orientation of the plane of polarization. Following Sekera (1956), the reference plane for these two defining quantities is taken as the vertical plane through the direction of observation, shown as the plane *ZPHO* in Figure 2.11. It is customary to express the polarization state of scattered light by means of the four Stokes parameters of polarization. Because elliptical polarization is not present significantly in atmospheric molecular scattering, the parameter V is equal to 0. The degree of polarization in terms of the remaining parameters I, Q and U is given by

$$\mathrm{P} = \frac{(Q^2 + U^2)^{1/2}}{I} = \frac{[(I_\perp - I_\parallel)^2 + U^2]^{1/2}}{I_\perp + I_\parallel} \tag{4-69}$$

The angle χ between the plane of polarization, that is, the plane containing the resultant electric vector, and the vertical plane through the direction of observation is defined by

$$\tan 2\chi = \frac{U}{Q} = \frac{U}{I_\perp - I_\parallel} \tag{4-70}$$

Figure 4.13 shows the distribution of the degree of polarization over the sky for an optical thickness of 0.15 and a solar zenith angle of 58.7 deg. Because the sun's vertical is a plane of symmetry, only one-half of the sky hemisphere is drawn. The band of maximum polarization, where P has a value of approximately 0.9, is quite evident.

Many readers may have examined the polarization of sky light qualitatively with nothing more elaborate than a small piece of sheet Polaroid. A pair of sun glasses made of this material is equally effective. When the sky is clear and the sun favorably situated, the band of maximum polarization indicated in Figure 4.13 can be found by rotating the polarizer which now is being used as an analyzer. When the analyzer is oriented to give minimum transmission for the direction of observation, the darkening of the sky and the resulting deep-blue color are quite impressive. A good appreciation of the depolarizing effects of scattering by particles can be obtained by repeating the observations when the sky is extremely hazy. Polarizing filters, which thus reject to varying degrees one component of the scattered light, are employed in color photography to produce a deep blue sky without sensibly altering other colors. Such a feat is not possible of course with spectral filters. Polarizing filters also are used in landscape photograpy to reject much of the atmospherically scattered light originating between the camera and the object, thereby improving the contrast rendition of distant scenes.

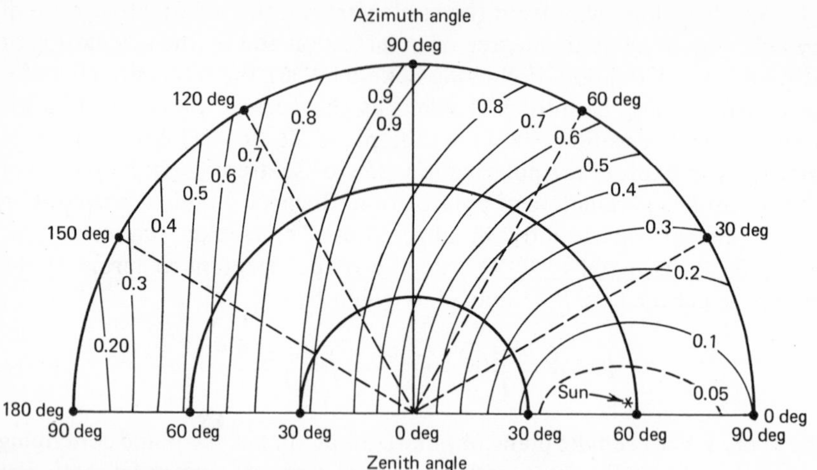

Figure 4.13 Distribution of the degree of polarization over the sky in a molecular atmosphere. One-half of the sky hemisphere is shown projected onto the horizontal plane. The sun is at an azimuth angle of 0 deg and a zenith angle of about 60 deg. The optical thickness of the atmosphere is 0.15. From Sekera (1956).

4.6.4 Studies of Sky Light Intensity and Polarization

The spectral intensity and polarization state of sky light from the molecular atmosphere have been investigated thoroughly. Most of the studies are based on the methods developed by Chandrasekhar (1950), in which he employed the Stokes parameters which oddly enough had lain unused for nearly a century. Calculations of intensity and polarization have been made by Sekera (1956, 1963), Sekera and Blanch (1952), Sekera and Ashburn (1953), Chandrasekhar and Elbert (1954), Coulson et al. (1960), and Dave and Furukawa (1966a). Further studies that variously take into account such factors as multiple scattering, turbidity, ground reflection, and ozone absorption have been made by Deirmendjian and Sekera (1953), Dave (1965), Dave and Furukawa (1966b), Brinkman et al. (1967), Coulson (1968), Fraser and Walker (1968), Kano (1968), Plass et al. (1973), and Blattner et al. (1974). Wolff (1964a,b) has computed the luminance profile of the molecular atmospheric horizon as seen from space.

Two of the works referenced above are reviewed at this point because they provide extensive tabulations of sky light intensity and polarization and take multiple scattering into account. Coulson et al. (1960) deal with 142,688 combinations of the following quantities:

- Optical thickness, 7 values from 0.02 to 1.00.
- Solar zenith angle, 7 values.
- Observation zenith angle, 16 values.
- Relative azimuth angle, 0 deg (30 deg) 180 deg.
- Ground reflectance, 0.00, 0.25, 0.80.

For such combinations the following quantities are tabulated for both upward and downward directions:

- Stokes parameters.
- Degree of polarization.
- Polarization orientation angle.

Dave and Furukawa (1966a) present tabulations of scattering by the molecular atmosphere at ozone absorption wavelenths for selected altitudes defined in terms of pressure levels. The computations were restricted to the solar and antisolar sides of the sun's vertical, hence the relative azimuth angles are 0 and 180 deg. All combinations of the following quantities were employed:

- Wavelengths, 16 values from 0.2875 to 0.6550.
- Solar zenith angle, 5 values.
- Observation zenith angle, 10 values on both the solar and antisolar sides.
- Pressure levels, 10 values from 0 to 1000 mb.

For such combinations the following quantities are tabulated for both upward and downward directions:

- Intensity of singly scattered flux.
- Intensity due to all orders of scattering.
- Degree of polarization.

5

Mie Scattering by Monodispersions

The sky always exhibits notable variations from the color, brightness, and polarization that would be produced by Rayleigh scattering alone. A nominally clear sky is not always a very blue sky, and even a blue sky is gray near the horizon. The rising and setting sun usually can be viewed directly, without discomfort, indicating that the overall path transmittance may be as little as about 10^{-6}. From the altitudes of jet aircraft the earth's actual horizon seldom can be seen, while the atmospheric haze horizon becomes quite evident. All such optical effects are due to scattering by the aerosol particles abounding near the earth's surface. The increasing obscuration as haze thickens into fog, and the almost complete opacity of lower-altitude clouds, attest to the efficacy of small water droplets as scatterers.

Such scattering by particles whose radius is greater than about 0.03 times the wavelength of the light is called *Mie scattering*. Starting with very small particles, as the particle size relative to the wavelength increases, there is a gradual transition from Rayleigh to Mie scattering, which is finally characterized by:

- A complicated dependence of scattered light intensity on the angle of observation, the complexity increasing with particle size relative to wavelength.
- An increasing ratio of forward scattering to backscattering as the particle size increases, with resulting growth of a forward lobe as in Figure 1.5.
- Little dependence of scattering on wavelength when particle size relative to wavelength is large, as may be inferred from the generally white appearance of clouds. Thus a white cloud and a blue sky epitomize the two types of scattering.

Since its inception more than half a century ago, Mie theory has been developed into many forms and expressions which best serve differing fields of application. This chapter presents the basic Mie theory, without derivation, and the concepts and functions that have found widest use in atmospheric optics. Because atmospheric particles cover such a broad size range, this chapter is restricted to monodispersions in order to present the essentials of the theory most simply. The complexities introduced by the usual polydispersions—haze, fog, cloud—are deferred to the following chapter. This chapter reviews several basic parameters and discusses the two intensity distribution functions that are the essence of Mie theory. The several angular scattering functions are then described. As in the case of Rayleigh theory, the study of angular scattering leads naturally to total scattering, which is treated in detail.

5.1 HISTORICAL DEVELOPMENT OF MIE THEORY

The present theory has been developed by many workers, principally from the basic researches by Mie (1908) on scattering by colloidal metal particles. Even before the time of Mie, however, much study had been devoted to the scattering by particles larger than gas molecules. Several of the later studies by Lord Rayleigh were concerned with this subject, and much of the present theory was anticipated by Lorentz during the period 1890 to 1900. Debye (1909) arrived at essentially the same results as Mie but through a somewhat different aptroach. These theoretical developments are recounted by Kerker (1969) and reviewed in an interesting manner by Logan (1965).

Mie theory found early applications in physical chemistry, where it has been developed into a powerful means for investigating colloids and macromolecules. Applications to atmospheric optics were much slower in coming about. In the 1930s the growing problems of visibility in aviation led Houghton (1931) and Stratton and Houghton (1931) to investigate the total scattering by fog and cloud droplets. A decade later studies of smoke screens in naval warfare, among other motivations, led to computation of angular scattering functions by La Mer and Sinclair (1943), later extended by Lowan (1949). All this early work required untold hours of manual calculation but laid the basis for the tables of functions now proliferated by electronic digital computers. The mathematical basis of the theory has been presented by Stratton (1941) in a treatment long regarded as classic.

In its complete formulation, Mie theory rigorously describes the scattering characteristics of particles for broad ranges of size and refractive index. When the particles are small or of molecular size the full theory reduces to the Rayleigh theory. While the derivation of that theory is a simple exercise in electromagnetics, this is not true of Mie theory. It is complex, and its derivation is beyond the scope of this book. We restrict ourselves therefore to explaining various aspects of the theory and to showing its applications to atmospheric optics. Even with this restriction there is no lack of subject matter. The problem rather is one of introducing the reader to the wealth of information that has been developed over many years and of guiding him in its use.

Readers whose interests go beyond these practical objectives are referred to the extended theoretical treatments by Stratton (1941), La Mer and Sinclair (1943), van de Hulst (1957), which is regarded by many as the standard work on the subject, Deirmendjian (1969), and Kerker (1969). The three last-named authors also give a synoptic view of the computational tasks and guidance in computing several required subfunctions. In addition, van de Hulst (1957), Deirmendjian (1960), and Kerker (1969) describe approximations that can be employed instead of the exact theory when the particle sizes are comparable to the wavelengths and larger. When a high degree of accuracy is not required, such methods can greatly reduce the labor of computing numerical values of the functions, and they also provide additional insights into scattering. One of these approximations is described in this chapter.

5.2 RATIONALE AND PHYSICAL PARAMETERS

Scattering by a particle composed of many molecules is an extension of scattering by a single molecule. Because many closely packed molecular oscillators are now involved, interferences among the scattered waves occur as a function of observation angle. The parameters that are important in the scattering by particles thus are the wavelength of the incident light and its state of polarization, the particle size and refractive index, and the angle at which the scattering is observed. In this section a brief rationale of scattering by particles is given and the geometry and radiometry of the scattering are discussed. The important role played by the ratio of particle size to wavelength is then emphasized, and the relevant aspects of refractive index are summarized.

5.2.1 Rationale and Radiometry of Particle Scattering

Although Mie scattering is complex in detail, the rationale or physical basis is readily seen by extending the concepts of Rayleigh scattering. The common bases for both types of scattering are of course the interactions between electromagnetic waves and the electric charges that constitute matter. Whereas with a gas molecule only a single dipole is involved, a particle consists of many closely packed, complex molecules, so that it may be considered an array of multipoles. These are excited by the primary or incident wave, thus creating oscillating multipoles. The multipoles give rise to secondary electric and magnetic waves, called *partial waves*, which combine in the far field to produce the scattered wave. The partial waves are represented in the theory by successive amplitude terms in a slowly converging series whose squared summation gives the scattered intensity at a particular observation angle. Because the size of the particle is comparable to a wavelength, the phase of the primary wave is not uniform over the particle, resulting in spatial and temporal phase differences between the various partial waves.

Where the partial waves combine to form the secondary or scattered wave, as it would be sensed by a detector, interferences caused by the phase differences occur between the partial waves. The interferences depend on the wavelength of the incident light, the size and refractive index of the particle, and the angular location of the detector. Thus sharp variations in scattered intensity are found when the detector is moved around the particle to various observation angles. It is understandable that the interferences causing variations in intensity become greater as the particle size increases. This may be appreciated from an inverse consideration that where only a single gas molecule is involved there is no interference and the scattered intensity varies smoothly as $1+\cos^2\theta$.

The geometry and radiometry of Mie and Rayleigh scattering are practically identical, although the effects of the former type are usually orders of magnitude greater. A representative geometry for Mie scattering is shown in Figure 5.1. The scattering center at point O may be either a single particle or a unit volume containing N particles. The light incident there is assumed to travel in the plus-X direction. The observation direction OD lies in the XZ plane and at angle θ from the forward direction of the incident light. Hence the plane of observation, always defined by these two directions except at $\theta = 0$ and 180 deg, coincides with the XZ plane. Occasionally in the literature, including several extensive tabulations of scattering functions, the observation angle is defined from the reverse direction of the incident light and is denoted by γ. In such cases, $\theta = 180\ \text{deg} - \gamma$. The distance OD from the scatterer to the detector, often called the *radiometric distance*, is assumed to be great enough that a unit volume of scatterers may be considered a point source for a detector at D.

The incident light may be either polarized or unpolarized. When it is plane-polarized, the orientation of the plane containing the electric vector is specified by angle ψ measured from the observation or XZ plane. Two rather standard cases are used for such incident light: (1) The plane containing the electric vector coincides with the XY plane, so that angle $\psi = 90$ deg and the vector is *perpendicular* to the plane of observation; and (2) the plane containing the electric vector coincides with the XZ

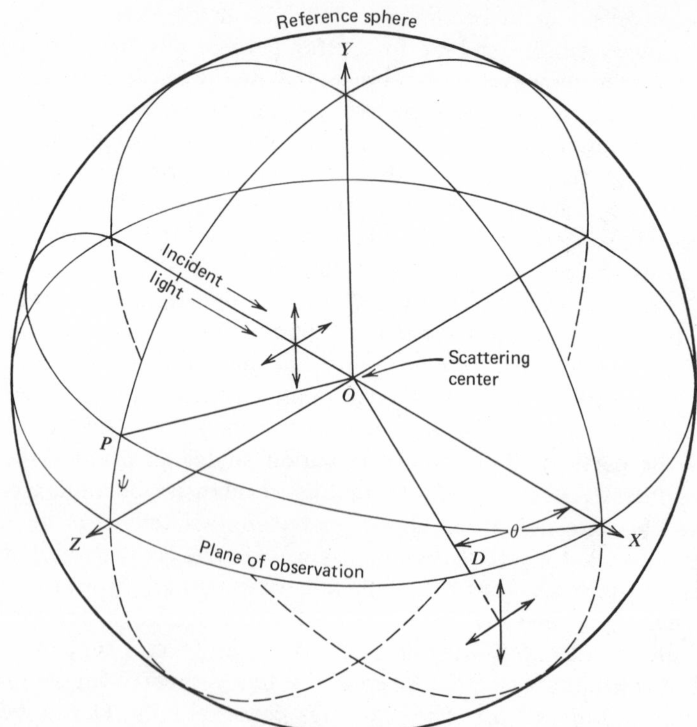

Figure 5.1 Geometry of Mie scattering. The two vectors correspond to incident unpolarized light. Line OD is the direction of observation, and θ is the angle of observation. When the incident light is polarized, its electric vector is assumed to lie in plane POX, at angle ψ to the plane of observation.

plane, so that angle $\psi = 0$ deg and the vector is *parallel* to the plane of observation. When the incident light is unpolarized, it may be helpful to employ the convention of (4-35) and the associated discussion. The observation plane then becomes the reference plane for defining the polarization directions of the two incoherent orthogonal components of the incident light.

Most of the radiometric concepts of Mie scattering are exact counterparts of those employed in Rayleigh scattering. Associated with each particle is an angular scattering cross section which defines the scattered intensity for a unit irradiance of incident light. Particularly important are the two components of the scattered light polarized perpendicular and parallel to the plane of observation. From the angular scattering cross section follows directly the volume angular coefficient, expressing the scattered intensity from a unit volume of aerosol. The total amount of light scattered in all directions by a single particle is given by its total scattering cross section. This cross section is related to the geometric cross section by a parameter, variously called the *efficiency factor* or *Mie coefficient*, which has no formal counterpart in Rayleigh scattering theory. Depending on the wavelength of the incident light and on the size and refractive index of the particle, the efficiency factor can take on values such that the particle scatters more flux than its geometric cross section would intercept from the incident flux. The volume total coefficient for a unit volume of aerosol follows directly from the total scattering cross section.

Although optical instrumentation is not covered in this book, it is convenient to call attention here to two general types of instruments for measuring the scattering by a relatively small volume of aerosol. The *angular* or *polar nephelometer*, devised by Waldram (1945a,b), incorporates the geometry of Figure 5.1 and measures the scattered intensity as a function of the observation angle. Middleton (1952) gives a description of this instrument, which originally was employed by Waldram in airborne measurements of atmospheric scattering. Since then it has been used in various forms for laboratory investigations of both aerosols and hydrosols. Information on this subject can be found in Johnson and La Mer (1947), Green and Lane (1957), Gucker and Egan (1961), Hodkinson (1966b), Holland and Draper (1967), and Holland and Gagne (1970). The *integrating nephelometer*, originated by Beuttell and Brewer (1949) and described by Middleton (1952), measures the scattered light as integrated over the approximate range 10 deg $< \theta <$ 170 deg. This allows a determination of the total scattering coefficient of the aerosol contained within the instrument. Further developments and applications of this nephelometer are discussed by Charlson et al. (1967) and Butcher and Charlson (1972). Its use has been noted in Section 3.2.4 along with additional references.

5.2.2 Particle Size Parameter and Shape

Qualitatively, the joint importance of wavelength and particle size may be appreciated from the fact that these two parameters determine the distribution of phase over the particle. It is assumed at this point that the particle is spherical, which is approximately true for a haze aerosol and even more so for fog and cloud. The ratio of particle size to wavelength is expressed by the *dimensionless size parameter* α, defined by

$$\alpha = \frac{2\pi r}{\lambda} \tag{5-1}$$

where λ is the wavelength in the medium surrounding the particle. It is seen that α represents the product of particle radius and the propagation constant $k = 2\pi/\lambda$ used in wave theory. Relationship (5-1) should be kept in mind, because the scattering is strongly dependent on the parameter α. Table 5.1 lists values of α for a range of particle sizes and a reference wavelength of 1.0 μm.

We should remember that the most important governing factor of Mie scattering is neither the absolute value of particle size nor the absolute value of wavelength. Rather it is the ratio of the two values, or *relative size*, as expressed by (5-1). The effect of increasing the wavelength is always to make the particle appear smaller. A haze particle having a radius of 1 μm can be considered a relatively large particle when illuminated by white light, which can be approximately represented by the wavelength 0.55 μm, but it becomes relatively small for an infrared wavelength of 10 μm. Further in this direction, a raindrop having a radius of 1 mm (1000 μm) appears large to this infrared wavelength but relatively small to a radar wavelength in the centimeter region.

Corresponding to this great range of relative sizes is an even greater range and diversity of scattering characteristics. The range and diversity can be appreciated from the trends of the scattering patterns illustrated in Figure 1.5. At the lower or molecular limit of relative size, scattering is explained by the prior Rayleigh theory, to which Mie theory reduces when the parameter α becomes very small. At the upper limit of relative size, as in dealing with raindrops and optical wavelengths, scattering can be explained satisfactorily from the principles of diffraction, reflection, and refraction. In the midrange of relative particle size, however, the only adequate theory is that of Mie. The distinctive feature of this theory lies in its ability to describe accurately and comprehensively the scattering by

TABLE 5.1 VALUES OF SIZE PARAMETER α FOR SMALL SPHERES HAVING THE INDICATED RADII, AT WAVELENGTH OF 1 μm

r (μm)	α	r (μm)	α	r (μm)	α
0.01	0.063	0.20	1.26	3.0	18.8
0.02	0.126	0.30	1.88	4.0	25.1
0.03	0.188	0.40	2.51	5.0	31.4
0.04	0.251	0.50	3.14	6.0	37.7
0.05	0.314	0.60	3.77	7.0	44.0
0.06	0.377	0.70	4.40	8.0	50.2
0.07	0.440	0.80	5.02	9.0	56.5
0.08	0.502	0.90	5.65	10.0	62.8
0.09	0.565	1.0	6.28	15.0	94.2
0.10	0.628	2.0	12.6	20.0	126.0

particles of *any* relative size. Agreement with its predictions is the test of any alternative theory or approximation.

First reminding the reader that Mie theory applies rigorously only to isotropic spheres, we note the matter of particle shape. From the standpoint of scattering it is usual to assume that atmospheric particles are spherical and isotropic. This assumption, employed here, is valid for "wet" haze particles, fog and cloud droplets, and small raindrops which are the important particles in a majority of scattering situations. The assumption is less valid for dust grains, and probably even less so for condensation nuclei that become crystalline at low relative humidity. Other kinds of non-spherical particles occur frequently. For example, large raindrops become distorted by air resistance as their falling speed increases, eventually dividing into smaller drops as discussed in Section 3.4.4. Ice crystals occurring in clouds have many spheroidal, prismatic, and columnar forms, and the variety of snowflake structures seems without limit.

Scattering by nonspherical particles is more complex than that by spheres and is outside the scope of this book although we do call attention to several investigations of such scattering. Van de Hulst (1957) gives thorough treatments for spheroids, ellipsoids, cylinders, and disks of both absorbing and nonabsorbing materials. Similar and very detailed treatments for all common geometric shapes can be found in Kerker (1969), who considers both homogeneous and stratified structures. He gives many references to the literature, particularly that of physical chemistry. Asano and Yamamoto (1975) and Barber and Yeh (1975) studied the scattering by spheroidal and ellipsoidal particles in terms of intensity distribution functions analogous to those discussed in Section 5.3.1.

We also note experimental work relevant to scattering by sharp dust grains and crystalline nuclei. Hodkinson (1963, 1966b) measured the angular scattering and extinction properties of particles having irregular shapes, random orientations, and radii in the approximate range 0.5 to 4 μm. Holland and Draper (1967) and Holland and Gagne (1970) investigated angular scattering by polydispersions of sharply irregular particles, whose sizes were deemed to be log-normally distributed with a mode radius of about 0.13 μm. Interesting comparisons were made between the scattering patterns of the irregular particles and the patterns predicted by Mie theory for isotropic spheres having the same log-normal distribution of size. In general the scattering at observation angles less than about 100 deg agreed well with the Mie predictions, but at larger angles the scattering was less. These results have been discussed by Plass and Kattawar (1971b) and Holland and Gagne (1971). The issues are whether another size distribution, expressed by (3-15), would have been more appropriate than the one used, and what effect that this would have on the predicted patterns. At very small values of the size parameter α, particle shape has little effect on the scattering pattern.

Finally we call attention to the large-scale phenomena produced in the sky by cloud droplets, raindrops, and multifarious ice crystals. The study of such phenomena is embraced by the subject usually called *meteorological optics*. Most persons have seen a *corona*, which consists of concentric rings of impure colors surrounding the sun or moon when that luminary is observed through a cloud of water droplets. Aircraft travelers and mountain hikers have observed the *glory*, or *anti-corona*, which consists of concentric rings of impure colors surrounding the shadow of the aircraft or the observer's head, when that shadow falls on a cloud deck. Both the corona and the glory can be explained either by Mie theory or by the principles of diffraction, reflection, and refraction. *The rainbow*, a sight of wonder and portent since the book of Genesis was written, and the *halo* produced by cloud ice crystals, are explained more easily by ray-optics than by electromagnetic theory. All the foregoing visual phenomena are treated by Minnaert (1940) in a delightfully descriptive manner, and in a technical manner by Humphreys (1940), Johnson (1954), and Tverskoi (1965). Tricker (1970) provides technical treatments with comprehensive descriptions and excellent photographs.

5.2.3 Applicable Values of Refractive Index

An atmospheric particle presents an optical discontinuity to a passing light wave because the refractive index of the particle is greater than that of the surrounding air. If the particle does not absorb any of the flux then only scattering occurs and the attenuation is expressed in terms of the total scattering coefficient. If the particle also absorbs a significant amount of the flux, the two processes are additive and produce an extinction coefficient according to Section 1.4.2. Not so simply defined are the effects of absorption on angular scattering; some of these are described later.

In considering refractive index, it is the relative and not the absolute value that is important. The relative index, as the name implies, is the ratio of the index of the particle to the index of the surrounding medium. However, the index of air is so near unity, about 1.0003, that the absolute index of the particle, from 1.33 to about 1.60, can be taken as the relative index. This would not be true for hydrosols where the index of the liquid medium is usually 1.33 or greater, so that the relative index of an immersed particle may be close to unity, or may even be less.

When absorption is not significant, the refractive index is expressed by a real number. For example, at visual wavelengths the nominal value

$n = 1.33$ applies to water. Values for the crystalline salts of condensation nuclei and for the quartz grains of dust are listed in Table 5.2. These values, which apply to the materials in dry form, decrease as the relative humidity increases above approximately 80% which causes water vapor droplet to condense on the nucleus and form a droplet. Depending on the size of the nucleus, a limiting value of $n = 1.33$ is approached when the droplet radius grows to several micrometers. The small amount of dissolved salt then has a negligible effect on the refractive index of the droplet. This matter is discussed by Bullrich (1964) who, in his extensive studies and measurements of scattering by haze, has selected the value $n = 1.5$ as an average for the crystalline materials in Table 5.2.

TABLE 5.2 REFRACTIVE INDICES OF AEROSOL PARTICLES HAVING CRYSTALLINE STRUCTURES

Particle composition	Refractive index	Particle composition	Refractive index
NH_4Cl	1.64	$CaSO_4$	1.57
NH_4NO_3	1.60	KCl	1.49
$(NH_4)_2SO_4$	1.52	Na_2SO_4	1.48
$MgCl_2$	1.54	SiO_2	1.49
$NaNO_3$	1.59	K_2SO_4	1.49

Source: Bullrich (1964).

When absorption is significant, the refractive index of the material must be represented by a complex number. Instead of denoting the index by n, we write

$$m(\lambda) = n(\lambda) - in_i(\lambda) \tag{5-2}$$

where m is the complex index, and n and n_i are the real and imaginary parts, respectively. The imaginary part, identical to a quantity called the *absorption index k* of the material, is related to the *Lambert absorption coefficient K* by

$$k \equiv n_i = \frac{K\lambda}{4\pi} \tag{5-3}$$

where K has the dimension L^{-1} and usually is expressed in cm^{-1}. The coefficient for liquid water has such high values in most of the infrared region that it can be measured only for very short paths in an absorption cell. At some wavelengths, path lengths less than 10 μm must be used. The resulting experimental difficulties are such that values found by different workers may vary by factors as great as two.

TABLE 5.3 OPTICAL CONSTANTS OF WATER: ABSORPTION INDEX AND REAL PART OF THE REFRACTIVE INDEX*

λ (μm)	k (λ)	n (λ)	λ (μm)	k (λ)	n (λ)	l (μm)	k (λ)	n (λ)
0.200	1.1×10^{-7}	1.396	3.40	0.0195	1.420	9.8	0.0479	1.229
0.225	4.9×10^{-8}	1.373	3.45	0.0132	1.410	10.0	0.0508	1.218
0.250	3.35×10^{-8}	1.362	3.50	0.0094	1.400	10.5	0.0662	1.185
0.275	2.35×10^{-8}	1.354	3.6	0.00515	1.385	11.0	0.0968	1.153
0.300	1.6×10^{-8}	1.349	3.7	0.00360	1.374	11.5	0.142	1.126
0.325	1.08×10^{-8}	1.346	3.8	0.00340	1.364	12.0	0.199	1.111
0.350	6.5×10^{-9}	1.343	3.9	0.00380	1.357	12.5	0.259	1.123
0.375	3.5×10^{-9}	1.341	4.0	0.00460	1.351	13.0	0.305	1.146
0.400	1.86×10^{-9}	1.339	4.1	0.00562	1.346	13.5	0.343	1.177
0.425	1.3×10^{-9}	1.338	4.2	0.00688	1.342	14.0	0.370	1.210
0.450	1.02×10^{-9}	1.337	4.3	0.00845	1.338	14.5	0.388	1.241
0.475	9.35×10^{-10}	1.336	4.4	0.0103	1.334	15.0	0.402	1.270
0.500	1.00×10^{-9}	1.335	4.5	0.0134	1.332	15.5	0.414	1.297
0.525	1.32×10^{-9}	1.334	4.6	0.0147	1.330	16.0	0.422	1.325
0.550	1.96×10^{-9}	1.333	4.7	0.0157	1.330	16.5	0.428	1.351
0.575	3.60×10^{-9}	1.333	4.8	0.0150	1.330	17.0	0.429	1.376
0.600	1.09×10^{-8}	1.332	4.9	0.0137	1.328	17.5	0.429	1.401
0.625	1.39×10^{-8}	1.332	5.0	0.0124	1.325	18.0	0.426	1.423
0.650	1.64×10^{-8}	1.331	5.1	0.0111	1.322	18.5	0.421	1.443
0.675	2.23×10^{-8}	1.331	5.2	0.0101	1.317	19.0	0.414	1.461
0.700	3.35×10^{-8}	1.331	5.3	0.0098	1.312	19.5	0.404	1.476
0.725	9.15×10^{-8}	1.330	5.4	0.0103	1.305	20.0	0.393	1.480
0.750	1.56×10^{-7}	1.330	5.5	0.0116	1.298	21.0	0.382	1.487
0.775	1.48×10^{-7}	1.330	5.6	0.0142	1.289	22	0.373	1.500
0.800	1.25×10^{-7}	1.329	5.7	0.0203	1.277	23	0.367	1.500
0.825	1.82×10^{-7}	1.329	5.8	0.0330	1.262	24	0.361	1.521
0.850	2.93×10^{-7}	1.329	5.9	0.0622	1.248	25	0.356	1.531
0.875	2.91×10^{-7}	1.328	6.0	0.107	1.265	26	0.350	1.530

0.900	4.86×10^{-7}	1.328	6.1	0.131	1.319	27	0.344	1.545
0.925	1.06×10^{-6}	1.328	6.2	0.0880	1.363	28	0.338	1.549
0.950	2.93×10^{-6}	1.327	6.3	0.0570	1.357	29	0.333	1.551
0.975	3.48×10^{-6}	1.327	6.4	0.0449	1.347	30	0.328	1.551
1.0	2.89×10^{-6}	1.327	6.5	0.0392	1.339	32	0.324	1.546
1.2	9.89×10^{-6}	1.324	6.6	0.0356	1.334	34	0.329	1.536
1.4	1.38×10^{-4}	1.321	6.7	0.0337	1.329	36	0.343	1.527
1.6	8.55×10^{-5}	1.317	6.8	0.0327	1.324	38	0.361	1.522
1.8	1.15×10^{-4}	1.312	6.9	0.0322	1.321	40	0.385	1.519
2.0	1.1×10^{-3}	1.306	7.0	0.0320	1.317	42	0.409	1.522
2.2	2.89×10^{-4}	1.296	7.1	0.0320	1.314	44	0.436	1.530
2.4	9.56×10^{-4}	1.279	7.2	0.0321	1.312	46	0.462	1.541
2.6	3.17×10^{-3}	1.242	7.3	0.0322	1.309	48	0.488	1.555
2.65	6.7×10^{-3}	1.219	7.4	0.0324	1.307	50	0.514	1.587
2.70	0.019	1.188	7.5	0.0326	1.304	60	0.587	1.703
2.75	0.059	1.157	7.6	0.0328	1.302	70	0.576	1.821
2.80	0.115	1.142	7.7	0.0331	1.299	80	0.547	1.886
2.85	0.185	1.149	7.8	0.0335	1.297	90	0.536	1.924
2.90	0.268	1.201	7.9	0.0339	1.294	100	0.532	1.957
2.95	0.298	1.292	8.0	0.0343	1.291	110	0.531	1.966
3.00	0.272	1.371	8.2	0.0351	1.286	120	0.526	2.004
3.05	0.240	1.426	8.4	0.0361	1.281	130	0.514	2.036
3.10	0.192	1.467	8.6	0.0372	1.275	140	0.500	2.056
3.15	0.135	1.483	8.8	0.0385	1.269	150	0.495	2.069
3.20	0.0924	1.478	9.0	0.0399	1.262	160	0.496	2.081
3.25	0.0610	1.467	9.2	0.0415	1.255	170	0.497	2.094
3.30	0.0368	1.450	9.4	0.0433	1.247	180	0.499	2.107
3.35	0.0261	1.432	9.6	0.0454	1.239	190	0.501	2.119
						200	0.504	2.130

Source: Hale and Querry (1973).
*The complex refractive index is related to these two quantities by Eqs. (5-2) and (5-3).

Liquid water is usually the principal material whose complex index is considered. This is because the condensation nuclei, whose real refractive indices are listed in Table 5.2, are so much smaller than infrared wavelengths that they scatter only inefficiently. In contrast, the water droplets of fog and cloud are large enough to be efficient scatterers throughout the currently used infrared region.

For many years the most complete set of values for the complex index of water were those determined by Centeno (1941). They cover the wavelength range 0.2 to 18.0 μm and have been frequently employed in computations of scattering. In recent years the complex index of water has been measured with very refined techniques by Draegert et al. (1966), Robertson and Williams (1971), and Rusk et al. (1971), to name but a few. Tabulated values have been published by Irvine and Pollack (1968), Querry et al. (1969), and Hale and Querry (1973), whose data are listed in Table 5.3. Tabulated values of the optical constants of ice are given by Irvine and Pollack (1968) and Schaaf and Williams (1973).

Recently the absorption characteristics of dust and haze particles, as expressed by the complex index, have come under scrutiny in studies of the earth's heat budget. To this end Eiden (1971) studied the use of polarized light techniques for measuring the optical constants of representative materials. Neumann (1973) and Palmer and Williams (1975) measured the absorption spectra of water solutions of sulfuric acid and of ammonium sulfate which appear to be the main constituents of stratospheric particles, as discussed in Section 3.1.2. Querry et al. (1972) tabulated the optical constants of water solutions of sodium chloride, an important constituent of maritime aerosols. Optical constants for other materials were measured by Grams et al. (1972), Volz (1972, 1973), Lin et al. (1973), Lindberg and Laude (1974), and Fischer (1975). Fischer and Hänel (1972) and Hänel (1976) provide much tabulated information on the physical and optical properties of particles, as functions of relative humidity, for urban, rural, and maritime aerosols.

5.3 THE INTENSITY DISTRIBUTION FUNCTIONS

The angular characteristics of Mie scattering for all particle sizes and wavelengths are expressed by two intensity distribution functions. These are basic to all subsequent definitions of scattering cross sections and volume coefficients; therefore their meaning is explained in detail here. Their formal definitions in mathematical terms are summarized, and operational definitions are discussed. The manner in which the functions control the scattered intensity, according to the polarization of the incident light, is described, and the effect of absorption by the particle is noted. Finally, attention is called to the tabulations of these functions that have been published.

5.3.1 Formal Definitions of the Functions

The light scattered at observation angle θ by a particle may be treated as consisting of two components having intensities $I_{\perp}(\theta)$ and $I_{\parallel}(\theta)$, polarized perpendicular and parallel to the plane of observation, respectively. The components are proportional to two intensity distribution functions i_1 and i_2, respectively. These functions are the essence of the Mie theory and depend on the size parameter defined by (5-1), on the refractive index defined by (5-2), and on the angle θ indicated in Figure 5.1. For a spherical, isotropic particle the functions are expressed by

$$i_1(\alpha, m, \theta) = |S_1|^2 = \left|\sum_{n=1}^{\infty} \frac{2n+1}{n(n+1)} (a_n \pi_n + b_n \tau_n)\right|^2$$
$$= |\mathrm{Re}\,(S_1) + \mathrm{Im}\,(S_1)|^2 \qquad (5\text{-}4)$$

$$i_2(\alpha, m, \theta) = |S_2|^2 = \left|\sum_{n=1}^{\infty} \frac{2n+1}{n(n+1)} (a_n \tau_n + b_n \pi_n)\right|^2$$
$$= |\mathrm{Re}\,(S_2) + \mathrm{Im}\,(S_2)|^2 \qquad (5\text{-}5)$$

where the n's are positive integers. Each intensity function thus is found as the sum of an infinite series. Each series converges slowly, and when α is greater than unity the number of terms required for satisfactory convergence is somewhat greater than the value of α. When $\alpha \ll 1$ and $m \approx 1$, the first term in each series corresponds to Rayleigh scattering.

In (5-4) and (5-5), S_1 and S_2 are the dimensionless complex amplitudes of the scattered waves observed as intensities $I_{\perp}(\theta)$ and $I_{\parallel}(\theta)$. The vertical lines indicate that the absolute values of the complex arguments are to be taken. As mentioned previously, the scattered waves consist of partial waves radiated by the multipoles formed by the electric charges constituting the particle. The first partial wave is considered to emanate from a dipole, the second partial wave from a quadrupole, and so on to higher orders. The amplitudes of the nth electric partial wave and the nth magnetic partial wave are given by the complex functions a_n and b_n, respectively. The values of a_n and b_n are found from Ricatti–Bessel functions, whose arguments are formed from the particle characteristics α and m but which are independent of the angle θ. Extensive tables of Ricatti–Bessel functions for scattering applications have been published by Gumprecht and Sliepcevich (1951a). Many values of a_n and b_n covering wide ranges of particle sizes and refractive indices are available in Gumprecht and Sliepcevich (1951b) and in Penndorf and Goldberg (1956).

The functions π_n and τ_n in (5-4) and (5-5) depend only on the angle θ and involve the first and second derivatives of Legendre polynomials having order n and argument cos θ. These functions are tabulated in MTP (1945), Gumprecht and Sliepcevich (1951c), and Shifrin and Zelmanovich (1966), reviewed by Penndorf (1967). The symbols Re and Im denote the real and imaginary parts of S_1 and S_2. Thus from data such as those cited above the functions $i_1(\alpha, m, \theta)$ and $i_2(\alpha, m, \theta)$ can be found. Manual calculations necessarily were employed in early years, although this method is laborious even for small values of α and becomes impractical at larger values. Prior to the availability of electronic digital computers, the functions were calculated only for coarse increments of α, m, and θ, and seldom for $\alpha > 6$. The use of digital computers in recent years has resulted in many tabulations of these functions for small increments and wide ranges of scattering parameters. Detailed instructions for computing the functions are given by van de Hulst (1957), Deirmendjian (1969), and Kerker (1969). Published tabulations are noted in Section 5.3.4.

5.3.2 Operational Definitions of the Functions

The intensity distribution functions, from now on indicated simply as i_1 and i_2, but with the dependence on α, m, and θ understood, express the intensity of each scattered component in units of $\lambda^2/4\pi^2$ when the incident light has a unit irradiance. Readers will recognize the quantity $\lambda^2/4\pi^2$ as the squared reciprocal of the wave propagation constant. The meanings of i_1 and i_2 are shown by the following examples, which should be read in the context of the geometry shown in Figure 5.1 and which can be demonstrated experimentally.

- When the particle is illuminated by plane-polarized light whose electric vector is *perpendicular* to the plane of observation, so that the polarization angle $\psi = 90$ deg, the intensity of the scattered light in this plane is given by

$$I_\perp(\theta) = E_\perp \frac{\lambda^2}{4\pi^2} i_1 \tag{5-6}$$

where $E_\perp$ is the irradiance of the incident light. With $\psi = 90$ deg, there is no i_2 component, hence no $I_\parallel(\theta)$ component, in this plane.

- When the particle is illuminated by plane-polarized light whose electric vector is *parallel* to the plane of observation, so that $\psi = 0$ deg, the intensity of the scattered light in this plane is given by

$$I_\parallel(\theta) = E_\parallel \frac{\lambda^2}{4\pi^2} i_2 \tag{5-7}$$

With $\psi = 0$ deg, there is no i_1 component, hence no $I_1(\theta)$ component, in this plane.

• When the particle is illuminated by plane-polarized light whose electric vector is neither perpendicular nor parallel to the plane of observation, so that 90 deg $> \psi > 0$ deg, the intensity of the scattered light in this plane is given by

$$I(\theta, \psi) = E_\psi \frac{\lambda^2}{4\pi^2}(i_1 \sin^2 \psi + i_2 \cos^2 \psi) \tag{5-8}$$

In general, this scattered light is elliptically polarized, because the two components are derived from the same scattering source but undergo different phase delays. The phase difference between the two components depends on the complex refractive index of the particle and on its size.

• When the particle is illuminated by unpolarized light represented by two electric vectors of equal magnitude perpendicular and parallel to the plane of observation but having no coherent relationship, the scattered light consists of two incoherent components according to

$$I(\theta) = \frac{I_\perp(\theta) + I_\parallel(\theta)}{2} = E\frac{\lambda^2}{4\pi^2}\left(\frac{i_1 + i_2}{2}\right) \tag{5-9}$$

In the last example, each of the two scattered components can be considered as being produced by the incident light component polarized in the same direction. The factor 2 in the denominator of (5-9) was inadvertently omitted from early scattering theory by several distinguished workers, as pointed out by Sinclair (1947). The requirement for this factor is due to the fact that i_1 and i_2 are each defined for incident polarized light of a unit irradiance. This is an important fact to keep in mind when working with numerical values of these functions.

5.3.3 Angular Dependence, Polarization, Absorption

The dependences of i_1 and i_2 on the observation angle at several values of the size parameter are shown in Figure 5.2 for small droplets of water. The indicated values of droplet radius correspond to an assumed wavelength of 0.55 μm, which means that the absorption is taken as zero and $m = n = 1.33$. The bottom pair of curves closely resembles the molecular scattering characteristics shown as polar plots in Figure 4.6. In particular, the function i_1 has but little dependence on angle θ, while i_2 varies somewhat as $\cos^2 \theta$ and indeed becomes practically zero at $\theta = 90$ deg. Actually, Rayleigh theory can be applied as an approximation to such small particles. According to Penndorf (1962), the error is not greater than 1% when $r < 0.03\ \lambda$.

At values of α larger than 0.5, the functions begin to depart markedly

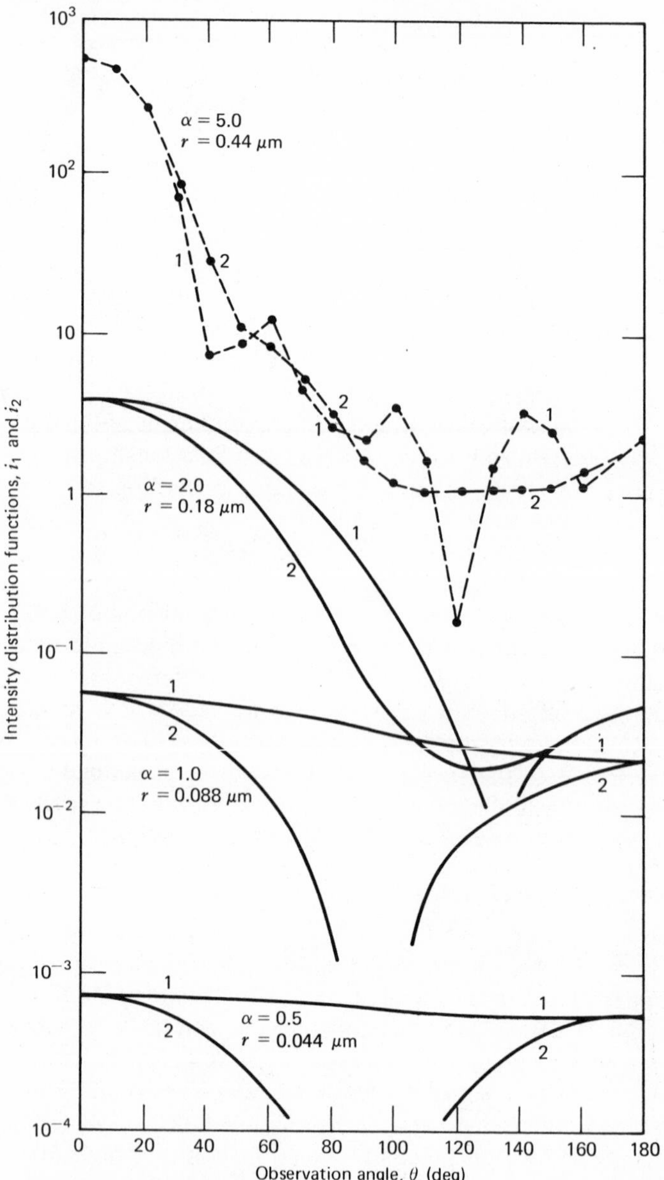

Figure 5.2 Intensity distribution functions for small nonabsorbing spheres having $n = 1.33$. Data from La Mer and Sinclair (1943).

from Rayleigh scattering characteristics and the values increase considerably, particularly at small values of θ. Thus forward scattering exceeds backscattering, whereas the two are equal in the Rayleigh case. Also, intermediate maxima and minima appear in each function, and their number increases with α. This characteristic of Mie scattering deserves emphasis. If all the inflection points of the functions are to be found, i_1 and i_2 must be computed for smaller and smaller increments of α and θ as α becomes larger. For example, the functions plotted in Figure 5.2 were computed in only 10-deg increments of θ. We should not assume that such large increments necessarily locate all the fluctuations of i_1 and i_2, even for values of α no greater than 5.0.

Reviewing each pair of curves in the figure, it is seen that i_1 and i_2 are always unequal except at $\theta = 0$ and 180 deg. Thus even when the incident light is unpolarized, the scattered light is partially polarized. Because each scattered component is proportional to its intensity distribution function, the degree of polarization is given by

$$P = \frac{i_1 - i_2}{i_1 + i_2} \tag{5-10}$$

analogously to (4-39). For $\alpha = 0.5$, the maximum value of P is nearly unity at $\theta = 90$ deg, in further resemblance to Rayleigh scattering. However, the polarization becomes weaker as α increases and even assumes negative values at some observation angles, as may be seen from the curves for $\alpha = 2.0$ and 5.0.

It is characteristic of scattering by an isotropic sphere that depolarization does not occur. Incident polarized light, as in the first two examples given in the preceding section, produces only a scattered component whose polarization is in the same direction as that of the incident light. Likewise, each conceptual component of incident unpolarized light produces only that scattered component whose polarization is in the same direction. In any of the cases, the scattered components can be isolated and observed separately by means of a polarization analyzer oriented perpendicular or parallel to the plane of observation. Significant depolarization does occur, however, with nonspherical particles, particularly anisotropic crystals such as are found in dry hazes and dust clouds. In all cases the i_1 and i_2 components are equal at $\theta = 0$ and 180 deg, because the plane of observation is not uniquely defined for these two directions, as can be visualized from Figure 5.1.

The effect of absorption by the particle on angular scattering has been studied by Deirmendjian (1963b), Deirmendjian et al. (1961), Deirmendjian and Clasen (1962), Plass (1966), and Kattawar and Plass (1967). Scattering theory takes absorption into account by employing the complex refractive index of the particle material in the computations of i_1 and i_2. The complex refractive index is defined by (5-2), and Table 5.3 lists the values of n and k for liquid water, which is the most important absorbing material to consider. The effect on i_1 and i_2 of increasing the value of n_i is shown in Figure 5.3 for small spheres having $\alpha = 5.0$, $n_r = 1.33$, and n_i as

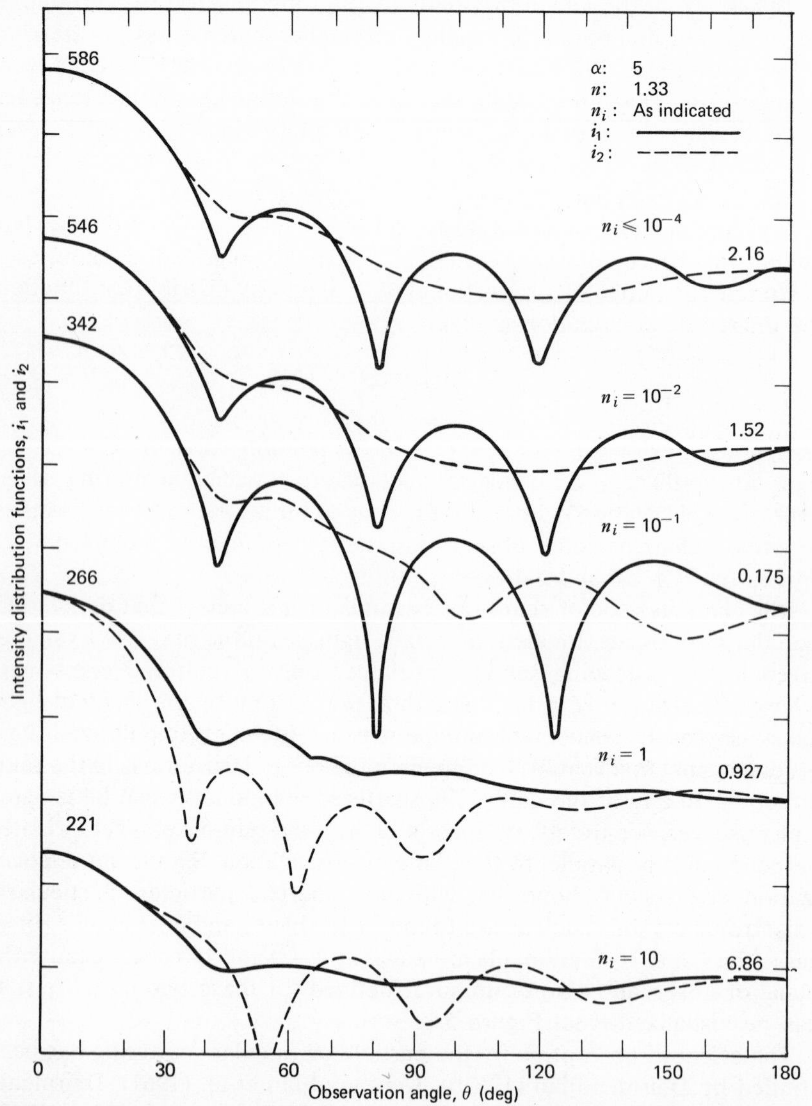

Figure 5.3 Intensity distribution functions for small absorbing spheres. The logarithm of the function is plotted so that one ordinate division equals a factor of 10. From Plass (1966).

indicated. Each ordinate division represents a factor of 10, and the curves have been displaced vertically to avoid overlapping. The anti-log values at $\theta = 0$ and 180 deg are indicated above each curve.

These curves should be compared with those for $\alpha = 5.0$ in Figure 5.2, where $n_i = 0$. When $n_i \leq 10^{-2}$, i_1 and i_2 are only slightly different from the nonabsorbing case. At $n_i = 10^{-1}$ the minima of i_1 have become sharper and deeper, while slight minima now appear in i_2. At $n_i \geq 1$ the oscillations in i_1 have practically disappeared, but pronounced minima have developed in i_2. According to Plass (1966), the ratio of backscattering (180 deg) to forward scattering (0 deg) always decreases at first as n_i increases from zero. The ratio reaches a minimum at $10^{-2} < n_i < 10^{-1}$ and then increases for larger values of n_i.

Hodkinson (1966b) also notes that the angular scattering by very small, moderately absorbing particles is about the same as that by nonabsorbing particles having the same value of n. When the particle size becomes greater than one-tenth of the wavelength, moderately absorbing particles show about the same scattering in the forward lobe as do nonabsorbing particles. Outside this lobe, however, the scattering by absorbing particles becomes less and oscillates more regularly and smoothly with the angle. Also, the backward maximum at or near $\theta = 180$ deg may be no greater than the other maxima away from the forward direction. These trends may be inferred from Figure 5.3.

5.3.4 Published Tabulations of Intensity Functions

Many computations of i_1 and i_2 have been made and published to meet the needs of atmospheric optics and physical chemistry. With the greater availability of electronic digital computers, which are essential for the task, the number of such computations has increased tremendously in recent years. Indeed, so many values have been computed for wide ranges and small increments of α and m that their tabulation in printed form has become inconvenient. Instead, as pointed out by Kerker (1969), the values of i_1 and i_2, or of equivalent quantities used to integrate the scattering function over a size distribution, may be stored on punched cards or magnetic tape at a computer facility. The tabulations most relevant to atmospheric optics, where the refractive index range is from about 1.3 to 1.6, are listed in Table 5.4 which, however, does not pretend to completeness. In some cases, the tabulated quantities may not be quite the same as those defined in this book, but the explanatory notes with the tabulations make the relationships clear. The listings range from the early work by Lowan (1949) to the monumental tabulations by Shifrin and Zelmanovich (1967) which, according to the review by Penndorf (1968), are the most comprehensive in the literature.

TABLE 5.4 PUBLISHED TABULATIONS OF INTENSITY DISTRIBUTION FUNCTIONS APPLICABLE TO ATMOSPHERIC PARTICLES

Author	Refractive index, m*	Size parameter, α	Observation angle, θ (deg)	Tabulated quantities
Lowan (1949)	1.33, 1.44, 1.55, 2.00	0.5 (various) 6.0	0(10)180	i
Gumprecht et al. (1952)	1.33	6, 8, 10(5)40	0(1)10(10)180	i
Penndorf and Goldberg (1956)	1.33, 1.40, 1.44, 1.486, 1.50	0.1(0.1)30	0(10)170(1)180	a_n, b_n, i
Havard (1960)	Complex values, $3.6 < \lambda < 13.5\ \mu m$	For particle $r =$ 1, 2, 4, 9, 12, 15 μm	0(10)180	i
Pangonis and Heller (1960)	1.05(0.05)1.30	0.2(0.2)7.0	0(5)180	i
Giese (1961)	1.33, 1.55	1(1)40	0(2)180	i
Giese et al. (1961)	1.5	0.2(0.2)159	0(1)10(10)180	i
Deirmendjian (1963b)	Complex values for $\lambda = 8.15$, 10.0, 11.5, 16.6 μm	0.5(0.5)15.0	0, 180	S
Denman et al. (1966)	1.05(0.05)1.30, 1.33	0.2(0.2)25.0	0(5)180	i
Shifrin and Zelmanovich (1966)	——	——	0 (152 values) 140	π_n, τ_n
McCormick (1967)	1.5	0.01(0.01)2.0 (0.1)181.9	180	i
Shifrin and Zelmanovich (1967)	Various complex values	0.5 (51 values) 100	0 (43 values) 180	i

5.4 ANGULAR SCATTERING BY MONODISPERSIONS

In this section the theoretical intensity distribution functions are extended to observable angular scattering by particles. The extension is made directly from the several examples of scattered intensity discussed in Section 5.3.2. Here the angular scattering cross sections of particles are defined, and those for very small particles are then compared to those found from Rayleigh theory. Next, the volume angular coefficient is derived for a monodispersion. The section concludes with a discussion of the multicolor effects seen when white light is scattered by a suspension of particles having a nearly uniform size.

5.4.1 Angular Scattering Cross Section

The angular scattering cross section of a particle is analogous to that of a molecule, as discussed in Section 4.3.1, and is defined as that cross section of an incident wave, acted on by the particle, having an area such that the power flowing across it is equal to the scattered power per steradian at an observation angle θ. In symbols, employing the subscript p for the general case of a particle, the angular cross section $\sigma_p(\theta)$ then is defined by

$$\sigma_p(\theta)E_0 = I(\theta) \qquad \sigma_p(\theta)\bar{S} = I(\theta) \tag{5-11}$$

where $\bar{S}$, the average value of the Poynting vector of the incident light, is equivalent to the irradiance E_0 of the incident light. Equation (5-11) shows that $\sigma_p(\theta)$ is the ratio of two measurable quantities, and that its value is normalized to a unit irradiance The dimensions of $\sigma_p(\theta)$ are L^2 corresponding to an area, and the units commonly used are $cm^2\ sr^{-1}$.

Such an angular cross section is implied in the preceding definitions of i_1 and i_2, which give the intensities of the two scattered components in units of $\lambda^2/4\pi^2$ for incident light having a unit irradiance. Thus when the incident light is plane-polarized with its electric vector perpendicular to the plane of observation, so that $\psi = 90$ deg, the angular cross section of the particle is

$$\sigma_{p,\perp}(\theta) = \frac{I_\perp(\theta)}{E_{0,\perp}} = \frac{\lambda^2}{4\pi^2} i_1 \tag{5-12}$$

When the incident light is plane-polarized with its electric vector parallel to the plane of observation, so that $\psi =$ deg, the cross section is

$$\sigma_{p,\parallel}(\theta) = \frac{I_\parallel(\theta)}{E_{0,\parallel}} = \frac{\lambda^2}{4\pi^2} i_2 \tag{5-13}$$

When the incident light is unpolarized, which is the prevailing natural condition in the atmosphere, the cross section is

$$\sigma_p(\theta) = \frac{I_\perp(\theta) + I_\parallel(\theta)}{2E_0} = \frac{I(\theta)}{E_0} = \frac{\lambda^2}{4\pi^2}\left(\frac{i_1 + i_2}{2}\right) \tag{5-14}$$

From the above definitions, it is clear that a plot of the angular cross section versus the observation angle is the pattern of scattered intensity in the plane of observation.

Thus the angular cross section is a function of the parameters α, m, and θ, through the functions i_1 and i_2, and depends on the polarization state of the incident light but is independent of its irradiance. The dependence on α and on θ are particularly strong, as may be seen in the cross sections plotted in Figure 5.4 according to (5-14). The data refer to water droplets for which the absorption is considered to be zero. This implies light at visual wavelengths, and the indicated values of droplet radius refer to $\lambda = 0.55\ \mu\text{m}$.

At smaller values of α scattering patterns of particles are scarcely distinguishable from a molecular scattering pattern in their angular dependence. In particular, the angular dissymmetry at small α is practically negligible, so that forward scattering and backscattering are almost equal. Even at $\alpha = 0.6$ the ratio of the scattered intensity at 0 and 180 deg is only about 1.4. As α increases above 0.6, however, the minimum of the pattern moves away from the 90-deg position and occurs near 105 deg when $\alpha = 1.0$. At still larger α, additional minima and maxima appear, but their general order is not easily defined when the refractive index has values appropriate to atmospheric particles. The pattern assumes a distinctive appearance for each value of α, as may be seen from the curves for $\alpha = 5$ and 25. The correlation becomes sufficiently strong that particle size in an approximately monodisperse suspension can be determined by comparing the measured pattern with the theoretical patterns for various values of α. It can be appreciated, however, that the overall pattern from a polydispersion, where the range of particle size is continuous, does not exhibit such sharp variations because of the overlapping of many different patterns.

The refractive index within the range of values applicable to atmospheric particles has a relatively modest influence on scattering compared to the other two parameters. This can be seen in Figure 5.5, where $\sigma_p(\theta)$ is plotted for $\alpha = 6.0$ and for three values of the index. Generalizations from a limited amount of data are unsafe, but in the examples shown the effect of increasing the index is to reduce the scattering in the forward lobe and to increase it nearly everywhere else, although in a rather erratic manner. Interestingly, the three cross sections are nearly equal at $\theta \approx 35$ deg and at $50\ \text{deg} < \theta < 60$ deg. Significantly, the shape of the forward lobe, out to

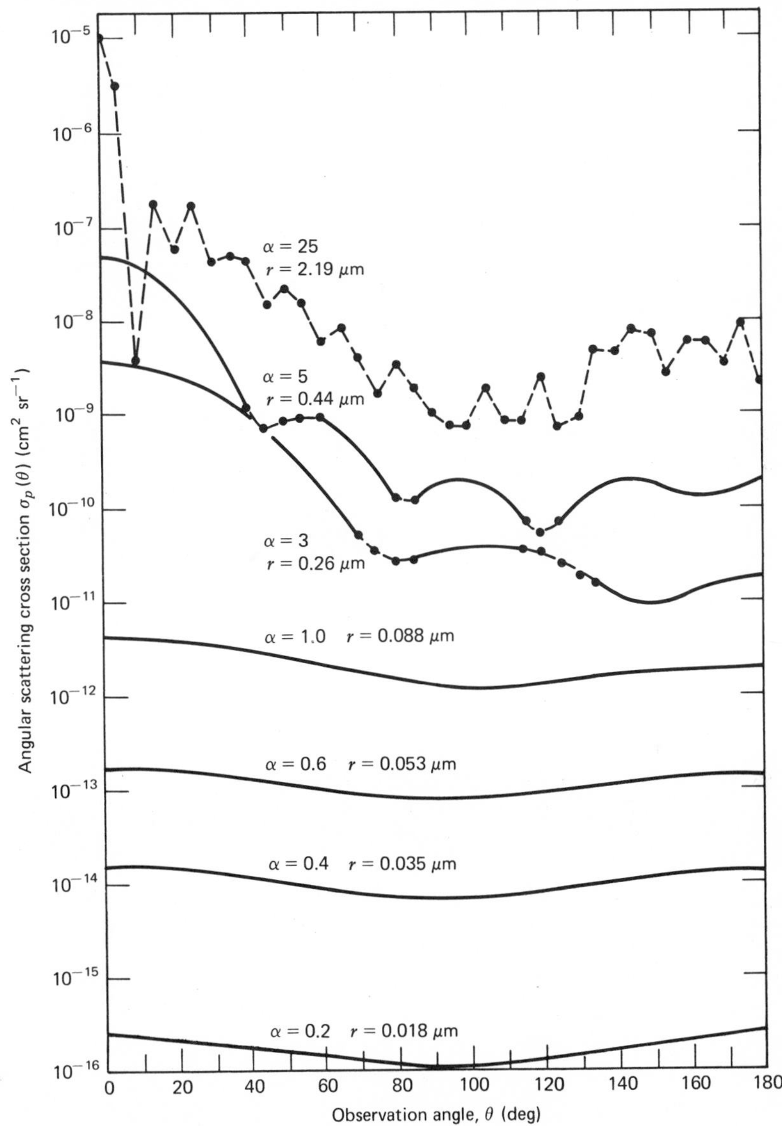

Figure 5.4 Angular scattering cross sections of water droplets illuminated by unpolarized light at $\lambda = 0.55\ \mu$m. Data from Denman et al. (1966).

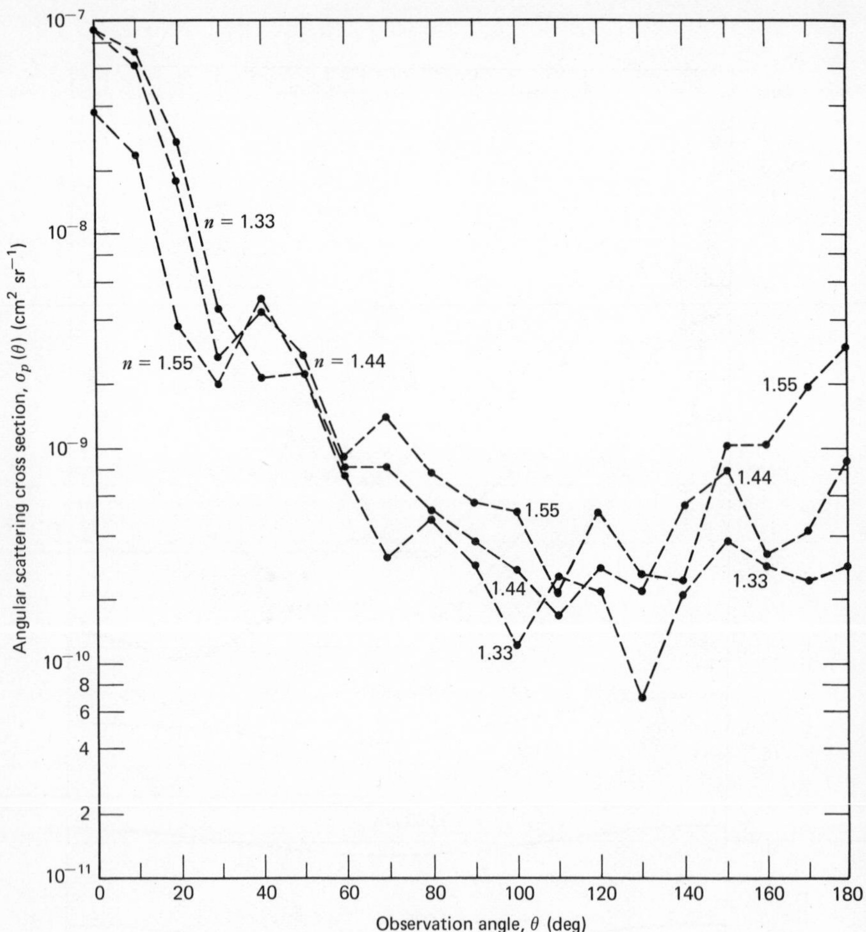

Figure 5.5 Angular scattering cross sections of particles having different refractive indices and $\alpha = 6.0$. Data from La Mer and Sinclair (1943).

about 30 deg is little affected by changes in the refractive index when the particle size is greater than a few wavelengths. This tendency can be seen even for $\alpha = 6.0$ in the figure, and the tendency becomes pronounced at larger values of α. This is because the forward lobe can be regarded mainly as the result of Fraunhofer diffraction by an opaque object. In such a view the forward lobe is independent of the refractive index of the object. The use of diffraction theory as an alternative to Mie theory in calculating the forward lobe has been investigated by Hodkinson (1966a,b) and Hodkinson and Greenleaves (1963).

The small amount of flux scattered into a very small solid angle by a typical single atmospheric particle can be appreciated from the ordinate values in Figure 5.4 and any of the relationships (5-12) through (5-14). Accurate measurement of such flux is difficult and has seldom been accomplished. The difficulties, cited here because they epitomize several aspects of scattering measurements, are of two types: (1) physical problems in isolating the particle from all other particles, measuring its diameter when required, and positioning it without mechanical means of support; and (2) optical problems in reducing stray light (which acts as a noise background) to a low level compared to that of the scattered light and collecting enough of this scattered light (at a fine angular resolution) to provide a good signal-to-noise ratio. Despite the difficulties, experiments have been reported wherein the angular scattering from single particles was measured and found to agree well with theory. Gucker and Egan (1961) employed droplets of dioctal phthalate having radius values near 1 μm, while Blau et al. (1970) employed droplets of this and other liquids with a radius in the range 3.75 to 55 μm. Phillips and Wyatt (1972) have measured the scattering by small droplets of substances that are common constituents of photochemical hazes (smogs).

The flux scattered into a large solid angle by a single particle, however, is readily detectable. In order to accomplish this, the particle is strongly illuminated and the detector optics are arranged so that the aperture subtends an appreciable angle, as seen from the particle. The creation and detection of such flux form the basis of particle counters which have found wide use in aerosol research and in industry for monitoring the quality of air in "clean rooms." Typical instrumentation is described by Hodkinson (1966b), while additional information is provided by Hodkinson and Greenfield (1965), Quenzel (1969), and Cooke and Kerker (1975).

5.4.2 Comparison of Rayleigh and Mie Cross Sections

It is informative to compare the scattering by very small particles as predicted by the Rayleigh and Mie theories. For the first theory we can see from (4-38) that the angular scattering cross section of a single small particle for unpolarized light varies as r^6. Thus in terms of the size parameter α, the angular cross section varies as α^6. Taking as examples the values $\alpha = 0.4$ and $\alpha = 0.2$, the sixth power of the ratio is

$$\text{Rayleigh theory: } \left(\frac{\alpha = 0.4}{\alpha = 0.2}\right)^6 = 64$$

In close agreement the ratio of the two cross sections, at the same values of α and at $\theta = 0$ deg, is found from the Mie data for Figure 5.4 to be

$$\text{Mie theory: } \frac{\sigma_p(\theta)(\alpha = 0.4)}{\sigma_p(\theta)(\alpha = 0.2)} = \frac{1.40 \times 10^{-14}\ \text{cm}^2\ \text{sr}^{-1}}{2.11 \times 10^{-16}\ \text{cm}^2\ \text{sr}^{-1}} = 67$$

which can be verified approximately from the plots. This ratio does not change very much over the full angular range.

Since scattering by very small particles thus varies nearly as α^6, it might seem that scattering also should vary as λ^6, for consistency with (5-1). However, (4-38) calls for scattering to vary as λ^{-4}, as in all cases in which Rayleigh theory is valid. This latter dependency is still the correct one for the above examples and is easily extracted from Mie theory by the following argument. As shown above, the angular cross section $\sigma_p(\theta)$ varies nearly as α^6 for $\alpha \leq 0.4$. But this means from (5-14) that, if the wavelength is held fixed, the quantity $i_1 + i_2$ varies as $(2\pi r/\lambda)^6$, where r is the variable. We thus have the scattering varying as the sixth power of particle radius, as called for by (4-38). Now let r be held fixed and let α be varied by varying λ. The quantity $i_1 + i_2$ now varies as λ^{-6}, but from (5-14) the cross section $\sigma_p(\theta)$ then must vary as λ^{-4}.

It is of interest to compare the numerical values of angular cross sections computed by the two theories. Assuming that $\alpha = 0.2$, $\lambda = 1.0\ \mu$m, and $n = 1.33$, we find from (4-38) and (5-14), along with values of i_1 and i_2 as tabulated in the literature, the values:

Rayleigh theory: σ_p (0 deg) $= 7.09 \times 10^{-16}$ cm^2 sr^{-1}
Mie theory: σ_p (0 deg) $= 6.99 \times 10^{-16}$ cm^2 sr^{-1}

Thus the two theories give nearly the same results for all $\alpha < 0.2$, which is equivalent to $r < 0.03\lambda$. Verifications from theory are better of course than deductions from numerical examples, and readers are referred to Stratton (1941) and Kerker (1969) for such a demonstration. They show that, when α and m are sufficiently small, only the terms a_1, π_1, and τ_1 in (5-4) and (5-5) need be considered. With unpolarized light this leads to the expression

$$\sigma_p(\theta) = \frac{8\pi^4 r^6}{\lambda^4}\left(\frac{m^2-1}{m^2+2}\right)(1+\cos^2\theta)$$

which differs from the Rayleigh expression (4-38) only in that m, which may be complex, replaces n. It is emphasized that the validity of this equivalence requires that both m and the product $m\alpha$ be small, including the real and imaginary parts of m. These are the criteria for Rayleigh scattering by small particles as distinguished from molecules.

5.4.3 Volume Angular Coefficient

Whereas the angular cross section of a particle represents scattering that is very difficult to observe, the volume angular coefficient represents scattering that is a principal observable in the outdoor world. This coefficient expresses the angular scattering characteristic of a unit volume of aerosol and is found directly from the angular cross sections of the particles. The random separations and motions of the particles preclude any interferences between the individually scattered fluxes, so that the scattering is incoherent, as in the molecular case, and the intensities of the waves are additive. When the particles are spherical and of the same size and refractive index, which are the assumptions here, all the scattering patterns or cross sections are identical. If there are N particles in the unit volume, the scattered intensity in any direction is just N times that from one particle.

The volume angular coefficient, denoted for the general case by $\beta_p(\theta)$, has the same form as the angular cross section $\sigma_p(\theta)$. When the incident light is plane-polarized with its electric vector perpendicular to the plane of observation, the coefficient is given by

$$\beta_{p,\perp}(\theta) = N\sigma_{p,\perp}(\theta) = N\frac{\lambda^2}{4\pi^2}\,i_\perp \tag{5-15}$$

When the incident light is plane-polarized with its electric vector parallel to the plane of observation, we have

$$\beta_{p,\parallel}(\theta) = N\sigma_{p,\parallel}(\theta) = N\frac{\lambda^2}{4\pi^2}\,i_\parallel \tag{5-16}$$

And when the incident light is unpolarized,

$$\beta_p(\theta) = N\sigma_p(\theta) = N\frac{\lambda^2}{4\pi^2}\left(\frac{i_1 + i_2}{2}\right) \tag{5-17}$$

Since the dimensions of $\sigma_p(\theta)$ are L^2 and the dimensions of the concentration N are L^{-3}, the dimension of $\beta_p(\theta)$ is L^{-1}. Frequently used units are $m^{-1}\ sr^{-1}$.

When the incident light has a unit irradiance, the numerical value of $\beta_p(\theta)$ represents the intensity of scattered light from the unit volume of aerosol, passing through an infinitesimal solid angle $d\omega$ at angle θ, in terms of either power or energy per steradian, per unit cross section of path, and per unit length (or per unit volume). If the particles are isotropic spheres, depolarization does not occur. When the incident light is plane-polarized, the polarization plane of the scattered light relative to the plane of observation is the same as that of the incident light. When the incident light is unpolarized, the scattered light is a mixture of polarized and unpolarized light, as may be seen from Figure 5.2. The degree of polarization is given by (5-10). Heller (1963) reviews the use of intensity and polarization measurements for determining particle sizes and concentrations and provides an extensive bibliography of this subject.

Although concerned largely with hydrosols, the experimental techniques described may well suggest various counterparts in the field of aerosols.

5.4.4 Higher-Order Tyndall Spectra

In this section we call attention to the preferential scattering of various wavelengths at certain angles when changes in wavelength produce variations in α. For example, when a monodispersion is illuminated by white light, very striking color effects can be seen. Historically, early studies of the color and polarization of the light scattered by small particles are associated with Tyndall, whose investigations dealt with particles "of every degree of fineness, from those visible to the eye to those far beyond the reach of the microscope." The term *Tyndall blue* is applied to the color of the light when the particles are small enough that $r \ll \lambda$. This is mostly Rayleigh scattering of course, and undoubtedly Tyndall's experiments prompted Rayleigh to find the explanation. Tyndall blue is characteristic of clean country hazes but also may be seen in exhausts from automobile engines consuming crankcase oil.

As the particles of a monodispersion illuminated by white light are made larger than those that produce Rayleigh scattering, additional colors having angular dependence appear in the scattered light. At intermediate sizes where $0.2\ \mu\text{m} < r < 0.8\ \mu\text{m}$, so that the radius is bracketed by the visual spectrum, various color bands appear in angular order. These were called *higher-order Tyndall spectra* by Johnson and La Mer (1947) who employed the phenomena to determine particle sizes of approximate monodispersions. For simplicity, and to obtain the most vivid colors, they observed the red bands at $\lambda = 0.629\ \mu\text{m}$ and the green bands at $\lambda = 0.524\ \mu\text{m}$, for which the wavelength ratio is 1.2, and measured the angular positions of the various maxima and minima. They dealt principally with the $I_\perp(\theta)$ or i_1 component which produced the most distinct colors. Particle size was then found by comparing the positions of the maxima, either with instrument calibrations for particles of known sizes, or with the positions (theoretical) as indicated by the intensity distribution function i_1. The latter method is facilitated by the format of the functions published by La Mer and Sinclair (1943) and Lowan (1949). Their tabulations give values of these functions for eight pairs of α values, from 0.5 to 6.0, whose ratios are 1.2. Consequently, each pair of i_1 or i_2 values yields the two intensities $I(\theta)_{\text{red}}$ and $I(\theta)_{\text{gr}}$, or the intensities at any two wavelengths having this same ratio. While the use of higher-order Tyndall spectra does not give a particle size determination as accurate as can be obtained from the angular location of maxima and minima in monochromatic light, the method is more rapid and simpler.

These spectra are mostly restricted, however, to actual and approximate monodispersions and to particles having a radius comparable in value to the wavelengths of visual light. Variations in the spectra with characteristics of the suspension are described in detail by Kerker et al. (1966). When the particles are much smaller than visual wavelengths, the scattered light is predominantly blue at all angles, as mentioned previously. In the size range $0.2\ \mu m < r < 0.8\ \mu m$, the number of color orders increases with particle radius, and it is an empirical rule that the radius in micrometers is approximately equal to one-tenth the number of maxima in $I_{\perp}(\theta)$. As the radius increases beyond $0.8\ \mu m$, the number of orders does not continue to increase. Instead, the colors begin to wash out and the orders become indistinct, and at $r > 1.2\ \mu m$ the scattered light is yellow in the sector $50\ \deg < \theta < 150\ \deg$. Similar effects are produced when a suspension is made polydisperse, even when the values of particle radii are not greater than visual wavelengths. In each instance the loss of color orders is due to the overlapping and merging of individual orders.

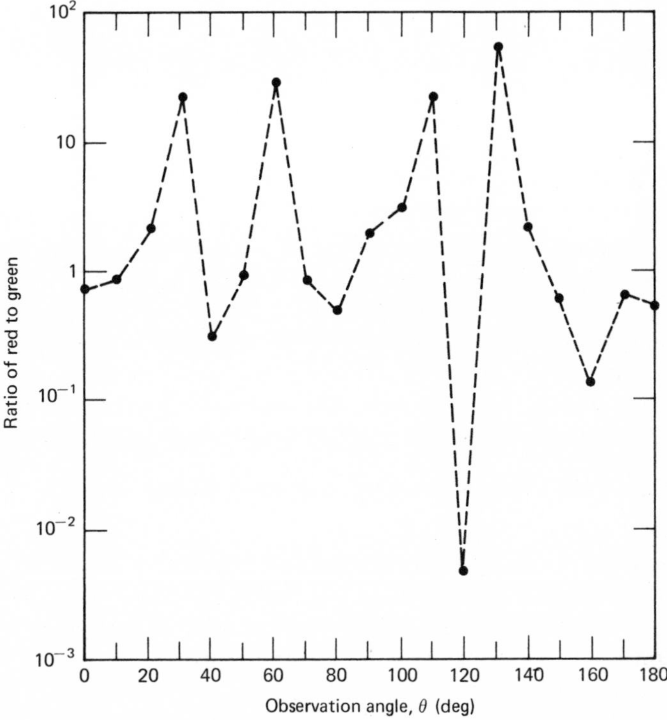

Figure 5.6 Intensity ratio of red light ($0.629\ \mu m$) to green light ($0.524\ \mu m$) as scattered by a small particle having $n = 1.44$ and $\alpha = 5.0$. Data from La Mer and Sinclair (1943).

The use of higher-order Tyndall spectra is an example of *multicolor detection*, which has undergone much development in recent years. Some of this work with respect to Tyndall spectra, including calculations of the angular positions of additional colors, is described and referenced by Kerker et al. (1966) and Kerker (1969). The large amount of information in even a simple two-color ratio may be seen in Figure 5.6, where the maxima define the positions of the red orders or centers of the color bands, and the minima do the same for the green orders. The pattern is a distinctive signature in the two colors. Although referred to as a ratio of intensities, because these are the quantities directly observed, the pattern is actually a ratio of angular scattering cross sections. This is because the irradiances and other common factors cancel out when the ratio is formed. Hence the experimental data can be made rather independent of fluctuations in the light source, variations in detector sensitivity, and uncertainties in particle concentration. The ratio $I_1(\theta)/I_2(\theta)$, where the subscripts indicate any selected pair of wavelengths, then depends only on particle size, refractive index, and wavelength.

5.5 TOTAL SCATTERING BY MONODISPERSIONS

As in molecular scattering, the concept of total scattering by particles refers to the total amount of flux scattered in all directions from the incident flux. This section treats in detail several measures of total scattering that are used. First, the total scattering cross section of a single particle is defined. Next, the concept of a scattering efficiency factor is developed and extended to take absorption into account. This results in efficiency factors for extinction and light pressure. An approximate but convenient method of calculating efficiency factors is reviewed. Single-particle concepts of scattering and extinction are extended to the volume coefficients, which are the measurable indices of all the processes. The section concludes with a list of published tabulations of efficiency factors.

5.5.1 Scattering Cross Section and Efficiency Factor

The total scattering cross section σ_p is defined as that cross section of an incident wave, acted on by the particle, having an area such that the power flowing across it is equal to the total power scattered in all directions. When this definition is compared with the one given for $\sigma_p(\theta)$ in Section 5.4.1, it is clear that σ_p is defined by

$$\sigma_p = \int_0^{4\pi} \sigma_p(\theta)\, d\omega \tag{5-18}$$

for unpolarized incident light. Analogous cross sections can be defined for polarized incident light. The geometry for the integration is the same as that shown in Figure 4.7 for the molecular case, and the relationship between the variables of integration is

$$d\omega = 2\pi \sin\theta \, d\theta$$

Hence we have

$$\sigma_p = 2\pi \int_0^{\pi} \sigma_p(\theta) \sin\theta \, d\theta \tag{5-19}$$

where σ_p represents the incident wavefront area applicable to the interaction with the particle. The dimensions of σ_p are L^2, and the usual units are $cm^2\ sr^{-1}$. Numerically, σ_p is equal to the total amount of power scattered in all directions by the given particle when the incident flux has a unit irradiance.

Values of the total scattering cross section cover a wide range greater than the corresponding range of geometric cross sections. The two are related by the *efficiency factor* Q_{sc}, often called the *Mie coefficient* K, which is the ratio of the scattering to the geometric cross section. In the present discussion the subscript sc is omitted for simplicity. Thus division of (5-19) by πr^2 gives

$$Q = \frac{\sigma_p}{\pi r^2} = \frac{2}{r^2} \int_0^{\pi} \sigma_p(\theta) \sin\theta \, d\theta \tag{5-20}$$

This ratio defines the efficiency (so called) with which the particle totally scatters the incident light. From (5-1) and (5-14), we write (5-20) as

$$Q = \frac{1}{\alpha^2} \int_0^{\pi} (i_1 + i_2) \sin\theta \, d\theta \tag{5-21}$$

The efficiency factor Q has a dual dependence on the size parameter α: implicitly in the intensity distribution functions i_1 and i_2, and explicitly in the denominator of (5-21). In addition, Q has the same dependence as i_1 and i_2 on the particle refractive index.

The integration indicated in the preceding equations is more illustrative than practical. For purposes of computation, σ_p and Q can be defined in terms of the complex functions a_n and b_n employed in (5-4) and (5-5) to define the functions i_1 and i_2. In such terms the total scattering cross section is expressed by

$$\sigma_p = \frac{\lambda^2}{2\pi} \sum_{n=1}^{\infty} (2n+1)(|a_n|^2 + |b_n|^2) \tag{5-22}$$

Division of this expression by πr^2 gives the scattering efficiency factor

$$Q = \frac{2}{\alpha^2} \sum_{n=1}^{\infty} (2n + 1)(|a_n|^2 + |b_n|^2) \tag{5-23}$$

as an alternative expression to (5-21) but one that is more suitable for computation. Techniques for thus computing σ_p and Q and associated functions are treated by Penndorf (1956), van de Hulst (1957), Deirmendjian (1969), and Kerker (1969). Many tabulations of Q for wide ranges of α and m are available in the literature; those having the greatest relevance to atmospheric scattering are cited in Section 5.5.6. A particularly applicable set of values from Penndorf (1956, 1957c) is reproduced in Appendix J.

Figure 5.7 shows Q for small spheres of water, as given by List (1966) from computations by Lowan (1949) and Houghton and Chalker (1949). When α is very small, the value of Q is much less than unity, and the particle scatters far less flux than would be intercepted by its geometric cross section. As α increases, Q rises to a maximum value of nearly 4 and then slowly converges to the value 2 in the manner of a damped oscillation. As the refractive index increases, the value of Q maximum also increases somewhat, becoming equal to about 4.4 when $n = 1.50$, and occurs at values of α less than that in the figure. Not shown in the plot of Q are the "ripples" of high frequency and small amplitude superposed on

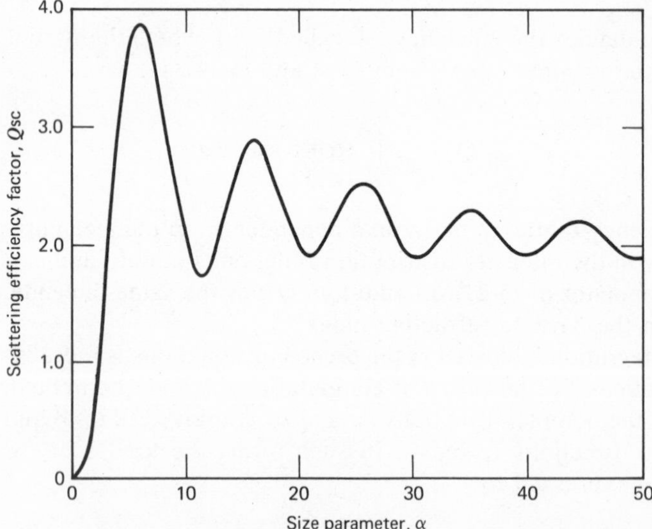

Figure 5.7 Scattering efficiency factor versus size parameter for water droplets. From List (1966).

the large-scale variations. These minor oscillations, while having theoretical interest for the study of dipole resonances, have no very discernible effects on atmospheric scattering.

Additional characteristics of the scattering efficiency factor are revealed by the log-log plot in Figure 5.8 from the values in Appendix J. The curve for $n = 1.50$ corresponds to a fairly dry haze particle, that is, one on

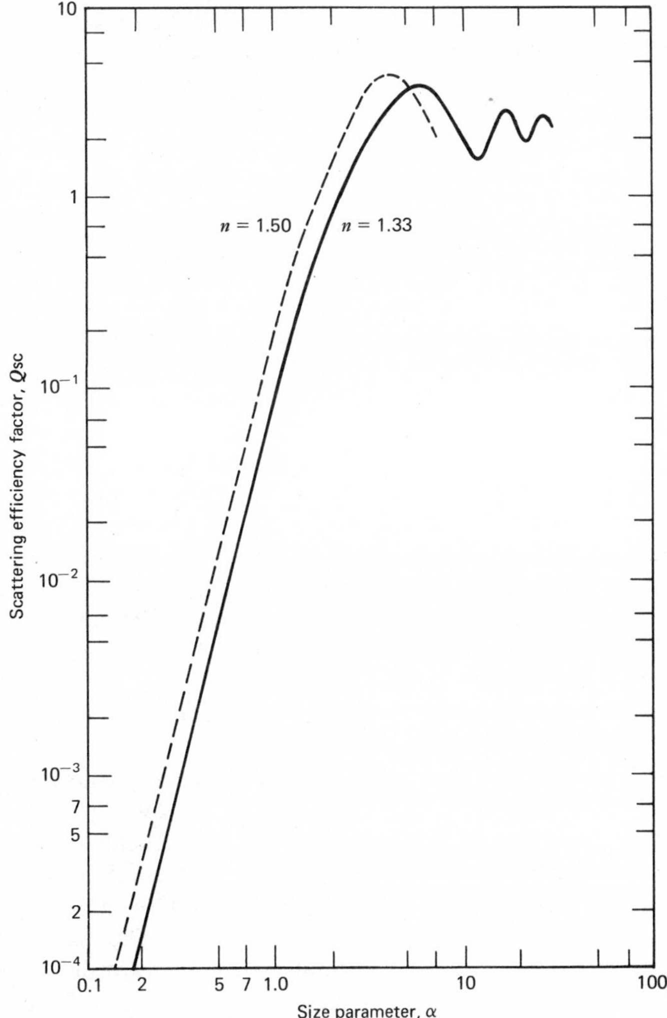

Figure 5.8 Log–log plot of scattering efficiency factor versus size parameter for two refractive indices. Data from Penndorf (1957c).

which little water has condensed. The curve for $n = 1.33$ represents the other extreme, where a condensation nucleus has grown to a droplet of water such as those in fog and cloud. For each curve the slope becomes equal to about 4 when $\alpha \ll 1$. Since α bears an inverse relationship to λ, such a slope indicates that the scattering varies as λ^{-4} in the manner of the Rayleigh expressions. In the tabulated data used to construct the curve for $n = 1.50$, the value of Q at $\alpha = 0.20$ is 3.70×10^{-4}. For comparison, the value $Q = 3.63 \times 10^{-4}$ was computed from Rayleigh theory in the text following (4-32). In Figure 5.8 the decreasing slope as α increases, whether from changing the particle size or the wavelength, shows the transition from Rayleigh to Mie scattering. For the curve $n = 1.33$, the maximum value $Q = 3.98$ is reached at $\alpha = 6.5$, which means that a water droplet scatters most efficiently when its radius is approximately equal to the wavelength.

5.5.2 Scattering Efficiency Factor Considered Further

From the definition of the efficiency factor, it follows that, for all $Q > 1$, the particle intercepts or interacts with an area of the incident wave front greater than its own geometric cross section. This means that the particle perturbs the electromagnetic field beyond the confines of the particle and that scattering may occur from the entire region. An explanation for the value $Q = 2$ can be found from the principles of reflection, refraction, and diffraction, as in van de Hulst (1957). Here we note that the particle is an obstacle in the path of the incident wave and intercepts a wavefront area equal to its own geometric cross section. If the particle is nonabsorbing, the flux intercepted is reflected and refracted in all directions. If the particle is partially or completely absorbing, a proportionate amount of the intercepted flux is expended in heating the particle. In any event, the particle directly removes from the optical beam an amount of power proportional to its geometric cross section; in this sense the scattering efficiency factor of the particle is unity.

According to Babinet's principle, however, the particle also diffracts an amount of the incident flux proportional to its own geometric cross section. In this sense also the scattering efficiency factor is unity. When this diffraction effect, which is independent of the refractive index, is combined with the effects of reflection and refraction, which are dependent on the refractive index, it is evident that $Q = 2$. The diffracted flux is sent into a very small angle centered about the forward direction of the incident flux, the angle varying inversely as the particle diameter. This fact hindered for many years the experimental verification that $Q = 2$ for $\alpha \gg 1$. Usually the total scattering is evaluated by placing a light detector in the path of the light beam and measuring the beam irradiance with the scatterers in and out of the path. The experimental requirement for demonstrating that $Q = 2$ is that the aperture of the detector be sufficiently small and far away enough from the scattering center that the aperture-subtended angle is much smaller than the diffraction angle. Unless this requirement is satisfied, some of the diffracted flux will be detected, and a lesser scattering will be inferred.

These ideas have been confirmed both theoretically and experimentally. Brillouin (1949) showed mathematically that the scattering cross section of a perfectly *reflecting* sphere having $\alpha = 160$ is πr^2 for light scattered (diffracted) into a small cone of half-angle $\theta = 7$ deg centered about the forward direction. The scattering cross section for light scattered (reflected) at angles $7\text{ deg} < \theta < 180\text{ deg}$ is also πr^2. Thus the total scattering cross section is $2\pi r^2$. Bar-Isaac and Hardy (1975) derived the same result directly from scalar diffraction theory at the limit where the wavelength goes to zero. Sinclair (1947) measured the attenuation of a layer of nearly monodisperse particles having $r = 15.0 \pm 1.0\ \mu\text{m}$. The light had a wavelength of $0.524\ \mu\text{m}$, so that α had the value 180. Using a photodetector with an aperture of about 2 in., he found by transmission measurements that the scattering cross section was πr^2 when the detector was 6 in. from the scatterers. When the distance was increased to 18 ft, however, the measurements indicated a scattering cross section of nearly $2\pi r^2$.

Further treatments of this subject can be found in Middleton (1952), Green and Lane (1957), and Kerker (1969). Hodkinson and Greenleaves (1963) provide graphs showing the reduction in an observed efficiency factor as a function of detector acceptance angle and give a helpful discussion of the factors important in transmission measurements. Latimer (1972) made analytic and experimental studies of the effects produced by incident beam convergence and detector acceptance angle on the observed transmission. These considerations regarding the experimental determination of the scattering efficiency factor are closely related to the matter of diffuse transmission discussed in Section 1.4.4.

5.5.3 Efficiency Factors of Absorption, Extinction, and Light Pressure

The difference between the flux removed from the incident beam and the totally scattered flux must be attributed to absorption by the particle. Absorption occurs when the refractive index is complex according to (5-2), and the absorbed energy goes into heating the particles. Insofar as attenuation or extinction of the flux in the beam is concerned, it makes no difference whether flux removal is caused by scattering or absorption. A simple transmission measurement does not distinguish between the two processes. Since the extinction effects of the two processes are additive, we can write

$$Q_{ex} = Q_{sc} + Q_{ab} \tag{5-24}$$

Here Q_{sc} is defined by (5-23), and Q_{ex} by

$$Q_{ex} = \frac{2}{\alpha^2} \sum_{n=1}^{\infty} (2n + 1)[\text{Re}\,(a_n + b_n)] \tag{5-25}$$

When needed, Q_{ab} can then be found from (5-24).

The overall manner in which absorption by the particle, expressed by its complex refractive index, affects Q_{ex} is shown in Figure 5.9. The refractive indices noted for the curves are representative of liquid water at various infrared wavelengths. When the absorption is practically zero, as for the curve $m = 1.29$, Q_{ex} is the same as Q_{sc}. As the absorption increases, the resonances that produce the primary maximum near $\alpha = 2\pi$ and the secondary maxima at larger values of α become highly damped. Their amplitudes thereby are greatly reduced, and the curve converges to the value 2 sooner than when the absorption is zero. For extreme absorption represented by $m = 1.29 - 0.472i$, the curve asymptotically reaches the value 2 after only a single mild overshoot.

The theory of the interactions between an incident wave and the absorbing material of a particle, which form the basis of (5-25), is beyond the scope of this book. Interested readers are referred to Johnson et al. (1954), Johnson and Terrell (1955), van de Hulst (1957), Deirmendjian et al. (1961), Irvine (1965a), and Kerker (1969). A good understanding of the behavior of the efficiency factors as m and α are varied can be obtained from Plass (1966) and Kattawar and Plass (1967). Relevant to this subject is a quantity known as the *assymmetry factor*, denoted by $\overline{\cos\theta}$. The assymmetry factor represents the mean value of $\cos\theta$ over the sphere weighted by the values of scattered intensity $I(\theta)$ and enters into the concept of radiation pressure described in the following two paragraphs.

An original study of the pressure exerted by an electromagnetic wave was made by Debye (1909). The wave carries momentum, so that the particle, by scattering and absorbing, removes momentum from the wave. The amount removed from an incident beam of unit irradiance is just equal to the extinction cross section, hence is proportional to Q_{ex}. Of the momentum removed by the scattering process, however, the part associated with the forward component of scattered intensity is restored to the incident beam. This part is proportional to the asymmetry factor $\overline{\cos\theta}$ mentioned above. Thus the efficiency factor of light pressure is

$$Q_{pr} = Q_{ex} - \overline{\cos\theta} \times Q_{sc} \tag{5-26}$$

which may be used for calculations in the same manner as Q_{sc} and Q_{ex}.

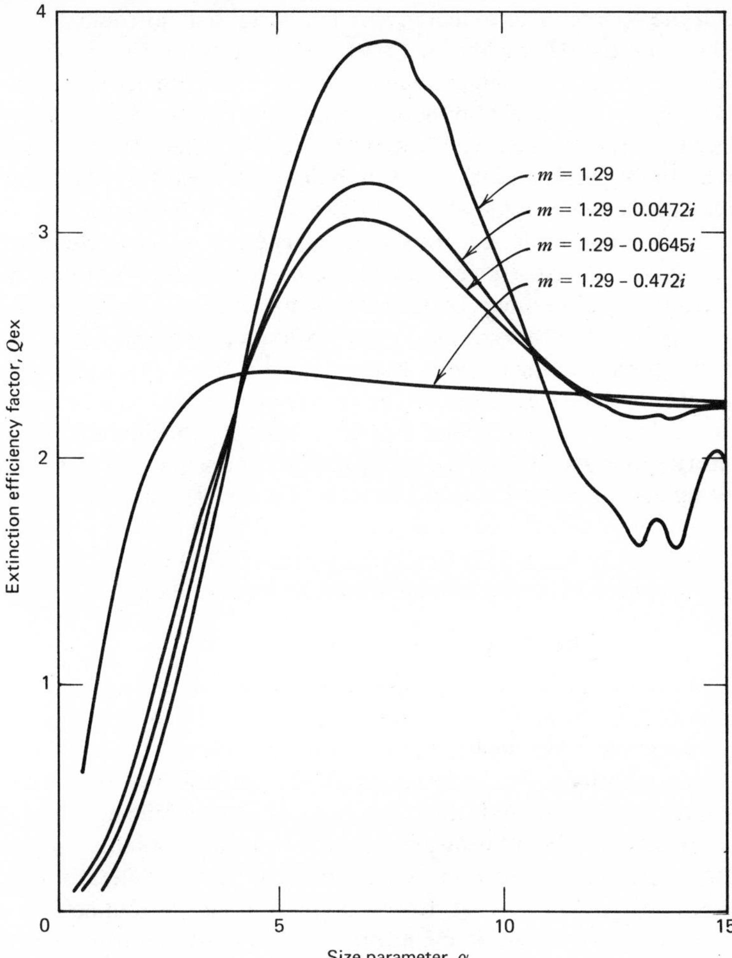

Figure 5.9 Extinction efficiency factor versus size parameter for complex refractive indices of water. From Kerker (1969).

For very small particles and molecules, the symmetry of the applicable Rayleigh scattering makes $\overline{\cos\theta}$ equal to zero. For this condition Q_{pr} is equal to Q_{ex}.

There is a close relationship between $\overline{\cos\theta}$ and m. In considering a large sphere, for example, when $m = 1$, there is neither reflection nor refraction, and all the flux incident on it passes through without deviation. The scattering is completely asymmetric, that is, $\overline{\cos\theta}$ equals unity, and Q_{pr} equals zero. At the other extreme, when $m = \infty$, no refracted flux can pass through the sphere. The reflected flux is distributed isotropically, consequently it makes no contribution to $\overline{\cos\theta}$. The only contribution is that by diffraction, which results in the value $\overline{\cos\theta} = 0.5$. Defining expressions for Q_{pr} and $\overline{\cos\theta}$ in terms of Mie theory parameters, along with graphical and tabulated data, are given by Irvine (1965) and Kerker (1969). Irvine and Pollack (1968) present extensive tabulations of $\overline{\cos\theta}$ for water droplets and ice spheres over wide ranges of radius and wavelength.

In this book only homogeneous spheres, each having a single refractive index, are considered. However, water-coated insoluble particles having two refractive indices may be found to some extent in haze aerosols. In heavily polluted atmospheres, for example, the plentiful carbon particles—perhaps tinged with hygroscopic material—acquire a water film whose thickness depends on the relative humidity. To a fair approximation, these two-component particles can be considered coated or concentric spheres, where the core has a decreasing influence on the scattering as the water coating becomes thicker. The angular scattering and extinction characteristics of coated spheres have been studied by Fenn and Oser (1965) and Pilat (1967). Kerker (1969) gives a comprehensive treatment with many graphical and numerical data.

5.5.4 Approximation from Optical Principles

Alternative treatments of total scattering, which do not employ the electromagnetic formalism of Mie theory, have been developed from optical principles. Originally such treatments offered the advantage of fewer computations than are required by Mie theory. When particle absorption is not involved, this advantage is less significant today since many tabulations of scattering efficiency factors are available for polydispersions as well as for discrete values of α. When particle absorption is involved, this advantage of fewer computations is retained because tabulations of extinction efficiency factors are less plentiful. The treatment of total scattering by optical principles affords an additional view which is not explicit in Mie theory, but the resulting expressions should be regarded as approximations. In this section such a treatment for the efficiency factors of scattering and extinction is summarized.

In the optical theory of total scattering developed by van de Hulst (1957), he assumed that

$$m - 1 \ll 1 \qquad \text{and} \qquad \alpha \gg 1$$

and proposed the name *anomalous diffraction* for the theory. The first assumption, initially imposed as a restriction or boundary limit, strictly means that $m \to 1$, but the theory has also been found generally applicable when $m \to 2$. The second assumption means that the particle radius must be several times greater than the wavelengths being considered. Hence for wavelengths in the visual and near infrared regions, fog and cloud droplets whose radii are greater than several micrometers exemplify the atmospheric particles to which the anomalous diffraction theory applies. For nonabsorbing particles the treatment leads to an expression for Q_{sc}, and for absorbing particles to an expression for Q_{ex}.

Basic to the theory is the *normalized size parameter* ρ, defined by

$$\rho = 2\alpha(m-1) = \frac{4\pi r}{\lambda}(m-1) \tag{5-27}$$

which thus links the three basic factors of scattering. In referring to Figure 5.10, when the droplet radius becomes notably greater than the

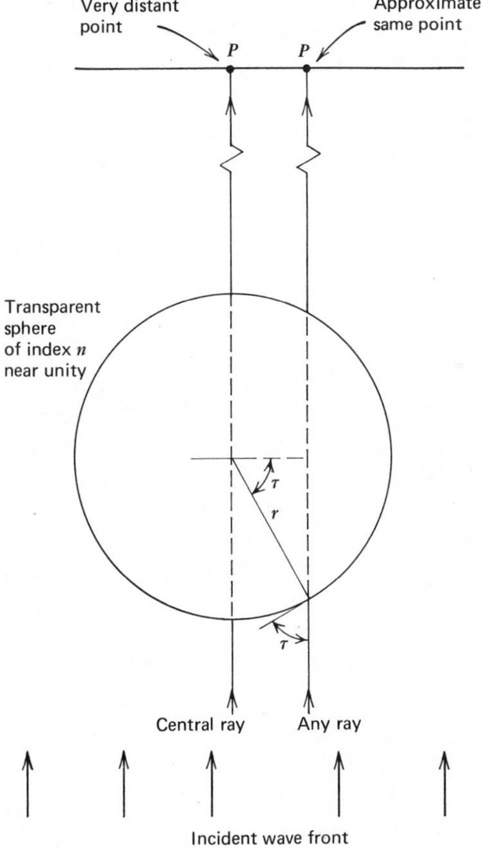

Figure 5.10 Path of light ray through a sphere having refractive index near unity. From van de Hulst (1957).

wavelength, an incident ray can be traced all the way through the droplet. Since $m - 1$ is assumed small, and real in this instance, the ray is deviated only slightly at the air–water interfaces and only a small fraction of the flux is reflected. Hence the amplitude of the field at a point P in the path of the wave is affected only slightly by the presence of the droplet. The phase of the wave at P, however, is delayed by the greater optical path length through the droplet whose refractive index is greater than unity. The geometric path length in the droplet is $2r \sin \tau$, so that the phase lag in radians at P is

$$\begin{aligned} \delta &= 2r \sin \tau (m - 1) \frac{2\pi}{\lambda} \\ &= \rho \sin \tau \end{aligned} \tag{5-28}$$

It is seen that ρ represents the phase lag for the central ray through the droplet and determines the kind and degree of interference that takes place between the transmitted light and the light diffracted to P by the droplet.

From such considerations and Huygens' principle of wave propagation, van de Hulst (1957) developed an expression for the scattering efficiency factor:

$$Q_{sc} = 2 - \frac{4}{\rho} \sin \rho + \frac{4}{\rho^2}(1 - \cos \rho) \tag{5-29}$$

This formula describes the principal features of the efficiency factor, not only for $m \to 1$ but also for $m \to 2$. It can be shown from (5-29) that the first maximum occurs at approximately $\rho = 4.08$, regardless of the refractive index. This result is also obtained from Mie theory, as shown in Figure 5.11, where the bottom scale in ρ is common to all the curves. Penndorf (1957b) employed (5-29) in detailed computations of Q_{sc} for values of m close to unity. Although such small values are not of interest here, empirical correction factors have been devised by Deirmendjian (1960) for indices of atmospheric particles. For example, the value found for the first maximum should be multiplied by the factor

$$1 + \frac{m - 1}{m}$$

for any value $1.5 \geq m > 1$. Thus from this factor and (5-29), we find that the values of Q_{sc} for $m = 1.33$ and $m = 1.50$ at $\rho = 4.08$ are, respectively, 3.96 and 4.18. In view of the simplicity of the computation, these values compare well with those in Appendix J for the corresponding arguments $\alpha = 6.2$ (for $m = 1.33$) and $\alpha = 4.08$ (for $m = 1.50$).

The case of absorbing particles is considered next. An additional

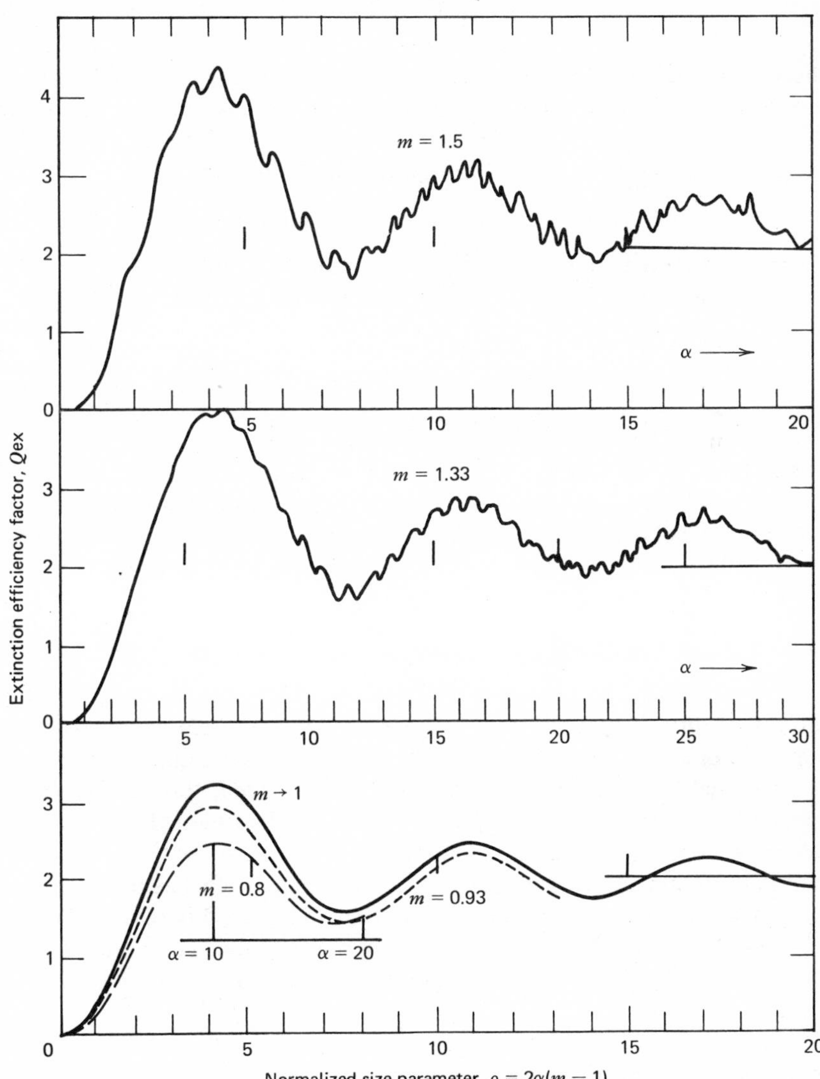

Figure 5.11 Extinction efficiency factor versus normalized size parameter, computed from Mie theory for refractive indices $m = 1.5$, 1.33, 0.93, 0.80, and $m \rightarrow 1$. The scales of α have been chosen so that the scale of $\rho = 2\alpha(m-1)$ is common to all the curves. From van de Hulst (1957).

parameter B is employed, defined by

$$\tan B = \frac{n_i}{n-1} \tag{5-30}$$

where n_i is the complex part of the refractive index. The initial requirement for the nonabsorbing case that $(m-1) \ll 1$ here means that both $m-1$ and n_i must be much less than unity. Analogously to (5-28), the phase lag for the central ray through an absorbing particle is

$$\rho^* = 2\alpha(m-1) = \rho(1 - i \tan B) \tag{5-31}$$

where the real part denotes an actual phase shift and the imaginary part a decay of amplitude. The extinction efficiency factor turns out to be

$$\begin{aligned} Q_{ex} = 2 &- 4 \exp(-\rho \tan B)(\rho^{-1} \cos B) \sin(\rho - B) \\ &- 4 \exp(-\rho \tan B)(\rho^{-1} \cos B)^2 \cos(\rho - 2B) \\ &+ 4(\rho^{-1} \cos B)^2 \cos 2B \end{aligned} \tag{5-32}$$

This reduces to (5-29) when $B = 0$, that is, when $n_i = 0$. Selected values of Q_{ex} for m close to 1 are listed in Table 5.5. These values, and any others that may be calculated from (5-32), can be corrected for various choices of ρ, n, and n_i in the ranges $1 < n < 1.50$ and $0 \le n_i \le 0.25$ by means of the correction factors given by Deirmendjian (1960).

5.5.5 Volume Scattering and Extinction Coefficients

The volume total scattering or extinction coefficient expresses the total amount of flux removed from a beam by a unit volume of particle suspension per unit irradiance of the volume. Assuming that the particles are spherical, and isotropic in other respects as well, this coefficient is independent of the polarization state of the incident light. Flux removal may be accomplished by pure scattering as indicated by Q_{sc}, or by any combination of scattering and absorption by the particles as indicated by Q_{ex}. Since the processes are incoherent, the amount of flux removed by N particles is just N times that removed by one particle. The volume total scattering coefficient for a monodispersion of particles thus is given by

$$\beta_{sc} = N\sigma_p \tag{5-33}$$

where N is the particle concentration, and σ_p is the scattering cross section defined by (5-19). Because Q_{sc} is the ratio of the scattering to the geometric cross section, (5-33) becomes

$$\beta_{sc} = N\pi r^2 Q_{sc} \tag{5-34}$$

In a similar manner we can write

$$\beta_{ab} = N\pi r^2 Q_{ab} \tag{5-35}$$

TABLE 5.5 EXTINCTION AND ABSORPTION EFFICIENCY FACTORS FOR PARTIALLY ABSORBING SPHERES, WITH m CLOSE TO 1

		Q_{ex}				Q_{ab}
	$\beta =$	0°	15°	30°	45°	45°
ρ	$\tan\beta =$	0	0.27	0.58	1.00	1.00
0		0.00	0.00	0.00	0.00	0.00
0.2		0.02	0.09	0.10	0.26	0.23
0.4		0.08	0.21	0.35	0.52	0.40
0.6		0.18	0.36	0.54	0.76	0.53
0.8		0.31	0.53	0.74	0.97	0.63
1.0		0.47	0.71	0.93	1.16	0.70
1.2		0.66	0.90	1.11	1.32	0.76
1.4		0.88	1.09	1.28	1.46	0.80
1.6		1.11	1.28	1.43	1.58	0.84
1.8		1.35	1.47	1.57	1.68	0.86
2.0		1.60	1.64	1.69	1.76	0.89
2.2		1.84	1.80	1.80	1.82	0.90
2.4		2.08	1.95	1.89	1.87	0.92
2.6		2.31	2.08	1.96	1.91	0.93
2.8		2.51	2.19	2.02	1.94	0.94
3.0		2.70	2.28	2.06	1.96	0.95
3.5		3.03	2.42	2.13	1.98	0.96
4.0		3.17	2.46	2.14	2.00	0.97
4.5		3.11	2.43	2.13	2.00	0.97
5.0		2.88	2.34	2.10	2.00	0.98
5.5		2.55	2.23	2.08	2.00	0.98
6.0		2.19	2.14	2.05	2.00	0.99
6.5		1.87	2.07	2.04	2.00	0.99
7.0		1.64	2.02	2.03	2.00	0.99
7.5		1.55	1.99	2.02	2.00	0.99
8.0		1.58	1.99	2.02	2.00	0.99
8.5		1.71	2.00	2.02	2.00	0.99
9.0		1.91	2.02	2.02	2.00	0.99
9.5		2.12	2.03	2.02	2.00	0.99
10.0		2.29	2.04	2.02	2.00	1.00

Source: van de Hulst (1957).

and, most importantly at ultraviolet and infrared wavelengths,

$$\beta_{ex} = N\pi r^2 Q_{ex} \tag{5-36}$$

As with the volume total coefficient β_m for molecular scattering, the dimension of β is L^{-1}, and customary units are m^{-1} or km^{-1}.

Numerical values of either β_{sc} or β_{ex} are easily found for a given wavelength, and for a monodispersion having a given particle size and concentration, when values of the corresponding efficiency factor are available. Assume that a haze is characterized by particles having an effective mean radius $r_e = 0.20\ \mu$m, a refractive index $n = 1.50$, and a concentration $N = 500\ \text{cm}^{-3}$. At the wavelength 0.55 μm the size parameter has the value 2.28 so that, from Appendix J, Q_{sc} equals 2.19. Then from (5-34) we find that

$$\beta_{sc} = 1.38 \times 10^{-6}\ \text{cm}^{-1} = 0.138\ \text{km}^{-1}$$

The resulting atmospheric transmittance, from (1-35) or (1-50) is

$$\tau = 0.87\ \text{km}^{-1}$$

The meteorological range, from (1-71), is

$$R_m = 28\ \text{km} \approx 15\ \text{nmi}$$

Reference to Table 1.8 discloses that this value of meteorological range corresponds to a *very clear* condition.

At the other extreme, consider a monodisperse cloud of droplets having a refractive index $n = 1.33$, an effective radius $r_e = 10.0\ \mu$m, and a concentration $N = 100\ \text{cm}^{-3}$. At the wavelength 0.55 μm, the absorption is negligible; the size parameter α equals 114; and Q_{sc} has the approximate value 2. For these conditions,

$$\beta_{sc} = 6.3 \times 10^{-4}\ \text{cm}^{-1} = 63\ \text{km}^{-1}$$

The transmittance is

$$\tau = 0.94\ \text{m}^{-1}$$

and the meteorological range is

$$R_m = 0.062\ \text{km} \approx 200\ \text{ft}$$

A concept known as the albedo of single scattering measures the effectiveness of scattering relative to extinction. For either a monodisperse or a polydisperse suspension, this albedo is defined by

$$\bar{\omega} = \frac{\beta_{sc}}{\beta_{ex}} = \frac{\beta_{sc}}{\beta_{sc} + \beta_{ab}} \tag{5-37}$$

or the ratio of the amount of flux scattered to that scattered and absorbed. The albedo of single scattering is much used in studies of cloud radiance. The definition given by (5-37) does not agree with common usage. There the albedo of a "cloud surface" often is taken as the ratio of the amount of flux scattered (whose component in any direction allows the cloud to be seen) to the amount of incident flux.

5.5.6 Published Tabulations of Efficiency Factors

Few aspects of scattering have greater importance or have attracted greater attention than atmospheric total scattering and extinction and the associated matter of visibility. Consequently, many computations of Q_{sc} and a somewhat fewer number of computations of Q_{ex} have been made and published. This latter situation reflects to some extent the generally greater difficulties of the required computations and the narrower interest in attenuation of the infrared. The increasing utilization of these

TABLE 5.6 PUBLISHED TABULATIONS OF EFFICIENCY FACTORS APPLICABLE TO ATMOSPHERIC PARTICLES

Reference	Refractive index*	Size parameter
Holl (1948)	1.33	4.8, 5.4, 6.0, 6.6 (24 values) 8.0
Lowan (1949)	1.33, 1.44, 1.55, 2.00, 1.50	0.5 (15 values) 6.0 0.5 (24 values) 12.0
Houghton and Chalker (1949)	1.33	7 (33 values) 24
Gumprecht and Sliepcevich (1951b)	1.2, 1.33, 1.4, 1.44, 1.5, 1.6	1 (many values) 8, 10(5)100(10)200 (50)400
Johnson et al. (1954)	Complex values, $3.6 < \lambda < 13.5\ \mu m$	1 (many values) 25
Pangonis et al. (1957)	1.05(0.05)1.30	0.2(0.2)7.0(1)15
Havard (1960)	Complex values, $3.6 < \lambda < 13.5\ \mu m$	For particle $r =$ 1, 2, 4, 9, 12, 15 μm
Deirmendjian (1963b)	Complex values, $8.15 < \lambda < 16.6\ \mu m$	0.5(0.5)7.0, 10.0, 15.0
Penndorf (1956, 1957c)	1.33; 1.40, 1.44, 1.486, 1.50	0.1(0.1)30.0
Giese et al. (1961)	1.5	0.2(0.2)159
McCormick (1967)	1.5	0.01(0.01)2.0(0.1) 181.9
Irvine and Pollack (1968)	Complex values, $0.7 < \lambda < 200\ \mu m$	For particle $r =$ 0.3, 1, 3, 10 μm
Zelmanovich and Shifrin (1968)	Various complex values	0.5 (many values) 100

*The complex values of refractive index refer to water.

wavelengths for various military and scientific purposes, however, renders important those computations that have been made and published. In Table 5.6 are listed some of the tabulations of Q_{sc} and Q_{ex} that are available in the literature. In most cases the definitions and symbols of the tabulated quantities agree with those used here. Where they do not, the explanatory notes accompanying the tabulations make the relationships clear. Of all the tabulations listed, those by Zelmanovich and Shifrin (1968) are the most comprehensive; the review of Penndorf (1969) will be found informative. A portion of the tabulations by Penndorf (1956, 1957c) are reproduced in Appendix J, for $n = 1.33$ and 1.50 and $0.1 < \alpha < 30.0$.

6

Mie Scattering by Atmospheric Polydispersions

This chapter treats the applications of Mie theory to the atmosphere, thus providing a real-world view of scattering by haze, fog, and clouds. Although the particle and droplet size distributions of these polydispersions extend across several orders of magnitude, Mie theory is valid for all these scatterers. In applying the theory to these polydispersions, it is necessary to integrate the appropriate functions over the size distributions. The essence of this chapter is the presentation and explanation of functions that have been developed in this manner. We consider first the angular and then the total scattering characteristics of haze aerosols as they are revealed by such functions. Similar treatments are then given for fog and cloud.

6.1 ANGULAR SCATTERING BY HAZE AEROSOLS

The angular scattering characteristics of the haze aerosols always present in the atmosphere have been studied and measured by many workers. This interest arises primarily from the fact that these characteristics govern the spectral and spatial distribution, and polarization, of skylight. They also influence to varying extents the clarity of distant scenes. This section examines these characteristics of hazes in terms of functions that have been developed for both exponential and power-law size distributions. The section concludes with a discussion of angular scattering by the entire haze atmosphere.

6.1.1 Phase Function and Stokes Parameters

In extending the scattering functions to a polydispersion, the size distribution function is substituted for the monodispersion concentration N. Consider first the volume angular coefficient $\beta_p(\theta)$ defined for a monodispersion by (5-15) through (5-17). Omitting for simplicity the subscripts used there, we write

$$\beta_j(\theta) = N\sigma_j(r, \lambda, m, \theta) = N\frac{\lambda^2}{4\pi^2} i_j(r, \lambda, m, \theta) \tag{6-1}$$

where i_j (with $j = 1, 2$) identifies the intensity distribution functions i_1 and i_2 discussed in Section 5.3.1. For a polydispersion, N is defined generally by (3-7). Making this substitution in (6-1) and omitting the dependency indicators r, λ, and m used above, we have

$$\beta_j(\theta) = \frac{\lambda^2}{4\pi^2} \int_{r_1}^{r_2} n(r) i_j(\theta)\, dr \tag{6.2}$$

Similarly, the volume total coefficient defined for a monodispersion by (5-34) becomes

$$\beta_{sc} = \pi \int_{r_1}^{r_2} r^2 Q_{sc} n(r)\, dr \tag{6-3}$$

Because many angular scattering studies are made in terms of the phase function, that concept is reviewed at this point. The phase function $P(\theta)$ for a suspension of particles is the counterpart of the molecular phase function discussed in Section 4.4.4. From the definition there, it is clear that

$$P_j(\theta) = \frac{\beta_j(\theta)}{\beta_{sc}/4\pi} \tag{6-4}$$

A detailed definition is obtained by substituting (6-2) into (6-4), which gives

$$P_j(\theta) = \frac{4\pi}{\beta_{sc}} \frac{\lambda^2}{4\pi^2} \int_{r_1}^{r_2} n(r) i_j(\theta)\, dr \tag{6-5}$$

Equation (6-4) or (6-5) means that numerical values of $\beta_j(\theta)$ can be obtained from tabulated values of $P_j(\theta)$ simply by multiplying them by $\beta_{sc}/4\pi$. This is a useful fact to remember. It is also evident from (6-4) that $P_j(\theta)$ is independent of the concentration N (assuming single scattering), that it is dimensionless, and that it meets the normalization requirement

$$\frac{1}{4\pi} \int_0^{4\pi} P_j(\theta)\, d\omega = 1 \tag{6-6}$$

Frequently scattering coefficients and phase functions are expressed in terms of the particle size parameter instead of the particle radius. The change in variable is easily made. From (5-1) we have

$$dr = \frac{\lambda}{2\pi}\, d\alpha \tag{6-7}$$

so that (6-3) becomes

$$\beta_{sc} = \frac{\pi}{k^3} \int_{\alpha_1}^{\alpha_2} \alpha^2 Q_{sc} n(\alpha)\, d\alpha \tag{6-8}$$

where the propagation constant $k = 2\pi/\lambda$. An analogous expression for β_{ex} is obtained by employing Q_{ex} instead of Q_{sc}. Similarly, (6-5) becomes

$$P_j(\theta) = \frac{4\pi}{k^3 \beta_{sc}} \int_{\alpha_1}^{\alpha_2} n(\alpha) i_j(\theta)\, d\alpha \tag{6-9}$$

Usually any scattering expression for a polydispersion can be treated in a similar manner.

Many studies of haze and cloud scattering have been based on the exponential size distribution defined by (3-15). Here we call attention to the works by Deirmendjian (1963a, 1964, 1969) in which the angular scattering information is expressed by four phase functions forming a scattering matrix which operates on the Stokes vector of the incident light. The resulting transformation gives the Stokes vector—hence all characteristics—of the scattered light. According to Deirmendjian (1969) this transformation depends on only two complex amplitude functions A_1 and A_2 from Mie theory. These functions are simply related to the complex amplitudes S_1 and S_2 defined by (5-4) and (5-5), according to

$$\begin{aligned} kA_1 &= S_1 \\ kA_2 &= S_2 \end{aligned} \tag{6-10}$$

The elements of the matrix are formed by extending the intensity distribution functions and cross sections discussed in Sections 5.3.1 and 5.4.1 to include

$$\begin{aligned} i_1(\theta)k^{-2} &= \sigma_1(\theta) = A_1 A_1^* \\ i_2(\theta)k^{-2} &= \sigma_2(\theta) = A_2 A_2^* \\ i_3(\theta)k^{-2} &= \sigma_3(\theta) = \mathrm{Re}\,\{A_1 A_2^*\} \\ i_4(\theta)k^{-2} &= \sigma_4(\theta) = -\mathrm{Im}\,\{A_2 A_1^*\} \end{aligned} \tag{6-11}$$

with the dependences on α and m understood.

The scattering transformation of the Stokes vector is found by means of the Mueller calculus described by Mueller (1948), Shurcliff (1962), and Shurcliff and Ballard (1964). For a single isotropic particle the representation is

$$\begin{bmatrix} I_s \\ Q_s \\ U_s \\ V_s \end{bmatrix} = \begin{bmatrix} \frac{1}{2}[\sigma_1(\theta)+\sigma_2(\theta)] & \frac{1}{2}[\sigma_1(\theta)-\sigma_2(\theta)] & 0 & 0 \\ \frac{1}{2}[\sigma_1(\theta)-\sigma_2(\theta)] & \frac{1}{2}[\sigma_1(\theta)+\sigma_2(\theta)] & 0 & 0 \\ 0 & 0 & \sigma_3(\theta) & \sigma_4(\theta) \\ 0 & 0 & -\sigma_4(\theta) & \sigma_3(\theta) \end{bmatrix} \begin{bmatrix} I_0 \\ Q_0 \\ U_0 \\ V_0 \end{bmatrix} \tag{6-12}$$

As indicated by the subscripts, the column of parameters on the left side is the Stokes vector of the scattered light, while the column on the right refers to the incident light. By the usual rules of matrix multiplication, the product of the operations is

$$\begin{aligned} I_s &= \tfrac{1}{2}I_0[\sigma_1(\theta)+\sigma_2(\theta)]+\tfrac{1}{2}Q_0[\sigma_1(\theta)-\sigma_2(\theta)] \\ Q_s &= \tfrac{1}{2}I_0[\sigma_1(\theta)-\sigma_2(\theta)]+\tfrac{1}{2}Q_0[\sigma_1(\theta)+\sigma_2(\theta)] \\ U_s &= U_0\sigma_3(\theta)+V_0\sigma_4(\theta) \\ V_s &= -U_0\sigma_4(\theta)+V_0\sigma_3(\theta) \end{aligned} \tag{6-13}$$

These relationships also apply to a monodisperse suspension of isotropic particles when the equations are multiplied by the concentration N.

The results of the transformation as shown in (6-13) are easily verified for common cases of scattering. As an example, when the incident light is unpolarized so that the normalized parameter $I_0 = 1$ and the remaining parameters are zero, we see that

$$\begin{aligned} I_s &= \tfrac{1}{2}[\sigma_1(\theta)+\sigma_2(\theta)] \\ Q_s &= \tfrac{1}{2}[\sigma_1(\theta)-\sigma_2(\theta)] \\ U_s &= 0 \\ V_s &= 0 \end{aligned} \tag{6-14}$$

Thus the scattered light has partial linear polarization directed either normally or parallel to the plane of observation accordingly as Q_s is positive or negative. When $\sigma_1(\theta) = \sigma_2(\theta)$, which occurs at $\theta = 0$ and 180 deg, the scattered light is completely unpolarized. Similarly, when the incident light has complete vertical polarization, with the parameters having the following values,

$$\begin{aligned} I_0 &= 1 \\ Q_0 &= 1 \\ U_0 &= 0 \\ V_0 &= 0 \end{aligned}$$

the parameters of the scattered light are

$$\begin{aligned} I_s &= \sigma_1(\theta) \\ Q_s &= \sigma_1(\theta) \\ U_s &= 0 \\ V_s &= 0 \end{aligned} \tag{6-15}$$

In this case the scattered light has complete linear polarization directed normally to the plane of observation at all values of θ. There is no scattered component parallel to this plane so long as the particles are isotropic. This result agrees with the statement following (5-6).

For a *polydispersion,* the scattering matrix must be written in terms of the volume angular scattering coefficient $\beta_j(\theta)$ defined by (6-2), but with $j = 1, 2, 3, 4$. As may be seen from (6-4), the coefficient is related to $P_j(\theta)$ by

$$\beta_j(\theta) = \frac{\beta_{sc}}{4\pi} P_j(\theta)$$

Analogously to (6-12), the scattering transformation of the Stokes vector is represented by

$$\begin{bmatrix} I_s \\ Q_s \\ U_s \\ V_s \end{bmatrix} = \frac{\beta_{sc}}{4\pi} \begin{bmatrix} \frac{1}{2}(P_1+P_2) & \frac{1}{2}(P_1-P_2) & 0 & 0 \\ \frac{1}{2}(P_1-P_2) & \frac{1}{2}(P_1+P_2) & 0 & 0 \\ 0 & 0 & P_3 & P_4 \\ 0 & 0 & -P_4 & P_3 \end{bmatrix} \begin{bmatrix} I_0 \\ Q_0 \\ U_0 \\ V_0 \end{bmatrix} \tag{6-16}$$

where for simplicity (θ) has been omitted from the phase function symbol. An instructive example of using the Mueller matrix to evaluate the scattering characteristics of a polydispersion is given by Holland and Gagne (1970).

6.1.2 Polarization Considered Further

From (6-9) and the meaning of $j = 1, 2$ it is clear that $P_1(\theta)$ and $P_2(\theta)$ refer to the scattered light components polarized perpendicular and parallel to the plane of observation. Hence from (5-10) the degree of polarization of this light is

$$\mathrm{P} = \frac{P_1(\theta) - P_2(\theta)}{P_1(\theta) + P_2(\theta)} \tag{6-17}$$

which is positive for $P_1(\theta)$ greater than $P_2(\theta)$, similarly to (4-39). In the matrix multiplication, $P_3(\theta)$ and $P_4(\theta)$ have to do with rotation of the polarization plane and the introduction of elliptical polarization when the incident light is linearly polarized. In such a case, if the plane of incident polarization lies at other than 0 or 180 deg to the plane of observation, the scattered light has varying amounts of elliptical polarization. The physical basis for this is noted in the statements following (5-8), and the analytical basis in the last two lines of (6-11). Harris (1972) has computed the polarization parameters for several aerosol size distributions found in the Los Angeles region. He shows many plots of the ratio of the perpendicular to the parallel component, the polarization in terms of the Stokes parameters, the ellipticity of the polarization ellipse, and the inclination angle of the ellipse major axis.

Few measurements have been made of the elliptical polarization of atmospherically scattered light. Such measurements can provide, at least in principle, information on the complex refractive index of the particles and may provide further insights into the nature of the aerosol. Elliptical polarization can be produced to a significant degree in laboratory experiments with special aerosols, as in the work by Holland and Gagne (1970). However, it is present only weakly in natural skylight, where the ambient incident light is mostly unpolarized. Eiden (1966) employed linearly polarized light to investigate the elliptical polarization of haze scattering. The increasing application of lasers to atmospheric optics problems may bring about greater interest in this subject which is still largely unexplored.

Deirmendjian (1969) has computed the four phase functions for the polydispersions listed in Table 3.2, and earlier information for a few cases can be found in Deirmendjian (1963a, 1964). The tabulations resulting from his computations for haze models are described in Section 6.2.4. One example of $P_1(\theta)$ and $P_2(\theta)$ for the haze model L is shown in Figure 6.1. Also shown for comparison are the Rayleigh phase functions for molecular scattering. At observation angles less than 40 deg the Mie components easily dominate the Rayleigh components and create the familiar solar aureole at small values of θ. The degree of polarization defined by (6-17) attains significant values over a broad angular range, but never reaches the value unity, as does molecular scattering at 90 deg. Deirmendjian (1963a, 1964) notes that the quantity

$$\frac{P_1(\theta)+P_2(\theta)}{8\pi}$$

at about $\theta = 40$ deg tends to remain constant regardless of the size distribution. This suggests that, in a polydispersion where the multiple scattering is small enough to be ignored, the particle concentration is directly proportional to the scattered intensity at this angle, *provided* the size distribution is of the type (3-15).

The warning sounded in Section 1.4.4 regarding the effect of forward scattering, that is, the *diffuse transmission*, applies to transmission measurements of haze. The effect is greater with haze particles than with molecules, as may be appreciated from the phase functions shown in Figure 6.1. Typically the haze scattering within a few degrees of the forward direction exceeds the molecular scattering by factors of 10 and more. Instruments for measuring direct transmission therefore should

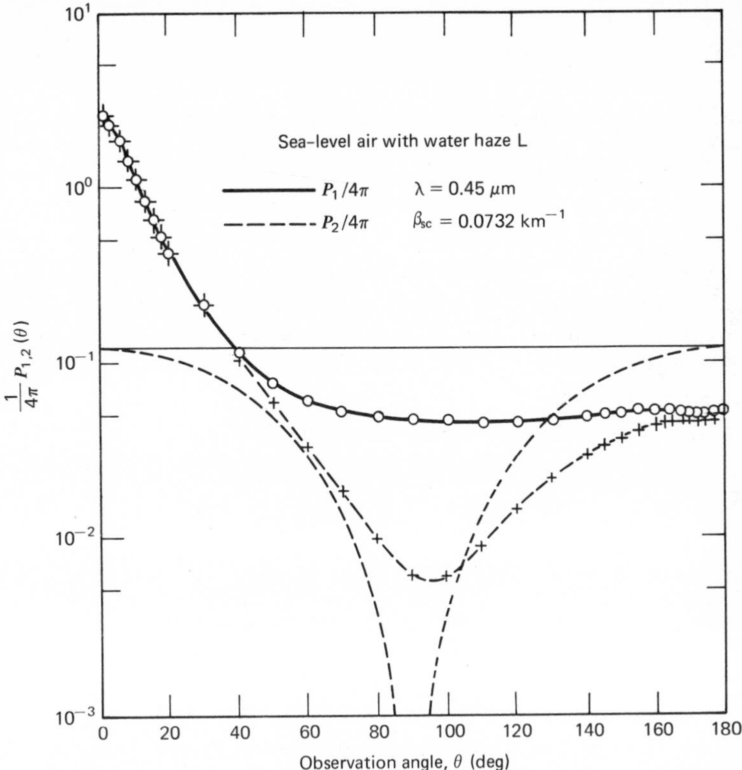

Figure 6.1 Normalized phase functions for haze model L of Figure 3.8. The corresponding phase functions for molecular scattering are shown by the thin traces for comparison. From Deirmendjian (1969).

have beam divergence angles no larger than necessary. Corrections may still be required for high accuracy.

6.1.3 Functions for the Power-Law Size Distribution

The power-law size distribution (3-20) has become almost basic in aerosol research. Its application to atmospheric scattering is developed in detail by Bullrich (1964), and the essentials of his treatment are summarized here. The volume angular coefficient for a monodispersion illuminated by unpolarized light can be written from (5-17) as

$$\beta(\theta) = N \frac{\lambda^2}{4\pi^2} i_t \tag{6-18}$$

where i_t represents the sum of the intensity distribution functions according to

$$i_t = \frac{i_1 + i_2}{2}$$

Equation (6-18) is extended to a polydispersion by substituting for N its value from (3-21), giving

$$\beta(\theta) = 0.434c \int_{r_1}^{r_2} \frac{\lambda^2}{4\pi} \frac{i_t}{r^{v+1}}\, dr \qquad (6\text{-}19)$$

where the value of c depends on the particle concentration. The value of v lies between 2.5 and 4.0 for most haze aerosols; smaller values mean a larger number of larger particles, as may be understood from (3-20).

As with the exponential size distribution, it is desirable to change the independent variable of the distribution from the particle radius to the size parameter α. From (5-1) we can write (6-19) as

$$\beta(\theta) = 0.434c \int_{r_1}^{r_2} \frac{i_t}{\alpha^2(\alpha\lambda/2\pi)^{v-1}}\, dr \qquad (6\text{-}20)$$

Substituting for dr its value from (6-7) gives

$$\beta(\theta) = 0.434c \left(\frac{2\pi}{\lambda}\right)^{v-2} \int_{\alpha_1}^{\alpha_2} \frac{i_t}{\alpha^{v+1}}\, d\alpha \qquad (6\text{-}21)$$

Since the dimension of c is L^{v-3}, as is evident from (3-20), and the integrand contains only dimensionless quantities, the dimension of $\beta(\theta)$ is L^{-1}.

Alternatively, angular scattering can be expressed in terms of the meteorological (or visual) range R_m instead of the size distribution constant c. The desired expression is obtained from (6-21) by substituting for c its value from (6-31), and for β_{sc} (thus introduced) its value from (6-34), where it is denoted by β_p to distinguish it from the molecular total scattering coefficient β_m. The result of these operations is

$$\beta(\theta) = \frac{1}{\pi K(0.55\ \mu\text{m})} \left[\frac{3.91}{R_m} - \beta_m\right] \left(\frac{0.55\ \mu\text{m}}{\lambda(\mu\text{m})}\right)^{v-2} \int_{\alpha_1}^{\alpha_2} \frac{i_t}{\alpha^{v-1}}\, d\alpha \qquad (6\text{-}22)$$

in which the factor K is defined by (6-29). Since R_m refers to the wavelength 0.55 μm, the value of β_m at this wavelength should be used. The first term in parentheses is, from (1-71), the *atmospheric* total scattering coefficient for haze particles plus air molecules. The indicated subtraction thus makes this expression equal to the total coefficient β_p for haze only.

The above expressions can be shortened by employing an angular function η for the integrand. This function is defined by

$$\eta(\alpha, m, \theta) = \int_{\alpha_1}^{\alpha_2} \frac{i_1 + i_2}{\alpha^{v+1}} \, d\alpha \tag{6-23}$$

Equation (6-21) thus becomes

$$\beta(\theta) = 0.434c \left(\frac{2\pi}{\lambda}\right)^{v-2} \frac{1}{2} \eta(\theta) \tag{6-24}$$

and (6-22) becomes

$$\beta(\theta) = \frac{1}{\pi} \left[\frac{3.91}{R_m} - \beta_m\right] \left(\frac{0.55 \ \mu\text{m}}{\lambda}\right)^{v-2} \frac{\eta(\theta)}{2K(0.55 \ \mu\text{m})} \tag{6-25}$$

Values of the angular function η are listed in Table 6.1 for various wavelengths and size distributions.

Several of the functions derived above have been employed by de Bary et al. (1965) in computations of angular scattering by the entire vertical thickness of the haze atmosphere. Their tabulations, from which $\beta(\theta)$ in km^{-1} can be obtained for many combinations of size distribution parameters, are described in Section 6.2.4. As part of his lidar studies, McCormick (1967) also employed these functions to compute the quantities K and $\eta(\theta)$ at $\theta = 180$ deg. His tabulations likewise are described in Section 6.2.4. Rensch and Long (1970) computed β (180 deg), often denoted by $\beta(\pi)$ in the literature, over the wavelength range 0.63 to 10.6 μm for model aerosols characterized by a power-law size distribution. Their values are plotted in Figure 6.2*a* for comparison with corresponding plots of β_{ex} in Figure 6.2*b*. McCormick et al. (1968) tabulated the average ratio $\beta(\pi)/\beta_{sc}$ for power-law aerosols having homogeneous spherical particles and a real refractive index of 1.50. Harrison et al. (1972) extended the computations to layered (onion) spherical particles having complex refractive indices.

We note the strong dependence of $\beta(\theta)$ on the altitude Z. As explained in Section 3.3.2, the size distributions of haze particles tend to remain unchanged with altitude, while the concentration decreases exponentially according to (3-24). A scale height $H_p = 1.25$ km has been used with (3-24) in many studies and tabulations of scattering functions. Table 3.3 lists additional values of H_p for various values of the surface meteorological range in a model haze atmosphere. Because $\beta(\theta)$ is directly proportional to particle concentration, its value at any altitude in such a model is readily found. We have from (3-24) that

$$\beta(\theta)_Z = \beta(\theta)_0 \exp - \frac{Z}{H_p} \tag{6-26}$$

where the subscript 0 indicates a selected reference altitude.

TABLE 6.1 VALUES OF THE ANGULAR FUNCTION $\eta(\alpha, m, \theta)$ FOR HAZE AEROSOLS CHARACTERIZED BY A POWER-LAW SIZE DISTRIBUTION HAVING $r_1 = 0.04\ \mu m$ and $r_2 = 10\ \mu m$ AND VALUES OF v AS SHOWN FOR WAVELENGTHS FROM 0.4 TO 1.2 μm

λ (μm) \ θ (deg)	1	4	7	10	30	60	80	90	110	120	130	150	180
						$v = 2.5$							
0.4	374.66	64.72	33.92	23.05	6.126	1.414	0.6266	0.4511	0.2816	0.2570	0.2626	0.4282	1.1831
0.45	360.50	64.63	33.85	23.01	6.108	1.412	0.6270	0.4517	0.2831	0.2586	0.2645	0.4306	1.1745
0.55	324.05	64.16	33.71	22.91	6.072	1.405	0.6263	0.4514	0.2843	0.2600	0.2662	0.4323	1.1613
0.65	286.09	63.92	33.62	22.83	6.038	1.398	0.6247	0.4503	0.2845	0.2604	0.2667	0.4329	1.1472
0.85	223.45	62.89	33.38	22.67	5.974	1.384	0.6209	0.4477	0.2841	0.2601	0.2666	0.4324	1.1243
1	188.78	61.96	33.12	22.54	5.930	1.374	0.6177	0.4455	0.2835	0.2596	0.2662	0.4316	1.1095
1.2	153.68	61.67	32.96	22.40	5.877	1.361	0.6133	0.4427	0.2826	0.2589	0.2656	0.4301	1.0873
						$v = 3$							
0.4	59.32	19.81	12.97	9.844	3.102	0.7520	0.3435	0.2505	0.1642	0.1514	0.1536	0.2151	0.4369
0.45	58.16	19.81	12.97	9.844	3.105	0.7545	0.3457	0.2526	0.1664	0.1538	0.1563	0.2183	0.4397
0.55	54.91	19.77	12.96	9.841	3.106	0.7568	0.3479	0.2547	0.1688	0.1565	0.1592	0.2219	0.4423
0.65	51.21	19.75	12.95	9.835	3.104	0.7575	0.3489	0.2556	0.1699	0.1577	0.1607	0.2236	0.4428
0.85	44.40	19.64	12.93	9.819	3.099	0.7570	0.3493	0.2562	0.1708	0.1588	0.1618	0.2250	0.4420
1	40.20	19.53	12.90	9.804	3.094	0.7562	0.3492	0.2562	0.1711	0.1591	0.1622	0.2254	0.4407
1.2	35.58	19.49	12.88	9.786	3.087	0.7547	0.3488	0.2560	0.1711	0.1592	0.1624	0.2255	0.4381

						$v = 3.5$							
0.4	12.93	7.397	5.730	4.745	1.790	0.4741	0.2243	0.1660	0.1127	0.1049	0.1058	0.1322	0.2111
0.45	12.84	7.402	5.736	4.751	1.796	0.4779	0.2272	0.1687	0.1156	0.1081	0.1093	0.1364	0.2157
0.55	12.56	7.406	5.742	4.758	1.802	0.4821	0.2306	0.1719	0.1190	0.1119	0.1135	0.1414	0.2213
0.65	12.20	7.407	5.745	4.760	1.805	0.4842	0.2323	0.1736	0.1208	0.1139	0.1157	0.1442	0.2243
0.85	11.46	7.398	5.746	5.762	1.807	0.4861	0.2339	0.1752	0.1225	0.1158	0.1179	0.1468	0.2270
1	10.96	7.385	5.743	4.761	1.808	0.4867	0.2345	0.1757	0.1231	0.1165	0.1187	0.1477	0.2279
1.2	10.35	7.381	5.741	4.759	1.807	0.4869	0.2348	0.1760	0.1236	0.1170	0.1192	0.1484	0.2283
						$v = 4$							
0.4	4.192	3.311	2.860	2.518	1.122	0.3324	0.1644	0.1239	0.08734	0.08242	0.08313	0.09701	0.1307
0.45	4.193	3.319	2.868	2.527	1.129	0.3373	0.1682	0.1275	0.09111	0.08652	0.08767	0.1024	0.1368
0.55	4.177	3.329	2.878	2.536	1.138	0.3433	0.1730	0.1320	0.09591	0.09179	0.09352	0.1095	0.1447
0.65	4.147	3.335	2.883	2.542	1.142	0.3467	0.1757	0.1346	0.09871	0.09488	0.09697	0.1137	0.1495
0.85	4.072	3.339	2.888	2.547	1.147	0.3502	0.1785	0.1373	0.10167	0.09816	0.10065	0.1182	0.1566
1	4.012	3.340	2.891	2.549	1.149	0.3515	0.1796	0.1384	0.10283	0.09945	0.10211	0.1200	0.1566
1.2	3.934	3.341	2.892	2.551	1.150	0.3526	0.1804	0.1392	0.10376	0.10048	0.10328	0.1215	0.1518

Source: Bullrich (1964).

6.1.4 Angular Scattering by the Haze Atmosphere

The angular scattering characteristics of the haze atmosphere are exhibited in the intensity, color, and polarization of the light from the daytime sky. These characteristics are governed chiefly by the particle size distribution, by the variation in particle concentration with altitude, and by functions such as those discussed in Sections 6.1.1 and 6.1.3. Additional but not usually controlling factors are the optical thickness of the atmosphere, the spectral reflectance of the ground, and the extent of multiple scattering. Readers will find that the work by Ivanov et al. (1969) is both a detailed treatment of many technical aspects of haze scattering and an excellent guide to the literature of the subject. In this section we survey several representative studies of angular scattering by haze. Several of these have produced tabulated values of various scattering functions, as described in Section 6.2.4.

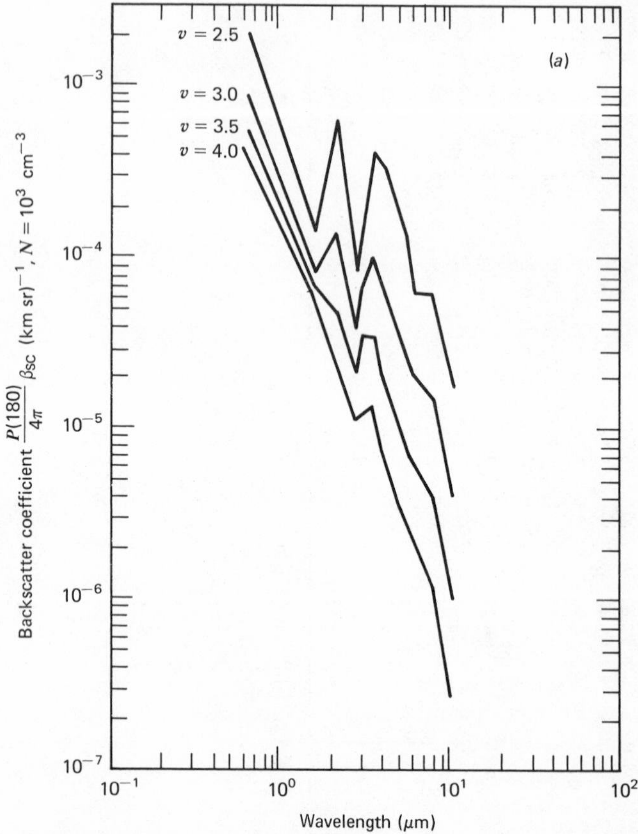

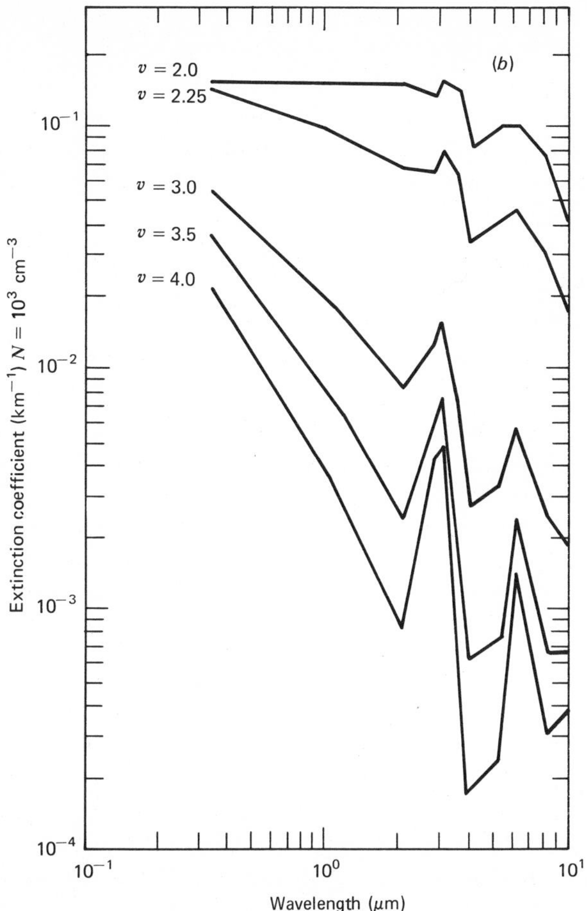

Figure 6.2 Backscattering (a) and extinction (b) coefficients for a continental aerosol model conforming to a power-law size distribution having values of the exponent ν as indicated on the figure. From Rensch and Long (1970).

An outstanding investigation is that by Bullrich (1964), who presents and compares much theoretical and experimental data. In his studies the haze particles are assumed to conform to the power-law size distribution, with $2.5 \leq v \leq 4.0$. The wavelengths employed are in the visual and very near infrared regions. Many charts showing the distribution of skylight intensity and polarization in the sun's vertical plane, as well as over the sky hemisphere, are given. Frequent comparisons with Rayleigh scattering are made for illustration. Although the computations deal only with single scattering, the effects of multiple scattering are estimated and discussed.

The numerical results of the computations appear in the tabulations by de Bary et al. (1965).

As an extension of the foregoing study, de Bary (1964) considers further the matter of multiple scattering and describes a semiempirical method for its evaluation. The essentials of the method are noted in the following discussion. First, it is evident that sky light has constituents from three processes:

1. Single and all higher orders of scattering by the molecular or Rayleigh atmosphere.
2. Single scattering by the haze atmosphere.
3. All higher-order scattering by the haze atmosphere.

The first constituent can be found from the tabulations by Coulson et al. (1960) for the molecular optical thickness T'_m corresponding to the given wavelength and for an estimated value of the ground reflectance. The second constituent can be found from the tabulations by de Bary et al. (1965) for estimated atmospheric turbidity at the given wavelength. The third constituent can be evaluated according to the following discussion.

It is well established that the number of small haze particles, having a radius less than 0.08 μm for example, is far greater than the number of larger particles, and it is assumed that the scattering likewise is greater. These small particles scatter (approximately) according to (4-38). The higher-order scattering by these particles for an optical thickness T'_p is equal to that of a Rayleigh atmosphere having $\mathrm{T}'_m = \mathrm{T}'_p$. These quantities and the turbidity T are related by (4-63). The higher-order scattering is then evaluated by subtracting the primary molecular scattering, which can be calculated with the volume angular coefficient given in Appendix H, from the value in Coulson et al. (1960) for all orders of molecular scattering. Such a decomposition of the total radiance is represented by

$$L_{\text{total}} = L_{msR} + L_{pspT} + L_{msR\mathrm{T}} - L_{psR\mathrm{T}}$$

where the subscripts have the following meanings:

msR: Multiple scattering (all orders) in a Rayleigh atmosphere.
$pspT$: Primary scattering in a haze atmosphere of turbidity T.
$msR\mathrm{T}$: Multiple scattering (all orders) in a Rayleigh atmosphere of optical thickness $\mathrm{T}'_m = \mathrm{T}'_p$ corresponding to the value of T.
$psR\mathrm{T}$: Primary scattering in a Rayleigh atmosphere having $\mathrm{T}'_m = \mathrm{T}'_p$.

Figure 6.3 shows values of the sky radiance computed from these relationships and the referenced tables, along with measured values. The visual range at the time was about 25 km, and the measurements justified the use of a turbidity factor $T = 2$. The measured values fall nicely

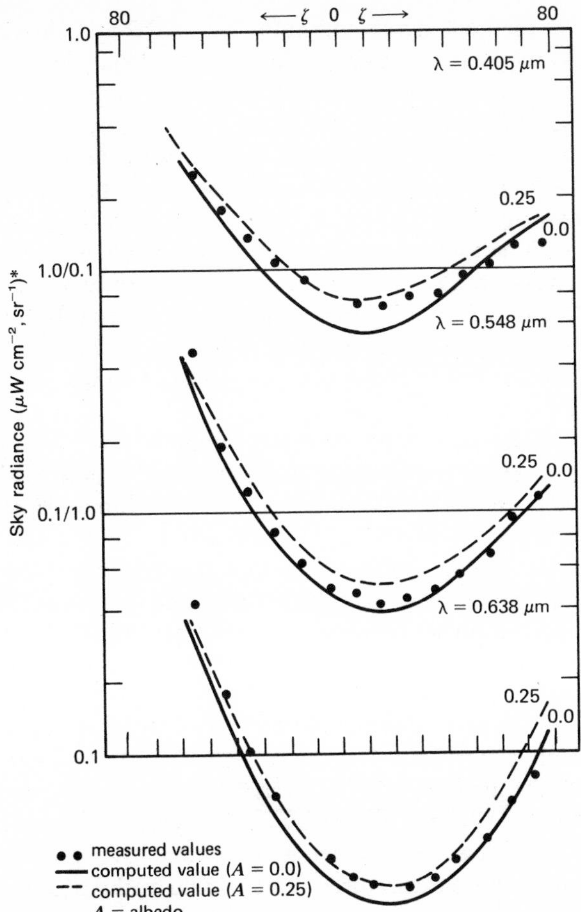

Figure 6.3 Comparison of computed and measured sky radiance versus zenith angle, in the plane of the sun's vertical, at three wavelengths for two values of the surface albedo. The zenith angle of the sun was 78 deg. From de Bary (1964).

*Spectral bandwidth not given.

between the values computed for a reflectance (albedo) of 0 and of 0.25. The immediate terrain at the measurement site consisted of grass, trees, and building roofs—all of low reflectance. The agreements seen in the figure indicate that the method accounts for multiple scattering in haze fairly well. Many similar comparisons of radiances and polarizations are given in the referenced paper. In a later work de Bary (1972) compares the results from this approximate method with rigorous calculations of multiple scattering and shows that at wavelengths of 0.55 and 0.65 μm and

$T = 4$ the two methods agree within factors of 1.2 and closer. At shorter wavelengths and greater turbidities, the agreements are less satisfactory.

Astronomers have studied multiple scattering by haze atmospheres in evaluating radiometric data of the planets. Generally the studies follow the approach developed by Chandrasekhar (1950) for a Rayleigh atmosphere. For a haze atmosphere, simplifying assumptions are employed almost necessarily. To cite but one factor, it is seen from Figure 6.1 that aerosol phase functions present greater analytic difficulties than Rayleigh functions. Thus van de Hulst (1948) in an early study assumed isotropic scattering (phase function invariant with angle) in order to bring out most clearly the essential nature of multiple scattering problems. Later studies are based on anisotropic phase functions, but of simplified form so that they can be expressed analytically.

Several workers have employed the *doubling principle* described by van de Hulst (1963). Studies along these lines, applicable to cloud as well as haze, have been made by Irvine (1965b, 1968), van de Hulst and Irvine (1963), and Hansen (1969, 1971a,b), to name but a few. Also, Dave (1970) and Dave and Gazdag (1970) show that the computer time ordinarily needed for multiple-scattering calculations can be reduced by representing the normalized phase function by a modified Fourier series. Readers wishing to pursue the analytic study of multiple scattering will find all the above-referenced papers helpful.

Multiple scattering is also treated by the Monte Carlo method. In this method, which can employ the exact Mie function having the strong forward lobe, the incident and scattered fluxes are considered photons. At each encounter between a photon and a scatterer, the scattering angle is selected by a random process from the cumulative distribution of the scattering function, and the three-dimensional paths of the photons are tracked through successive encounters. The method can be used either with a molecular or a haze atmosphere, or with both concurrently. It is applicable to any vertical distribution of scatterers, so it is well suited to nonhomogeneities as dissimilar as exponential atmospheres and cloud layers. Since individual photons and random events are dealt with, many case histories must be compiled to avoid large statistical fluctuations in the computed radiance. In general this requires a large amount of computer time, but this disadvantage may be compensated for by the great versatility of the method. Monte Carlo calculations of the radiance and polarization of skylight have been made by Collins (1968), Collins et al. (1972), and Plass and Kattawar (1972a,b).

6.2 TOTAL SCATTERING AND EXTINCTION BY HAZE AEROSOLS

Because of their relationship to atmospheric transmission and visibility, the total scattering and extinction characteristics of haze have been investigated thoroughly. This section presents and discusses the applicable functions, emphasizing particularly the wavelength dependence of attenuation. This dependence is compared with the inverse fourth-power relationship that obtains for Rayleigh scattering. The concepts of optical thickness and turbidity of the haze atmosphere, considered as a totality, are then described. Attention is called to the many published tabulations of scattering functions for haze. Readers wishing to pursue further the subject matter of this section, particularly with respect to the scattering efficiency factor as a parameter, may profitably consult the treatments by Penndorf (1965).

6.2.1 Functions for the Power-Law Size Distribution

The extension of the volume total scattering coefficient, defined for a monodispersion by (5-34), to a polydispersion requires integration over the selected size distribution. The integration may be performed analytically by computer or by numerical and graphical methods. The general form of the expression to be integrated is given by (6-3) or by (6-8). For the exponential size distribution (3-15), Deirmendjian (1969) has employed (6-8) in his computations of β_{sc}, β_{ab}, and β_{ex}. For the power-law distribution the expressions developed by Bullrich (1964) for β_{sc} are presented in the following paragraphs. A numerical method of finding β_{sc} for any given size distribution is then described. Finally the dependence of β_{sc} on altitude is discussed.

The basic expression for β_{sc} is given by (5-34). Substituting for N its value from (3-21) gives

$$\beta_{sc} = 0.434c\pi \int_{r_1}^{r_2} Q_{sc} r^2 \frac{1}{r^{v+1}}\, dr \tag{6-27}$$

where the dependency indicators r, λ, and m have been omitted. As in the previous integrations over a size distribution, it is convenient to change the variable from r to α. Substitutions from (5-1) and (6-7) into (6-27) yield

$$\beta_{sc} = 0.434c\pi \left(\frac{2\pi}{\lambda}\right)^{v-2} \int_{\alpha_1}^{\alpha_2} \frac{Q_{sc}}{\alpha^{v-1}}\, d\alpha \tag{6-28}$$

This expression is shortened by letting

$$K = \int_{\alpha}^{\alpha_2} \frac{Q_{sc}}{\alpha^{v-1}}\, d\alpha \tag{6-29}$$

Equation (6-28) then becomes

$$\beta_{sc} = 0.434c\pi \left(\frac{2\pi}{\lambda}\right)^{v-2} K \tag{6-30}$$

It is helpful to consider the concentration factor c of the power-law distribution a little further. From (6-30) this factor is defined by

$$c = 2.304 \left(\frac{\lambda}{2\pi}\right)^{v-2} \frac{\beta_{sc}}{\pi K} \tag{6-31}$$

The role of c in the distribution function may be seen from (3-20), where it is explained that the dimensions of c are L^{v-3}, and that its value depends on the particle concentration. Hence c is closely related to the observable scattering properties of an aerosol, that is, to the meteorological range R_m and the atmospheric turbidity T. Table 6.2 lists values of c for $v = 2.5$ and 4.0, and for various values of R_m and T, along with corresponding values of β_{sc} and β_p. The difference in the values of these two coefficients is, from (1-38), just equal to β_m which has the constant value 0.0124 for all the meteorological conditions listed in the table. The values of c correspond to r expressed in centimeters.

A definition of β_{sc} in terms of R_m can be derived as an alternative to (6-30). Rewriting (1-71) in the light of (1-38), we have

$$\beta_p\ (0.55\ \mu\text{m}) = \frac{3.91}{R_m\ (0.55\ \mu\text{m})} - \beta_m\ (0.55\ \mu\text{m}) \tag{6-32}$$

where the subtraction removes the effect of molecular scattering. The parenthetical terms are reminders that the concept of meteorological range refers to a wavelength of 0.55 μm. The sense of (6-32) is now extended to other wavelengths. From (6-30) it is easy to show that

$$\beta_p\ (\lambda) = \beta_p\ (0.55\ \mu\text{m}) \left[\frac{0.55\ \mu\text{m}}{\lambda\ (\mu\text{m})}\right]^{v-2} \frac{K\ (\lambda)}{K\ (0.55\ \mu\text{m})} \tag{6-33}$$

where the values at 0.55 μm serve as reference levels. Substitution for β_p (0.55 μm) from (6-32) gives

$$\beta_p\ (\lambda) = \left[\frac{3.91}{R_m\ (0.55\ \mu\text{m})} - \beta_m\ (0.55\ \mu\text{m})\right]\left(\frac{0.55\ \mu\text{m}}{\lambda\ (\mu\text{m})}\right)^{v-2} \frac{K(\lambda)}{K\ (0.55\ \mu\text{m})} \tag{6-34}$$

which is useful when values of R_m can be measured or estimated.

The coefficient β can also be found by numerical methods for a given size distribution. From (3-6) the number of particles having radii between r and $r + dr$ is

$$dN = n(r)\ dr = N_i$$

TABLE 6.2 CONVERSION FACTORS BETWEEN METEOROLOGICAL RANGE, ATMOSPHERIC TURBIDITY, AND PARAMETERS OF THE POWER-LAW SIZE DISTRIBUTION HAVING $r_1 = 0.04\ \mu m$ AND $r_2 = 10\ \mu m$ AT $\lambda = 0.55\ \mu m$.

	When meteorological range R_m is given:							
R_m (km)	40	20	15	10	5	2	1	0.5
β_{sc} (km^{-1})	0.0978	0.1956	0.2608	0.3912	0.7823	1.9559	3.9117	7.8234
β_p (km^{-1})	0.0854	0.1832	0.2484	0.3788	0.7700	1.9435	3.8993	7.8110
$10^9 c$ ($v = 2.5$)	0.481	1.032	1.398	2.133	4.335	10.987	21.954	43.977
$10^{16} c$ ($v = 4$)	0.708	1.519	2.059	3.140	6.383	16.110	32.333	64.748

	When atmospheric turbidity T is given:						
T	2	3	4	5	6	10	15
$T_p + T_m$	0.1975	0.2962	0.3950	0.4937	0.5924	0.9874	1.4811
T_p	0.0987	0.1975	0.2962	0.3950	0.4937	0.8887	0.3824
$10^9 c$ ($v = 2.5$)	0.445	0.889	1.334	1.779	2.224	4.003	6.226
$10^{16} c$ ($v = 4$)	0.665	1.310	1.964	2.619	3.274	5.893	9.167

Source: Bullrich (1964).

where N_i is the number within the size class whose width is dr. Then β is given by a summation over i size classes:

$$\beta = \sum_{i=0} N_i Q_i \pi r^2 \tag{6-35}$$

where Q_i is an average value for each size class. The values listed in Appendix J may be used in computations of this type. Evidently the accuracy and the labor increase as the size classes are made narrower. Figure 6.4*a* shows β_{sc} as computed by Pueschel and Noll (1967) for haze having a particle size distribution obeying

$$n(r) \propto n^{-3}$$

as shown in Figure 6.4b. The coefficient for the entire size range is proportional to the area under the curve. It is apparent that the total scattering for this distribution, which is very common in hazes, is governed mostly by particles in the size range 0.2 to 0.4 μm.

The total coefficient, whether β_{sc} or β_{ex} has the same dependence on altitude Z as does the angular coefficient $\beta(\theta)$. On the average it decreases exponentially as altitude increases, correspondingly to the

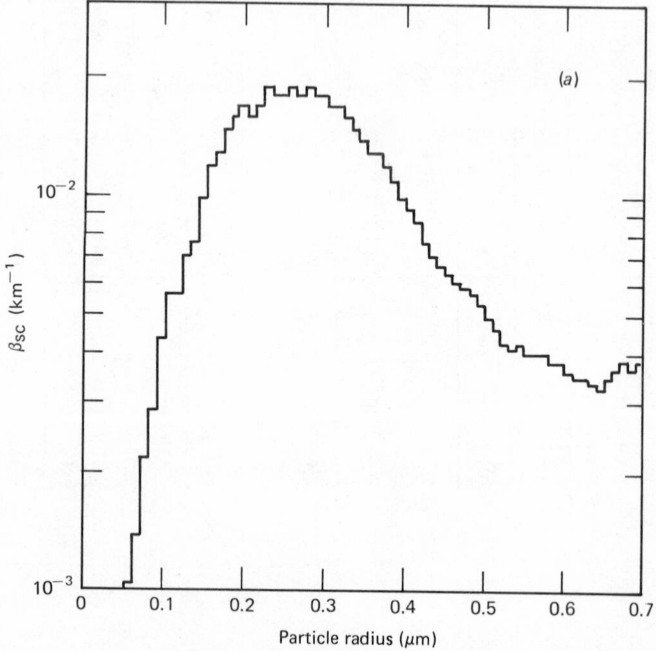

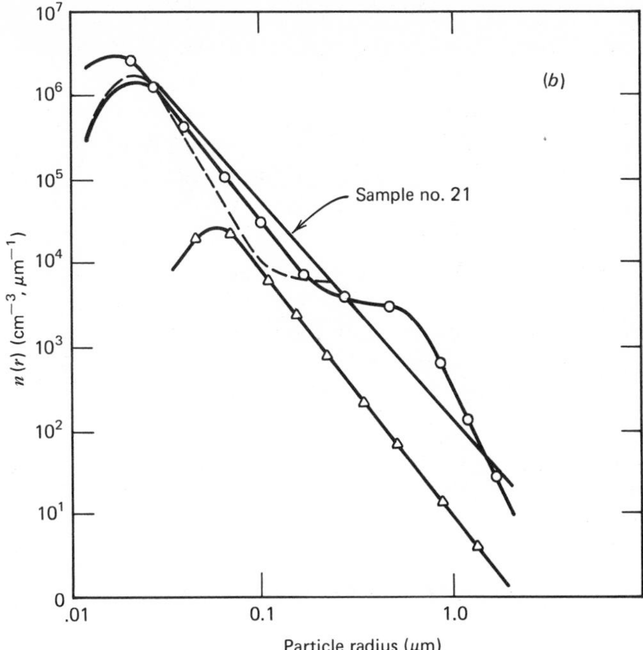

Figure 6.4 Total scattering coefficient for a sampled size distribution of a haze aerosol in the Seattle region. In (*a*) is shown the coefficient, computed from an expression equivalent to (6-35) for sample no. 21 at $\lambda = 0.55\ \mu$m. In (*b*) are shown four of the measured distributions. From Pueschel and Noll (1967).

decrease of particle concentration expressed by (3-24). Thus we write

$$(\beta_{sc} \text{ or } \beta_{ex})_Z = (\beta_{sc} \text{ or } \beta_{ex})_0 \exp\left(-\frac{Z}{H_p}\right) \tag{6-36}$$

where the subscript 0 indicates a selected reference altitude.

6.2.2 Wavelength Dependence of Attenuation

Many workers have studied the wavelength dependence of attenuation by haze; much of the earlier work has been reviewed by Middleton (1952). The dependence can be generally described, at least for a narrow wavelength range, by

$$\beta_{sc} = \text{constant} \times \frac{1}{\lambda^{\gamma}} \tag{6-37}$$

analogously to the Rayleigh expression (4-44). The exponent γ may vary from about 4, for Rayleigh scattering by *very small* particles, to practically 0 for visual and near-infrared wavelengths in fog. Between these extremes lies the wavelength dependence of scattering by customary haze. If the haze particles were of a uniform, known size, the appropriate value of γ could be found from the plots of Q_{sc} in Figure 5.8. That is, the value of γ would be equal to the slope of the curve at the given value of α. For the actual polydispersed aerosol, the complexity of the dependence can be appreciated from Figure 6.5, which shows β_{sc} as a function of wavelength for various particle radii. Each of the curves is a left-right transposition of the curve for $n = 1.33$ in Figure 5.8. It is clear that only for a suspension of very small or very large particles can the dependence be a simple function for a broad range of wavelengths.

First, we ask what predictions of the dependence can be made from theory, for example, from the definition of β_{sc} given by (6-28). There part of the dependence is expressed by the parenthetical term, while the remaining dependence is contained in the integrand. Van de Hulst (1957) and Bullrich (1964) point out that the integrand of (6-28) becomes independent of wavelength if the limits are taken as $\alpha_1 = 0$ and $\alpha_2 = \infty$. In such a case the dependence can be simply expressed, from (6-28), by

$$\beta_{sc} = \text{constant} \times \frac{1}{\lambda^{v-2}} \tag{6-38}$$

which is the same as (6-37) with $\gamma = v - 2$. Thus, in principle, spectral transmission measurements of haze can determine the value of the exponent of a power-law size distribution. In real cases, where the limits α_1 and α_2 are finite, the relationship (6-38) is modified but the expressed tendency still holds. This is shown by Figure 6.6, taken from the work by Curcio et al. (1961) discussed below. It is seen that the curves for both refractive indexes approximately satisfy the relationship $\gamma = v - 2$ for $2 < v < 5$. There remains the question, not easily answered, of the wavelength range over which any such fixed relationship is valid.

Second, we should note that the dependence is affected by the form of the particle size distribution. This is evident from the curves in Figure 6.2b, which show β_{ex} as a function of λ for several values of the exponent v. These curves were plotted by Rensch and Long (1970) from computations of the backscattering and extinction of laser light by haze, fog, and rain. In the visual and near infrared to about 2 μm, the curves generally conform to (6-38), especially at lower values of v. For example, $v = 2$ characterizes a dense haze merging into fog, which is known experimentally to exhibit little selective attenuation at these wavelengths. In confirmation, the slope of the curve for $v = 2$ is almost zero out to 2 μm. At higher values of v, (6-38) is still followed, although not so closely. The minima of the curves near 2, 4, and 10 μm, and the associated maxima, are due to variations in the imaginary part of the refractive index of water.

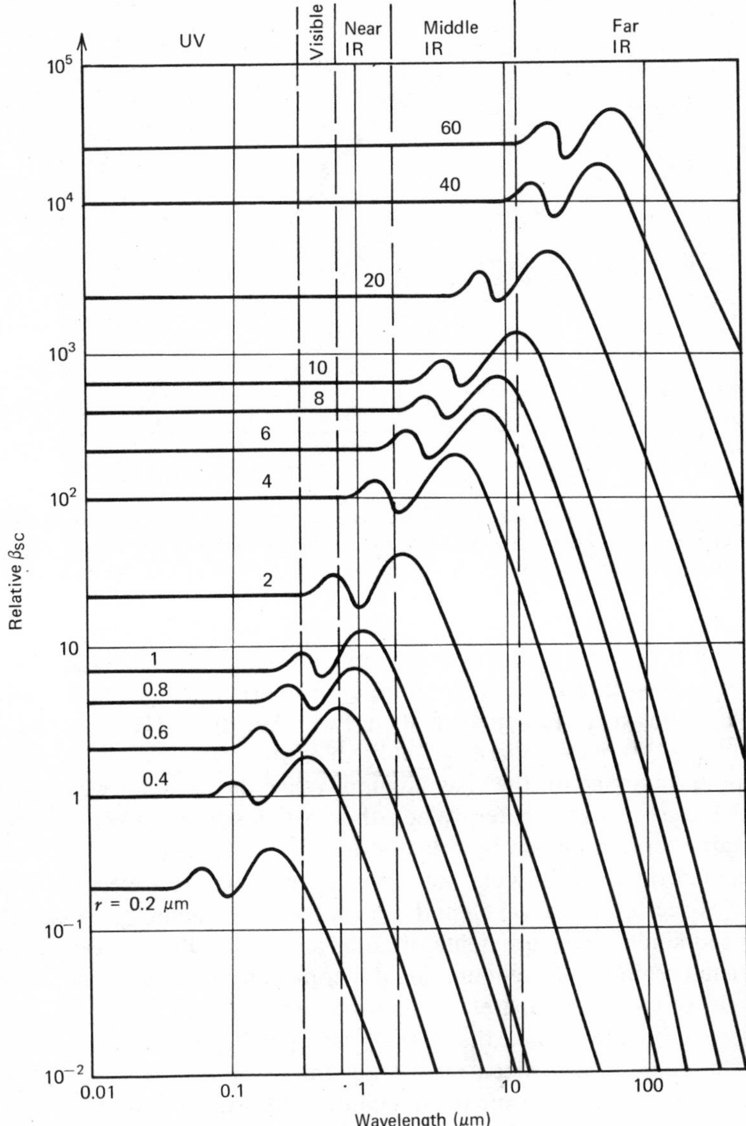

Figure 6.5 Relative values of total scattering coefficients for different wavelengths and various particle radii. From Gaertner (1947).

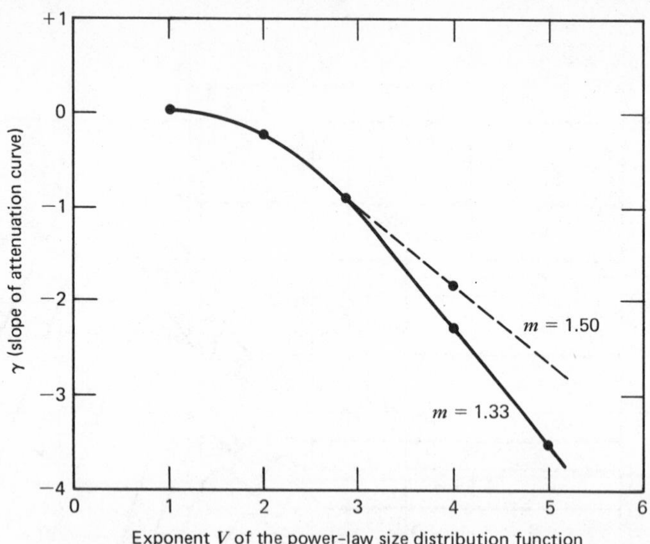

Figure 6.6 Slope of spectral attenuation coefficient curve versus exponent of the power-law size distribution function, for two refractive indices. The ordinate γ is the exponent in (6.37), while the abscissa v is the exponent in (6.28). From Curcio et al. (1961).

We turn now to experimental work, considering first those spectral measurements that have utilized the entire vertical extent of the atmosphere. From measurements of solar flux, Angström (1951) found that formula (6-37) was generally applicable, the value $\gamma = 1.3$ being an acceptable average for the visual region. When the particles were smaller, the value of γ was greater. When the particles were larger, as after a volcanic eruption or during a dust storm over a desert, γ had values as low as 0.5. In the ultraviolet and infrared regions, however, the dependence of scattering on wavelength was not so easily definable. Volz (1956) analyzed similar measurements of solar and stellar fluxes, principally in the visual region. After eliminating the contribution of molecular scattering, which is small compared to the usual amounts and variations in haze scattering, he determined that (6-37) was applicable and that $0.12 < \gamma < 2.3$. The value 1.2 occurred most frequently.

Spectral transmission measurements over horizontal paths are considered next. Curcio and Durbin (1959) found that (6-37) was valid in the visual region, with $\gamma = 1.3 \pm 0.6$ for meteorological ranges of 2.7 to 120 km. The comprehensive measurements by Yates and Taylor (1960) are frequently cited in the literature, and small portions of their summarized data regarding scattering coefficients are shown in Figure 6.7. Their

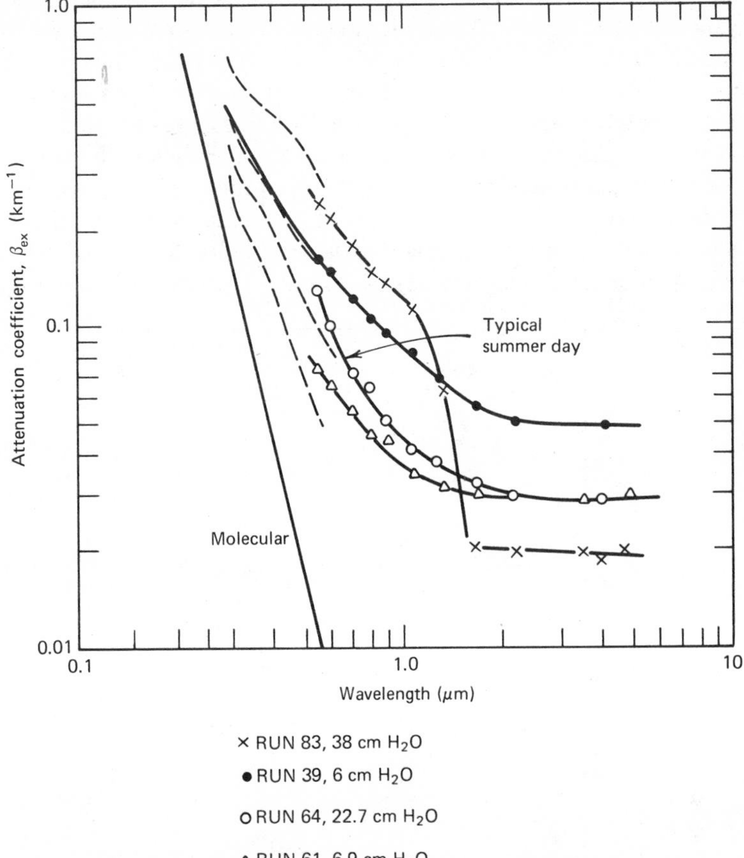

Figure 6.7 Attenuation coefficient versus wavelength for a 16.25-km path at sea level with values of precipitable water in centimeters as indicated. From Yates and Taylor (1960).

results demonstrate that, while the wavelength dependence is generally defined by (6-37), γ does not have a constant value in the visual and infrared regions but decreases in going into the infrared, as shown by the decreasing slopes of the curves. There has been conjecture that this is caused by a small excess of larger particles in the aerosol—beyond the number called for by a single power-law size distribution. According to Plass and Yates (1965), however, the tendency of such curves to flatten in the infrared cannot be explained by any hypothetical size distribution. They reason that the flattening is probably due to absorption by the wings of neighboring gaseous absorption bands.

Curcio et al. (1961) and Curcio (1961) made an extensive series of transmission measurements over horizontal sea-level paths in the Chesapeake Bay region. They employed 10 narrow-wavelength bands between 0.40 and 2.27 μm selected to be as free as possible from gaseous absorption. The physical conditions included relative humidities of 26 to 100% and meteorological ranges of 5.5 to 97 km. The results indicated that (6-37) was occasionally obeyed over the entire wavelength range; such agreements usually occurred during periods of relatively high attenuation. More commonly, however, the curves of β_{sc} versus λ exhibited decreasing slopes with increasing wavelength. In Figure 6.8, which shows a portion of their results, the values of β_{sc} have been grouped according to

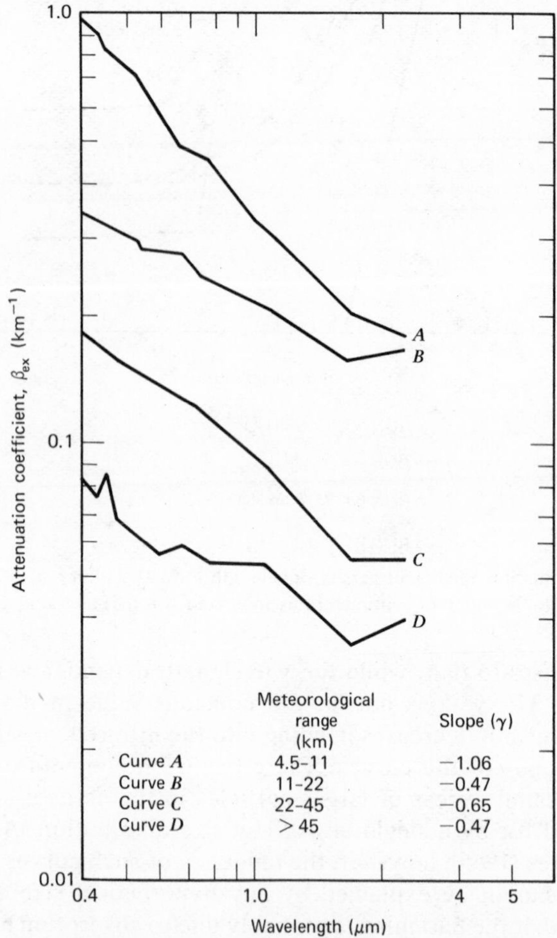

Figure 6.8 Values of attenuation coefficient for haze versus wavelength, grouped by meteorological range. From Curcio et al. (1961).

meteorological range. The curve for each group represents the average of about nine sets of observations, from which the effects of molecular scattering have been removed. For each curve the listed value of slope γ is an average over the wavelength range. In each case, however, the value of γ decreases somewhat at the longer wavelengths, similarly to the behavior in Figure 6.7.

Curcio et al. (1961) explained their experimental data by postulating various mixtures of "two-component" aerosols. The major component was a continental aerosol obeying (3-20) over the size range 0.10 to 20 μm, the distribution exponent having values near 3. Concentrations were variously assumed as several hundred particles per cubic centimeter. The minor component was taken as a maritime aerosol with a size distribution similar to that shown in Figure 3.9*b* over the range 1.0 to 20 μm and having concentrations of only a few particles per cubic centimeter. Attenuation coefficients computed from the above assumptions were in good agreement with the experimental data. Several combinations of the two-component aerosols could be formed, however, that gave nearly equivalent results. This points up a lack of uniqueness in the synthesized representations and indicates the need for concurrent measurements of aerosol characteristics and attenuation coefficients. Elterman (1968, 1970a) has employed the spectral attenuation data from Curcio et al. (1961) to fix the sea-level values of β_{sc} in his atmospheric attenuation model. These values are listed in Table 6.3, from the tabulations identified in Section 6.2.4.

Data from the classic investigations of Gebbie et al. (1951) of transmission of a horizontal path over water have been employed by many workers. A small portion of their data is shown in Figure 6.9, as presented by Sanderson (1955). Each curve applies to a 2000-yd path in several absorption region windows, that is, in a spectral region where the absorption is a minimum. In all cases the path was characterized by 1.7 cm of precipitable water. The state of haze was defined by the transmittance of red light at the wavelength 0.61 μm. Extrapolation of the transmittances to 0.55 μm resulted in the values of R_m indicated for the curves. The dashed lines connecting the plotted points represent reasonably well the course of haze scattering over these spectral ranges. In some cases the transmittance at 10.01 μm is slightly less than that at 3.61 μm; this indicates stronger water vapor absorption, not greater scattering.

Ahlquist and Charlson (1969) employed an integrating nephelometer to investigate the wavelength dependence of scattering by haze aerosols in the Seattle region, as described at the end of Section 3.2.4. Barnhart and Streete (1970) postulated several model hazes consisting of continental and maritime components and having size distributions similar to those of

TABLE 6.3 SPECTRAL ATTENUATION (SCATTERING) COEFFICIENT VERSUS METEOROLOGICAL RANGE

	β_p		
λ	$R_m = 4$ km	$R_m = 6$ km	$R_m = 10$ km
0.27	2.00	1.33	7.85×10^{-1}
0.28	1.89	1.25	7.42×10^{-1}
0.30	1.78	1.18	6.98×10^{-1}
0.32	1.67	1.11	6.55×10^{-1}
0.34	1.56	1.03	6.12×10^{-1}
0.36	1.45	9.61×10^{-1}	5.69×10^{-1}
0.38	1.40	9.28×10^{-1}	5.49×10^{-1}
0.40	1.30	8.61×10^{-1}	5.10×10^{-1}
0.45	1.15	7.62×10^{-1}	4.51×10^{-1}
0.50	1.05	6.96×10^{-1}	4.12×10^{-1}
0.55	9.66×10^{-1}	6.40×10^{-1}	3.79×10^{-1}
0.60	8.60×10^{-1}	5.70×10^{-1}	3.37×10^{-1}
0.65	7.80×10^{-1}	5.17×10^{-1}	3.06×10^{-1}
0.70	7.30×10^{-1}	4.84×10^{-1}	2.86×10^{-1}
0.80	6.40×10^{-1}	4.24×10^{-1}	2.51×10^{-1}
0.90	5.80×10^{-1}	3.84×10^{-1}	2.28×10^{-1}
1.06	5.20×10^{-1}	3.45×10^{-1}	2.04×10^{-1}
1.26	4.70×10^{-1}	3.11×10^{-1}	1.84×10^{-1}
1.67	4.00×10^{-1}	2.65×10^{-1}	1.57×10^{-1}
2.17	3.60×10^{-1}	2.39×10^{-1}	1.41×10^{-1}

Source: Elterman (1970a,b).

Figure 3.9. They then computed scattering coefficients at 20 wavelengths from 0.5 to 15.0 μm for 7 relative humidities from 50 to 98.8%. Presenting their results graphically, they found good agreement with the coefficients measured by Knestrick et al. (1962) in the Chesapeake Bay region. Postulating similar hazes, Hodges (1972) considered the infrared absorption by the particles as well as the scattering. Employing the meteorological data recorded by Yates and Taylor (1960) during their transmission measurements, he computed extinction coefficients which are in fair agreement with their measured values. Additional data on the wavelength dependence of haze scattering are given by Rangarajan (1972). Certain instances of anomalous extinction, that is, increasing attenuation with increasing wavelength over a limited spectral range, apparently produced by log-normal size distributions, have been noted in Section 3.2.4.

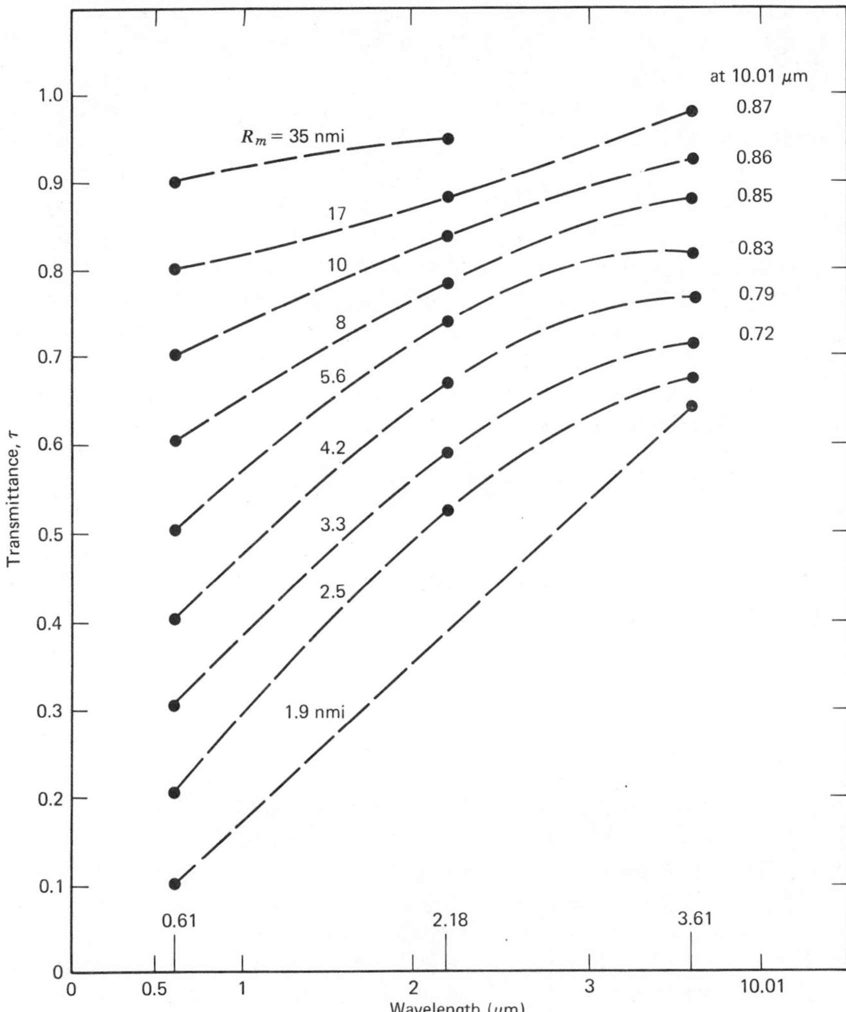

Figure 6.9 Transmittance of a 2000-yd path in haze at visual and infrared wavelengths for various values of the meteorological range R_m as indicated. From Sanderson (1955), in *Principles of Guided Missile Design—Guidance* by A. Locke. © 1955 by Litton Educational Publishing Co. Reprinted by permission of Van Nostrand Reinhold Co.

6.2.3 Optical Thickness of the Haze Atmosphere

The optical thickness T_p of a path in a homogeneous haze atmosphere expresses the combined total scattering of all the particles along the path. Horizontal paths are usually considered homogeneous with respect to haze, or can be divided into sections that are approximately so. For such a condition, T_p is defined by (1-51), which shows that it is analogous to the molecular optical thickness. In this section the subscript sc is not used with the coefficient β, it being understood that the concepts apply to either scattering or extinction. When the path is slant or vertical, β is a function of altitude according to (6-36), hence of position along the path. In principle, for all such cases T_p must be found by an integration such as that indicated by (4-61). When only single scattering is concerned, however, T_p is independent of the spatial distribution of the particles; only their total number and scattering characteristics need be considered.

The optical thickness of the overhead haze atmosphere is found by combining the volume total scattering coefficient and the thickness of the homogeneous haze atmosphere discussed in Section 3.3.2. This thickness is equal to the aerosol scale height H_p. Hence the atmospheric optical thickness for haze is given by

$$T'_p = \beta_0 \int_0^\infty \exp\left(-\frac{Z}{H_p}\right) dZ \tag{6-39}$$

where the prime signifies that the entire vertical extent of the atmosphere is involved, and β_0 is the value of the coefficient at ground level. The indicated integration yields

$$T'_p = \beta_0 H_p \tag{6-40}$$

which is analogous to (4-62). In using the above expressions, values of β_0 may be selected from either Table 3.3 or 6.3, and values of H_p from Table 3.3. The use of altitude limits other than 0 and ∞ with (6-39) gives the value of T_p for a vertical path between the selected altitudes. For a slant path, the value of T_p for the vertical path should be multiplied by $\sec \zeta$, or by the OAM when the zenith angle is greater than about 75 deg. Elterman (1968, 1970a) provides tabulations of T_p for many altitude intervals, wavelengths, and meteorological ranges as described in Section 6.2.4.

Tabulations of T'_p have been made by de Bary et al. (1965), employing the scattering functions discussed in Section 6.2.1 and an aerosol scale height of 1.25 km. The tabulations are described in Section 6.2.4; one set of values is given in Table 6.4. The values apply to an aerosol characterized by a power-law size distribution having the exponent v equal to 3, values of the cutoff radii as indicated, and a turbidity of 2. This turbidity value means that the haze total scattering is just equal to the molecular total scattering, as may be seen from (4-63). The values of T'_p in Table 6.4 can be converted for other turbidities by multiplying them by the factor $T-1$. Inspection of the listed values discloses that T'_p varies inversely as the wavelength to a close approximation, as expected from (6-38) and the condition $v = 3$. The approximation is closest when the size range is greatest, which agrees with the discussion centered around (6-38).

TABLE 6.4 OPTICAL THICKNESS OF THE HAZE ATMOSPHERE FOR A POWER-LAW SIZE DISTRIBUTION HAVING $v = 3$, CUTOFF RADII r_1 AND r_2 AS INDICATED, AT WAVELENGTHS OF 0.40, 0.65, 0.85, AND 1.20 μm

	Optical thickness, T'_p		
r_1 (μm)	$r_2 = 3.0$ μm	$r_2 = 5.0$ μm	$r_2 = 10.0$ μm
$\lambda = 0.40$ μm			
0.04	0.1362	0.1356	0.1352
0.06	0.1341	0.1336	0.1332
0.08	0.1307	0.1302	0.1298
$\lambda = 0.65$ μm			
0.04	0.08314	0.08337	0.08353
0.06	0.08364	0.08369	0.08385
0.08	0.08408	0.08431	0.08447
$\lambda = 0.85$ μm			
0.04	0.06280	0.06335	0.06373
0.06	0.06326	0.06381	0.06418
0.08	0.06415	0.06470	0.06507
$\lambda = 1.20$ μm			
0.04	0.04341	0.04430	0.04488
0.06	0.04381	0.04469	0.04528
0.08	0.04457	0.04546	0.04605

Source: de Bary et al. (1965).

6.2.4 Published Tabulations of Functions for Haze

In this section are described several tabulations of angular and total scattering functions and coefficients for haze aerosols. Wide ranges of wavelengths, particle sizes, and meteorological conditions are covered. The tabulated quantities are either identical to or very similar to the quantities discussed in the preceding sections. Where differences exist, the explanatory text accompanying a tabulation makes the relationship clear. This listing by no means includes all the tabulations in the literature, but consists of those we have found useful and easy to obtain.

1. DeBary et al. (1965) provide tabulations of:

- T'_p at seven wavelengths from 0.4 to 1.2 μm for power-law distributions having $v = 2.5$, 3.0, and 4.0, with $r_1 = 0.04$, 0.06, and 0.08 μm and $r_2 = 3.0$, 5.0, and 10.0 μm. Values of β_{sc} in km^{-1} may be obtained by dividing the tabulated values by 1.25 km.

• The angular scattering function for the entire vertical extent of the haze atmosphere. This function, which is the angular counterpart of T'_p, is tabulated at the wavelength 0.55 μm for the size distribution parameters employed for T'_p. Values of $\beta(\theta)$ in km^{-1} may be obtained by dividing the tabulated values by 1.25 km.

• Intensities of the scattered flux from the solar-irradiated atmosphere (entire vertical extent) for the same size distribution employed for T'_p and for $T = 2, 4$, and 6. The tabulations cover various zenith angles of the sun and various zenith and relative azimuth angles of observation.

2. Deirmendjian (1960, 1963a) provides values of β_{ex} and $\bar{\omega}$ at several wavelengths from 2.6 to 14.0 μm for hazes having the size distributions defined therein.

3. Deirmendjian (1969) tabulated the values of the four phase functions and $\bar{\omega}$ for each of the suspensions listed in Table 3.2 at 15 wavelengths from 0.45 to 16.6 μm in some cases, and at a lesser number in others. The tabulations for rain and hail include values at centimeter wavelengths.

4. Elterman (1968) gives values of β_p (Z), T_p (0 to Z), and T_p (Z to ∞) at altitude intervals of 1 km from 0 to 50 km at 22 wavelengths from 0.27 to 4.00 μm for a surface meteorological range R_m of 25 km. The tabulated quantities are the Mie counterparts of similar quantities tabulated for molecular scattering and ozone absorption. Descriptive material can also be found in Elterman (1964) and Elterman et al. (1969).

5. Elterman (1970a) extends the data in item 4 to cover eight values of R_m from 2 to 13 km at 20 wavelengths from 0.27 to 2.17 μm. Descriptive material can also be found in Elterman (1970b).

6. McCormick (1967) tabulated several scattering quantities important in lidar theory. Included are a back-scattering function η (180 deg), which is a special case of the function η defined by (6-23), and the function K defined by (6-29). These are tabulated for power-law size distributions having $v = 2.5$, 3.0, 3.5, and 4.0, with $r_1 = 0.04$, 0.06, and 0.08 μm and $r_2 = 3$, 5, and 10 μm, at five laser wavelengths from 0.3472 to 1.06 μm.

7. McCormick et al. (1968) tabulate the average angular cross section $\sigma_p(\pi)$ for backscattering and the total cross section σ_p. Values are given for the same size distributions as in item 6, but with $r_1 = 0.04$ and 0.08 μm and $r_2 = 3.0$ and 10.0 μm, at four wavelengths from 0.3472 to 1.06 μm.

8. McClatchey et al. (1971) tabulate β_{sc} and β_{ab}, along with molecular scattering and absorption coefficients, for five model atmospheres: tropical, midlatitude summer and winter, subarctic summer and winter. Values are given at altitude intervals of 1 km from 0 to 25 km, and at larger intervals to 100 km. The tabulations cover 12 laser wavelengths from 0.3371 to 337 μm for surface meteorological ranges of 23 km (clear) and 5 km (hazy).

9. Zelmanovich and Shifrin (1971) provide tabulations of total and angular scattering functions for the power-law distribution and the gamma distribution, which is similar to (3-15). For the power-law type, three values of the exponent v are used, along with $r_1 = 0.04$, 0.06, and 0.08 μm and $r_2 = 3$, 5, and 10 μm. For the gamma type, seven values of average radius from 1 to 15 μm are used. Total cross sections for scattering, attenuation, and light pressure are given at 124 wavelengths from 0.6 to 40 μm, including 13 laser wavelengths. Angular cross section and degree of polarization are given at 37 angles from 0 to 180 deg at 12 wavelengths from 0.714 to 12.1 μm. These tabulations are reviewed by Penndorf (1971).

6.3 TOTAL SCATTERING AND EXTINCTION BY FOG

The physical and optical properties of fog are of considerable interest to workers in both meteorology and atmospheric optics. The peculiar transition from haze to fog as the relative humidity increases is described in Section 3.4.2. This transition is marked by profound changes in scattering characteristics, manifested principally by the severe reduction in visibility. In this section we review several empirical and theoretical relationships between the meteorological range and the liquid water content of fogs. The extinction characteristic of fog and its wavelength dependence are then discussed in detail.

6.3.1 Meteorological Range versus Liquid Water Content

In the theoretical evaluation of scattering and the interpretation of data, fog presents more difficulties than haze. A serious obstacle to generalizing is created merely by the wide ranges of optical and meteorological conditions encompassed by the term *fog*. Table 1.8 shows that β_{sc} increases from 1.96 to 78.2 km^{-1} in going from *thin* to *thick fog*, while R_m decreases from 2 to 0.05 km. In terms of transmittance, τ decreases from 0.14 km^{-1} to about 10^{-34} km^{-1}. Correspondingly, the droplet size distribution may change from a simple power-law function characteristic of a dense haze to a bimodal form characteristic of a thick fog, as illustrated in Figure 3.17. The wavelength dependence of the scattering is very sensitive to features of the distribution, particularly the bimodal singularity. Also, two fogs that produce equal values of R_m may vary in composition, depending on their origin, history, and proximity to sources of pollution.

A fair correlation usually exists between the liquid water content of fog (and cloud) and R_m. The majority of fog droplets are sufficiently large that for visual wavelengths we can take $Q_{sc} = 2$ as a good approximation from

Figure 5.7. The coefficient β_{sc} defined by (5-34) then can be written as

$$\beta_{sc} = 2N\pi r_e^2 \tag{6-41}$$

where r_e is the effective mean radius defined by (3-13). The total volume of droplets per unit volume of space, or the liquid water volume content lw, is given by

$$lw = N\tfrac{4}{3}\pi r_e^3 \tag{6-42}$$

which is the counterpart of (3-12) and likewise is dimensionless. Combining (6-41) and (6-42), we see that

$$\beta_{sc} = \frac{3}{2}\frac{lw}{r_e} \tag{6-43}$$

The meteorological range is now introduced. Substituting (6-43) into (1-71) gives

$$R_m = \frac{2.62 r_e}{lw} \tag{6-44}$$

where R_m has the unit of length used for r_e and for the volume to which lw is referred. A convenient unit is the meter. As derived above (6-44) is dimensionally correct but, if lw is expressed in grams per cubic meter, as is usually done, the dimensions do not check. Equation (6-44) is Trabert's (1901) law whose application to clouds has been examined by aufm Kampe and Weickman (1952, 1957). They found it necessary to use a correction factor (multiplier) on the right-hand side to produce agreement between values of R_m as computed from (6-44) and as measured. The correction factor varied from about 1 for fair-weather cumulus having narrow size distributions, to 5 and more for cumulonimbus having broad size distributions which included large droplets. Thus the correction factor is a measure of how well the effective mean radius represents a size distribution. The basis of the problem is that the plentiful but smaller droplets have the major effect on R_m, while the fewer but larger droplets make the major contribution to liquid water content.

It is clear from the derivation of (6-44) that, if the droplet sizes remain constant but their numbers vary as meteorological conditions change, R_m varies inversely as lw. If the numbers remain constant but the sizes change, r_e is proportional to $(lw)^{1/3}$. For this reasonable assumption, (6-44) can be written in the convenient form

$$R_m = c\,(lw)^{-0.67} \tag{6-45}$$

The value of the term c depends on the form of the size distribution and the measurement units.

The relationship between R_m and lw found by Houghton and Radford (1938) for coastal and mountain-slope fogs is shown by the dashed curve A in Figure 6.10. Reviewing the slope and position of this curve in the light of (6-45), we deduce that the exponent of lw has the value -0.7 and that c is about 0.055 for R_m in km and lw in g m^{-3}. As discussed in Section 3.4.2, Eldridge (1966) synthesized the droplet size distributions that would produce the spectral attenuations measured by Arnulf et al. (1957) for three classes of fog. He computed R_m and lw for each of the synthetic distributions and determined that

$$R_m = 0.024(lw)^{-0.65} \tag{6-46}$$

for evolving and stable fogs, as shown by the dashed curve B in Figure 6.10. For selective fogs, in which the scattering varied significantly with wavelength, the value 0.024 of the constant was replaced by 0.017.

Platt (1970) called attention to the appreciable difference between the two sets of data represented by curves A and B in Figure 6.10, and suggested that different types of fog having dissimilar droplet distributions could be responsible. Eldridge (1971) reviewed the two investigations. He shows that the results are in better agreement when corrections are made for the different values of droplet radii limits employed. Application of the correction to the data for curve A produced the solid curve C, and to curve B produced the solid curve D. A further correction

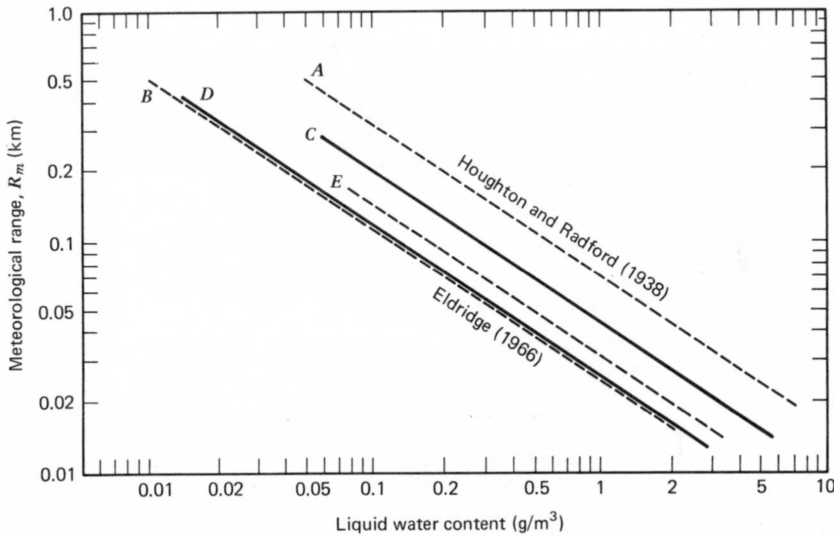

Figure 6.10 Meteorological range as a function of liquid water content in stable fogs. From Eldridge (1971).

for the admitted uncertainties of the lower size limit of droplets in the Houghton and Radford investigation yielded dashed curve E from curve C. The discrepancy between curves D and E is small for data of this sort.

Additional investigations of the relationship between R_m and lw are noted. As part of his investigation of arctic fogs, Kumai (1973) computed values of lw and R_m from measured droplet distributions, employing values of 0.02 and 0.5 for the luminance contrast threshold ϵ described in Section 1.5.3. Measurements of R_m were made concurrently with the droplet sampling. Figure 6.11 shows the data points for the relationship between measured R_m and lw for two types of fog designated long-lasting (L) and short-period (S). These fogs, being remote from civilization, are perhaps cleaner than those at lower latitudes. Also shown in the figure, as a solid and a dashed line, are the corresponding computed relationships. Agreement with the measured data points is good. It is evident that the dashed-line plot is described by

$$R_m \approx 0.020(lw)^{-1} \tag{6-47}$$

The solid line also has a slope of -1, but the value of the constant is about 0.032. Zabrodskii and Morachevskii (1959) provide many data which

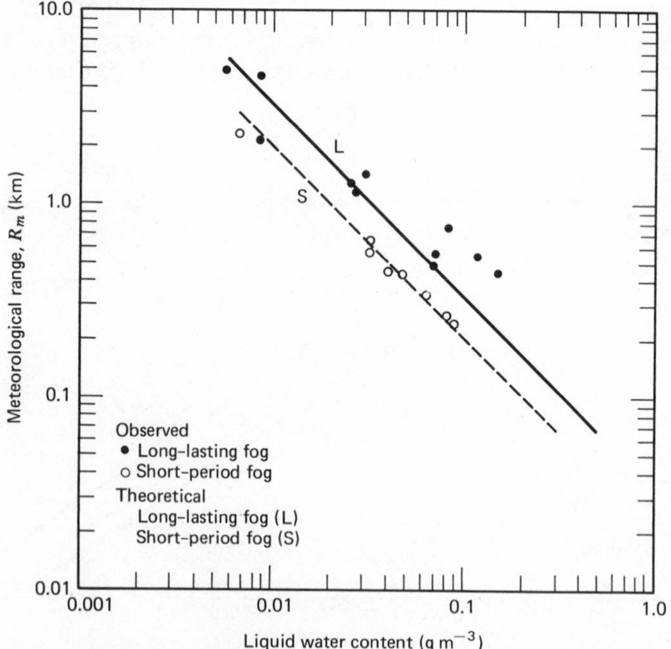

Figure 6.11 Meteorological range versus liquid water content of arctic fogs. From Kumai (1973).

allow the relationship between R_m and lw to be studied further. In the context of (6-45), a study of their data for "high fog" merging into stratocumulus cloud indicates that the exponent of lw is about 0.8 and that c is about 0.02.

6.3.2 Extinction and Its Wavelength Dependence

The wavelength dependence of extinction by fog can be best described only by general statements. Any fog (or cloud) scatters selectively in one or another part of the optical spectrum, depending on the values of α over the size distribution and the given spectral range, and on the relationship of Q_{ex} to α. When all $\alpha \ll 1$, as in the extreme left-hand side of Figure 5.8, near-Rayleigh scattering obtains and (6-37) becomes

$$\beta_{ex} = \text{constant} \times \frac{1}{\lambda^{\gamma \to 4}} \tag{6-48}$$

This condition is realized at radar and extreme infrared wavelengths in fog and cloud. When the wavelengths are comparable to the droplet sizes, as in the middle and far-infrared regions, the dependence on λ fluctuates strongly. This can be seen from the behavior of Q_{sc} in the region of the early maxima and minima, as shown in Figure 5.7. When all $\alpha \gg 1$, which is true for fog and visual wavelengths, Q_{sc} has a value of about 2, and

$$\beta_{sc} = \text{constant} \times \frac{1}{\lambda^{\gamma \to 0}} \tag{6-49}$$

showing that the dependence is practically zero at these relatively short wavelengths.

The decreasing dependence of total scattering on wavelength in the visual region as haze thickens into fog is shown by the work of Foitzik (1938), widely quoted in the literature. He measured the spectral attenuation of atmospheric paths during conditions such that $1.3\ \text{km}^{-1} \leq \beta_{sc} \leq 130\ \text{km}^{-1}$, corresponding to $3\ \text{km} \geq R_m \geq 0.03\ \text{km}$. From that work, Figure 6.12 shows the relative attenuation found for blue (0.483 μm), green (0.565 μm), and red (0.675 μm) light as a function of the *mean* attenuation coefficient for these wavelengths. Near the value $\beta_{sc} = 1.3\ \text{km}^{-1}$ or $R_m = 3.0\ \text{km}$, attenuation at the blue and red wavelengths varied about $\pm 13\%$ from that at the green. The sudden decrease in selectivity near $\beta_{sc} = 4\ \text{km}^{-1}$ marks, according to Middleton (1952), "the change from haze to fog, the point where hygroscopic nuclei may be supposed to start increasing in radius in an unstable manner." The general tendencies for such unstable growth at values of relative humidity near saturation may be seen from the droplet behavior illustrated in Figure 2.1. Beyond $\beta_{sc} = 6\ \text{km}^{-1}$ the relative attenuation for blue and red light remains

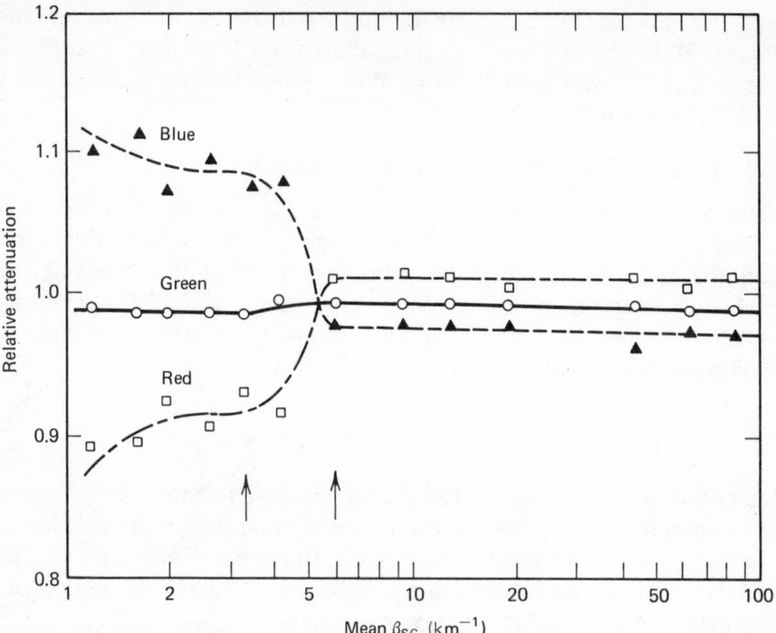

Figure 6.12 Relative attenuations of blue, green, and red light in fog. The arrows mark the transition condition between haze and fog known as mist. From Eldridge (1969) and Foitzik (1938).

constant within about +2% of the value for green, which means that the attenuation is no longer spectrally selective in the visual region as the fog thickens. Eldridge (1969) points out that the transition region between haze and fog, marked by arrows in the figure, corresponds to the condition called *mist*.

Fogs vary considerably in their spectral scattering characteristics, particularly over a broad wavelength range, because of different distributions of droplet sizes. This is shown by the spectral attenuation measurements made by Arnulf et al. (1957). Measurements were made in a variety of hazes and fogs, with the latter generally classified as selective fog, evolving fog, stable fog (first type), or stable fog (second type). Many sets of measurements were made for each class over the spectral range 0.35 to 10.0 μm and are presented graphically in the cited paper. Figure 6.13 shows a typical spectral attenuation for each class, along with values of the associated meteorological range and the liquid water content. All the attenuations decrease at wavelengths greater than about 4 μm, and each class would be selective in the far- and extreme-infrared regions. How-

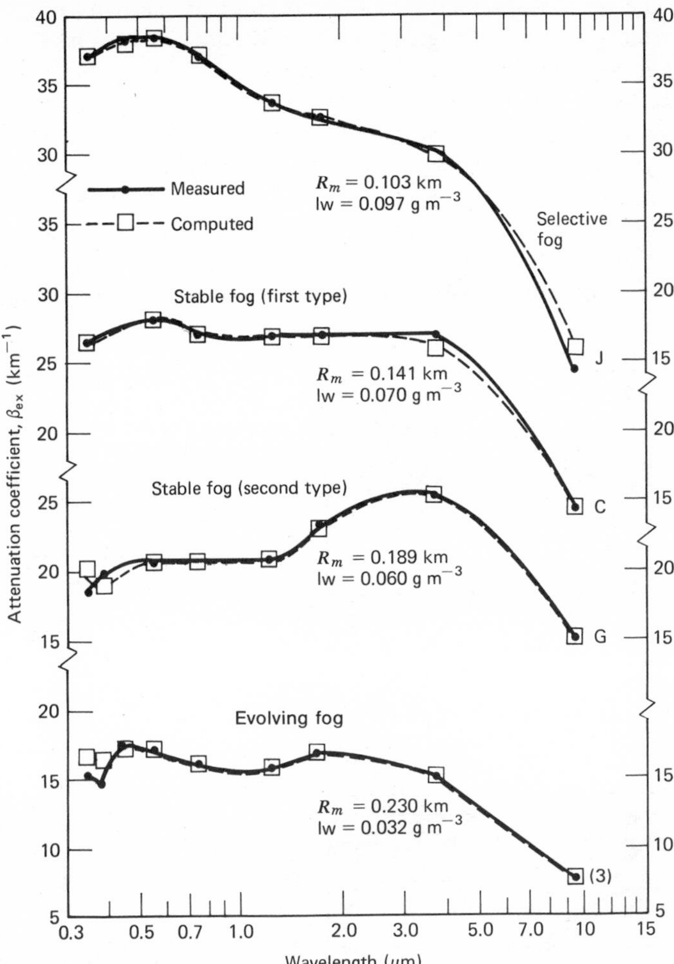

Figure 6.13 Spectral attenuations of four types of fog. From Arnulf et al. (1957) and Eldridge (1966).

ever, as pointed out in the cited paper, even at 10 μm the resulting increase in distance corresponding to a given threshold of detection is too small to be of practical interest. Eldridge (1966) has analyzed these measurements in detail, as described previously. Figure 6.13 is from his work.

Chu and Hogg (1968) calculated the extinction for the fog models shown in Figure 3.18 at wavelengths of 0.63, 3.5, and 10.6 μm. Similar calculations at additional wavelengths were made by Rensch and Long

(1970). The results obtained by Chu and Hogg are shown in Figure 6.14, where attention should be given to the ordinate and abscissa units. The extinction is presented as decibels per kilometer per milligram of liquid water per cubic meter. From (1-43) the number of decibels may be converted to β_{ex} by multiplying by 0.2303, while the liquid water content of fogs typically may vary between 10 and 500 mg m^{-3}. The abscissa unit is the mode radius r_c, that is, the droplet radius at the maximum of the distribution curve. The similarity of Fig. 6.14a, b, and c is attributed to the similarity of the slopes at the right of the maxima of the corresponding distributions in Figure 3.18. At smaller values of r_c the extinction is

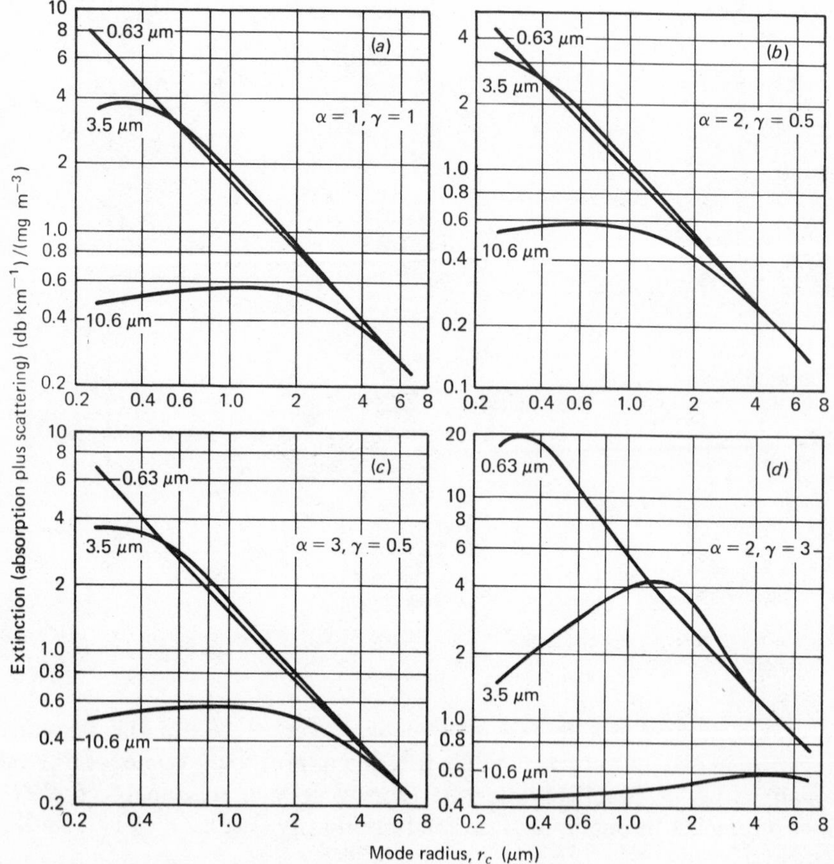

Figure 6.14 Extinction by the fog models of Figure 3.18. From Chu and Hogg (1968). (Reprinted with permission from *The Bell System Technical Journal*, Copyright 1968, The American Telephone and Telegraph Company.)

strongly wavelength-dependent, as would be expected for a thin fog (thick haze) made up of small droplets. At values of r_c greater than 4 μm, however, the wavelength dependence has disappeared except in Fig. 6.14*d* which corresponds to the narrow distribution in Figure 3.18. In all cases the extinction (per unit content of liquid water) versus r_c becomes asymptotic to a slope of approximately -1.

6.4 SCATTERING AND EXTINCTION BY CLOUDS

The scattering by clouds profoundly affects the spatial distribution of skylight and the spectral distribution and amount of solar flux reaching the earth's surface. This section considers first the angular scattering characteristics of major tropospheric cloud types and then their total scattering characteristics. Next, cloud reflectance is discussed for both steady-state and pulsed laser sources of irradiation. Attention is called to the published tabulations of scattering functions for model clouds and representative major cloud classes.

6.4.1 Angular Scattering by Clouds

The scattering characteristics of clouds have been studied much more thoroughly than those of fogs, probably because fogs are infrequent and temporary occurrences in most parts of the world but clouds in varying amounts are permanent features of the terrestrial environment. Their scattering characteristics are basic factors in the transfer of optical energy at all wavelengths through the atmosphere. In general, the scattering by low- and medium-altitude clouds, consisting principally of water droplets and called *water clouds*, is similar to that by fog. The scattering by high-altitude clouds, consisting principally of ice crystals and called *ice clouds*, is considerably different from that by water clouds. The differences, manifested in the angular distributions of intensities and polarizations, have been studied for criteria to distinguish between the liquid and ice phases of cloud particles. This section reviews the angular scattering characteristics of clouds as disclosed by the numerous studies cited herein.

The general subject of scattering by water clouds has been surveyed briefly by Bauer (1964). He employed the size distribution (3-15) with several choices of parameter values to produce different mode radii and distribution widths. Harris (1969) studied the changes in polarization and intensity of scattered light during the growth of the model cloud established by Neiburger and Chien (1960), illustrated in Figure 3.16. He found marked variations in polarization and intensity as the droplet size distribution changed with time. His study indicates that measurement of these quantities may be a useful method of studying cloud formation processes.

The angular scattering characteristics of a cloud are expressed completely by the phase function $P_j(\theta)$, with $j = 1, 2, 3, 4$. This function is defined by (6-9) and (6-11), while the scattering matrix and transformation of the Stokes vector are shown by (6-16). Deirmendjian (1969) tabulated the four phase functions for the cloud models listed in Table 3.2 at many wavelengths in the visual and infrared. Of the cloud models listed, the cumulus model C.1 best represents a tropospheric cloud, and the functions $P_1(\theta)/4\pi$ and $P_2(\theta)/4\pi$ for that model are plotted in Figure 6.15. Earlier, limited information for a few wavelengths can be found in Deirmendjian (1963a, 1964).

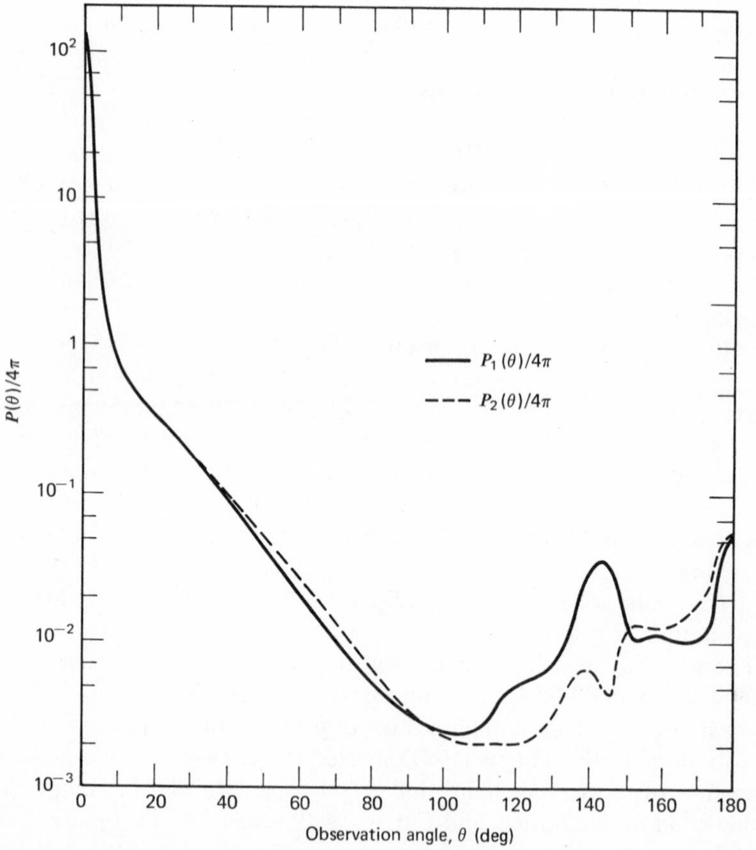

Figure 6.15 Phase functions for cumulus cloud model C.1 of Figure 3.21 at $\lambda = 0.70\ \mu$m. Data from Deirmendjian (1969).

Yamamoto et al. (1971) have computed an average phase function defined by

$$P(\theta) = \frac{1}{2}[P_1(\theta) + P_2(\theta)] \tag{6-50}$$

They tabulated this function for the three cloud classes, whose droplet size distributions are shown in Figure 3.23, at 38 wavelengths from 5.00 to 40.00 μm. Figure 6.16 shows $P(\theta)$ for altostratus clouds at several of

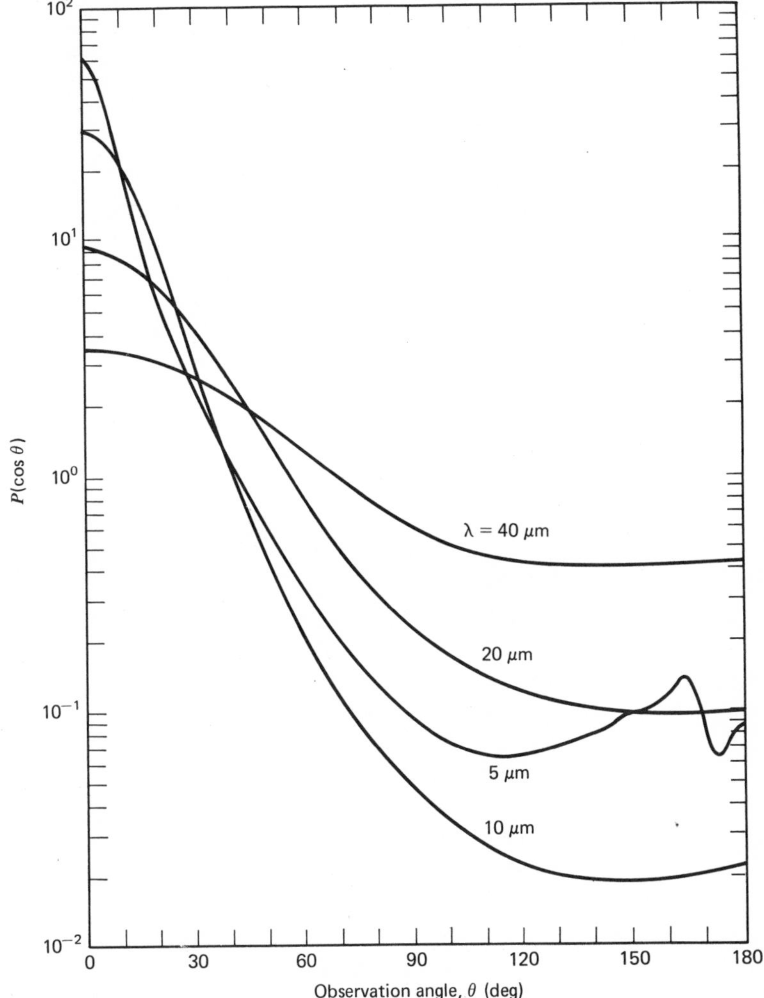

Figure 6.16 Phase functions for the altostratus cloud of Figure 3.23 at four infrared wavelengths. From Yamamoto et al. (1971).

these wavelengths. Outstanding is the decrease in the predominantly forward scattering as the wavelength increases. This increase makes the droplets appear smaller, thus reducing and broadening the diffraction lobe. Although the trace of a glory may be seen in the backscattering curve for 5.0 μm at 165 deg $< \theta <$ 180 deg, the remaining curves are smooth. This smoothness is the result of integrating over the size distribution, thereby averaging out the otherwise sharp variations with angle, and to the small values of the size parameter α at these wavelengths.

The use of lidar in cloud research involves the backscattering and total scattering or extinction functions. Carrier et al. (1967) have tabulated $\beta(\theta)$ at $\theta = 180$ deg, designated $\beta(\pi)$, and β_{ex} for the cloud models of Figure 3.22 at five wavelengths in the visual and infrared. They also give information on the variation in these coefficients with cloud thickness, height above base, and geographical location. Their values of $\beta(\pi)$ are listed in Table 6.5. Liou and Schotland (1971) show the effects of a bimodal size distribution and multiple scattering on the phase functions of cumulus clouds. Anderson and Browell (1972), dealing with the cloud models of Figures 3.21 and 3.23, show graphically first- and second-order backscattering, taking into account the field of view and other geometric factors of a lidar transceiver.

Cloud scattering has been investigated as a means of distinguishing between water and ice clouds. Harris (1971) provides an extensive bibliography of this work, which in recent years has benefited from the use of polarized laser beams and lidar instrumentation. In one such study, Harris (1971) employed the model cumulus C.1 cloud of Table 3.2, assuming that the particles were spherical in the ice phase as well as in the liquid phase and that only single scattering occurred. Scattering computations for the water and ice phases were made at five infrared wavelengths where the refractive indices for the two phases are notably different. He then presents graphically the intensity distribution functions i_1 and i_2 (in effect), the polarization ratio i_1/i_2, the degree of polarization defined by (5-10), and the ellipticity and orientation of the polarization ellipse. The corresponding plots for the water and ice phases show theoretical differences which should be detectable experimentally.

Using a different approach, Liou (1972a) considers that ice cloud particles are crystals whose forms can be approximated by long, circular cylinders of polydisperse sizes. He presents graphically the phase functions and degree of polarization for such cylinders, and for polydisperse ice spheres, at wavelengths of 0.7, 3.0, 3.5, and 6.05 μm. Compared to the spheres, the cylinders show greater scattering near $\theta = 90$ deg, at the expense of scattering in the forward and backward

TABLE 6.5 BACKSCATTERING COEFFICIENTS FOR THE MAJOR CLOUD MODELS OF FIGURE 3.22

	$\beta(\pi)$ (m^{-1} sr^{-1})				
	Wavelength (μm)				
Cloud type	0.488	0.694	1.06	4.0	10.6
Nimbostratus	7.16×10^{-3}	6.03×10^{-3}	6.45×10^{-3}	3.96×10^{-3}	1.54×10^{-4}
Altostratus	6.77×10^{-3}	4.52×10^{-3}	4.99×10^{-3}	3.20×10^{-3}	1.25×10^{-4}
Stratus II	6.04×10^{-3}	4.76×10^{-3}	4.62×10^{-3}	2.87×10^{-3}	1.31×10^{-4}
Cumulus congestus	3.97×10^{-3}	3.01×10^{-3}	3.66×10^{-3}	2.43×10^{-3}	7.88×10^{-5}
Stratus I	3.13×10^{-3}	2.88×10^{-3}	3.08×10^{-3}	1.47×10^{-3}	7.42×10^{-5}
Cumulonimbus	2.40×10^{-3}	2.21×10^{-3}	2.19×10^{-3}	9.13×10^{-4}	1.16×10^{-4}
Stratocumulus	2.44×10^{-3}	1.91×10^{-3}	2.08×10^{-3}	8.91×10^{-4}	5.95×10^{-5}
Fair-weather cumulus	1.18×10^{-3}	8.68×10^{-4}	1.00×10^{-3}	4.17×10^{-4}	2.31×10^{-5}

Source: Carrier et al. (1967).

directions. The glory, a characteristic feature of scattering by spheres, is absent from scattering by cylinders. For randomly oriented cylinders, the degree of polarization is notably different from that for spheres. In experimental work devoted to cloud scattering at far-infrared wavelengths, account must be taken of emission by the cloud. For example, the study by Hall (1968) indicates that the radiance of cirrus clouds due to thermal emission in the spectral band 8 to 13 μm exceeds the radiance due to scattering of the flux from the ground and lower atmosphere.

Because of their great optical thickness, except for short paths, clouds present multiple-scattering problems more often than do hazes. Readers wishing to pursue this subject will find generally helpful the references cited in the last two paragraphs of Section 6.1.4. Additional studies specifically devoted to clouds have been made by Hansen and Pollack (1970), Yamamoto et al. (1970), Heggestad (1971), and Liou and Schotland (1971). Monte Carlo calculations at various infrared wavelengths are described by Kattawar and Plass (1971, 1972) and Plass and Kattawar (1971a, b, c). These studies are parallel to those by the same investigators cited in Section 6.1.4, and the results likewise are shown by many detailed graphs. Danielson et al. (1969) also have made Monte Carlo calculations for visual wavelengths and provide many graphs and a tutorial discussion.

With any form of particle suspension, the tendency of multiple scattering is to average out the sharp variations in scattered intensity with observation angle, as described in Section 1.3.3. This tendency can often be seen by observing the sun through a uniform layer of dense stratus cloud. Usually the strongly peaked forward scattering, such as that shown in Figure 6.14, produces a very bright aureole when the sun shines through a thin cloud of any type. According to Deirmendjian (1969), however, when the cloud layer has the proper density and thickness, the aureole is suppressed by multiple scattering. The sun's disk then can be seen sharply defined, not much brighter than the surrounding cloud field. He estimates that the sun's apparent radiance in such cases has been reduced by the order of 10^{-7}, which means, from (1-52), that the optical thickness of the slant path through the cloud is about 16. This phenomenon, called the filtered *sun effect*, is quite striking as the sharp-edged solar disk has the hue of molten silver although not so bright. Persons who live in coastal regions where stratus clouds drifting inland are a common occurrence are encouraged to look for this sight.

6.4.2 Total Scattering and Extinction by Clouds

The total scattering and extinction properties of water clouds are similar to those of fogs, but they may differ considerably in detail. The general relationship between meteorological range and liquid water content of fog, expressed by Trabert's law (6-44), also holds for clouds. As explained there, however, a correction factor based on droplet size distribution and concentration is usually required. Hence the relationship can be expected to vary between fogs and clouds and between clouds of different classes. Since the liquid water content of a cloud may be from 1 to 6 g m^{-3}, while that of a fog is seldom greater than 0.1 g m^{-3}, visibility tends to be more restricted in clouds. When an aeroplane is passing through a cloud, for example, the wing tips frequently and momentarily disappear from view.

In his extensive studies of scattering by polydispersions, Deirmendjian (1969) has computed the three attenuation coefficients, β_{sc}, β_{ab}, and β_{ex}, and the albedo $\bar{\omega}$ for the three cloud models of Table 3.2 and Figure 3.21. The values of β_{ex} and $\bar{\omega}$ at several wavelengths are listed in Table 6.6, from which the values of β_{sc} and β_{ab} can be found through (5-37). In Figure 6.17 β_{ex} is plotted, along with β_{sc}, in regions of significant absorption. Shown for comparison are the corresponding coefficients for the three haze models of Table 3.2 and Figure 3.8 and the molecular

TABLE 6.6 EXTINCTION COEFFICIENT AND SINGLE SCATTERING ALBEDO FOR THE THREE CLOUD MODELS OF FIGURE 3.21

	Model C.1		Model C.2		Model C.3	
λ (μm)	β_{ex} (km^{-1})	$\bar{\omega}$	β_{ex} (km^{-1})	$\bar{\omega}$	β_{ex} (km^{-1})	$\bar{\omega}$
0.45	16.33	1.0	11.18	1.0	2.906	1.0
0.70	16.73	1.0	11.43	1.0	3.021	1.0
1.19	17.29	0.9994	—	—	—	—
1.45	17.63	0.9849	—	—	—	—
1.61	17.58	1.0	12.56	1.0	4.126	1.0
1.94	18.05	0.9395	—	—	—	—
2.25	18.36	0.9894	—	—	4.549	0.9975
3.00	17.98	0.4923	12.39	0.4809	3.245	0.4653
3.90	20.64	0.9140	17.76	0.9489	3.241	0.9660
5.30	23.87	0.8848	—	—	1.619	0.8927
6.05	19.86	0.5433	13.06	0.5591	1.836	0.4546
8.15	18.75	0.7465	—	—	0.7290	0.5713
10.0	11.18	0.6014	4.944	0.5262	0.4298	0.3118
11.5	10.10	0.2886	—	—	—	—
16.6	16.79	0.3949	9.753	0.3385	1.179	0.1544

Source: Deirmendjian (1969).

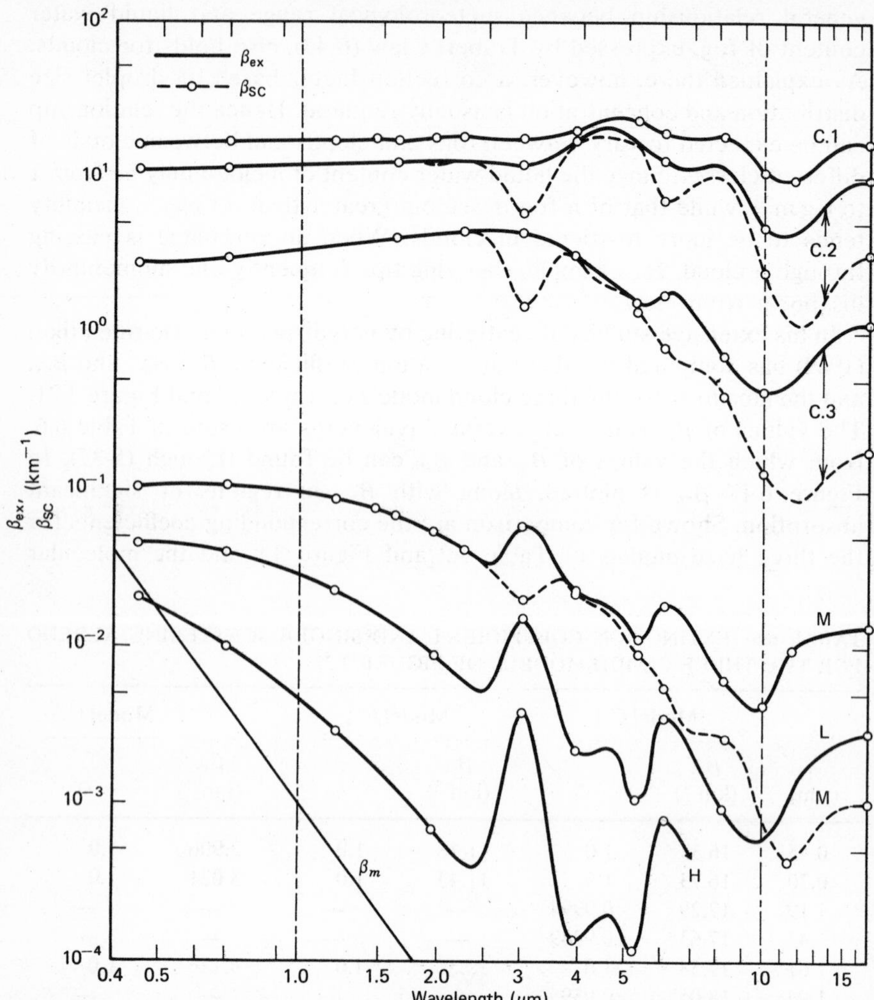

Figure 6.17 Extinction and scattering coefficients versus wavelength for haze and cloud models of Figures 3.8 and 3.21. From Deirmendjian (1969).

scattering coefficient β_m. The reader is reminded that the droplet concentration for each haze and cloud model is 100 cm^{-3}, a value selected for ease of scaling. When the concentration is changed by a given factor, the curve is shifted vertically by that factor, without change in shape. The cited paper discusses cloud scattering in detail and in practical terms.

Yamamoto et al. (1971) have computed the three attenuation coefficients and the albedo for the three cloud classes of Figure 3.23. The values of β_{sc} and β_{ab} at 43 wavelengths are listed in Table 6.7, while β_{ex}

TABLE 6.7 TOTAL SCATTERING AND ABSORPTION COEFFICIENTS FOR THE THREE CLOUD CLASSES OF FIGURE 3.23

	Altostratus		Stratocumulus		Nimbostratus	
λ (μm)	β_{sc} (km^{-1})	β_{ab} (km^{-1})	β_{sc} (km^{-1})	β_{ab} (km^{-1})	β_{sc} (km^{-1})	β_{ab} (km^{-1})
5.00	107.9	23.38	52.57	8.632	111.4	39.84
5.40	—	—	54.32	6.709	122.7	32.97
5.48	117.5	19.87	—	—	—	—
5.80	97.77	32.12	40.38	12.15	105.2	52.27
6.00	66.89	49.65	27.97	20.33	79.14	69.30
6.16	72.28	52.10	—	—	—	—
6.20	—	—	35.64	20.69	81.71	70.30
6.35	84.92	45.02	—	—	—	—
6.60	94.62	38.15	41.23	14.65	98.43	59.35
6.80	97.62	35.56	—	—	—	—
7.00	98.46	34.24	40.69	12.96	105.3	55.19
7.28	97.42	33.42	—	—	—	—
7.55	95.55	32.88	—	—	—	—
7.60	—	—	36.99	12.33	108.9	53.67
8.00	90.48	32.09	—	—	110.0	53.07
8.50	83.27	30.89	29.89	11.55	111.1	51.46
9.00	73.87	29.96	—	—	—	—
9.50	61.60	29.56	20.55	11.03	101.8	49.66
10.00	46.68	30.63	15.15	11.51	85.85	50.80
10.50	31.79	33.29	10.16	12.69	63.48	53.68
11.00	22.66	36.88	—	—	—	—
11.50	20.32	41.91	6.624	16.71	40.59	61.99
12.00	21.97	46.81	—	—	—	—
13.00	28.08	53.94	9.859	22.85	48.56	72.22
14.00	32.56	57.86	—	—	—	—
15.00	35.64	60.16	12.90	25.78	57.36	78.51
16.00	37.96	61.58	—	—	—	—
17.00	39.22	62.25	—	—	—	—
18.00	39.82	62.45	—	—	—	—
19.00	39.96	62.30	—	—	—	—
20.00	39.87	61.90	13.96	25.53	64.12	83.46
22.00	39.28	60.49	—	—	—	—
24.00	38.03	58.15	—	—	—	—
25.00	—	—	12.02	22.13	65.54	82.11
26.00	36.03	55.21	—	—	—	—
28.00	33.04	52.79	—	—	—	—
30.00	29.37	50.72	8.817	19.07	59.86	78.28

TABLE 6.7 (*Continued*)

	Altostratus		Stratocumulus		Nimbostratus	
λ (μm)	β_{sc} (km^{-1})	β_{ab} (km^{-1})	β_{sc} (km^{-1})	β_{ab} (km^{-1})	β_{sc} (km^{-1})	β_{ab} (km^{-1})
32.00	25.57	48.99	—	—	—	—
34.00	21.96	47.61	—	—	—	—
35.00	—	—	5.742	17.51	47.26	74.92
36.00	18.52	46.65	—	—	—	—
38.00	15.24	46.04	—	—	—	—
40.00	12.64	45.83	3.459	17.22	32.71	72.67

Source: Yamamoto et al. (1971).

and β_{sc} are shown in Figure 6.18. Noteworthy are the minor minima near the wavelength 6 μm and the pronounced minima near 12 μm. These regions can scarcely be considered "transmission windows," however, when the corresponding values of ordinate are noted. Outstanding is the rapid roll-off of values beyond approximately 25 μm. It is interesting to see the slopes of β_{sc} approach the value -4 at wavelengths greater than 30 μm, where the droplets begin to resemble Rayleigh scatterers.

Attention is called to additional studies of extinction by clouds. Of great utility to lidar workers are the computations of β_{ex} at five laser wavelengths by Carrier et al. (1967). The computations were made for the cloud models of Figure 3.22, and the resulting values are listed in Table 6.8. These are complementary to the values of $\beta(\pi)$ in Table 6.5. The cited paper gives information on the variation in the coefficients with cloud altitude and geographical location. Gates and Shaw (1960) computed β_{ex} for fair-weather cumulus clouds having $N = 300\ \text{cm}^{-3}$ and a distribution maximum at 7 μm, and for nimbostratus clouds having $N = 80\ \text{cm}^{-3}$ and a broad maximum extending from 5 to 12 μm. Platt (1970) computed attenuations for clouds and fogs at submillimeter wavelengths and compared them with the values measured at 337 and 1200 μm by Burroughs et al. (1966) and Bastin et al. (1964), respectively. Deirmendjian (1975) presents a very comprehensive study of the attenuation at far-infrared and submillimeter wavelengths by clouds and rainfall.

Zabrodskii and Morachevskii (1961) employed airborne instrumentation to measure the droplet size spectra and liquid water content of low clouds and high fogs. Concurrent measurements of transmittance at the wavelength 0.52 μm were made over the short path provided by the

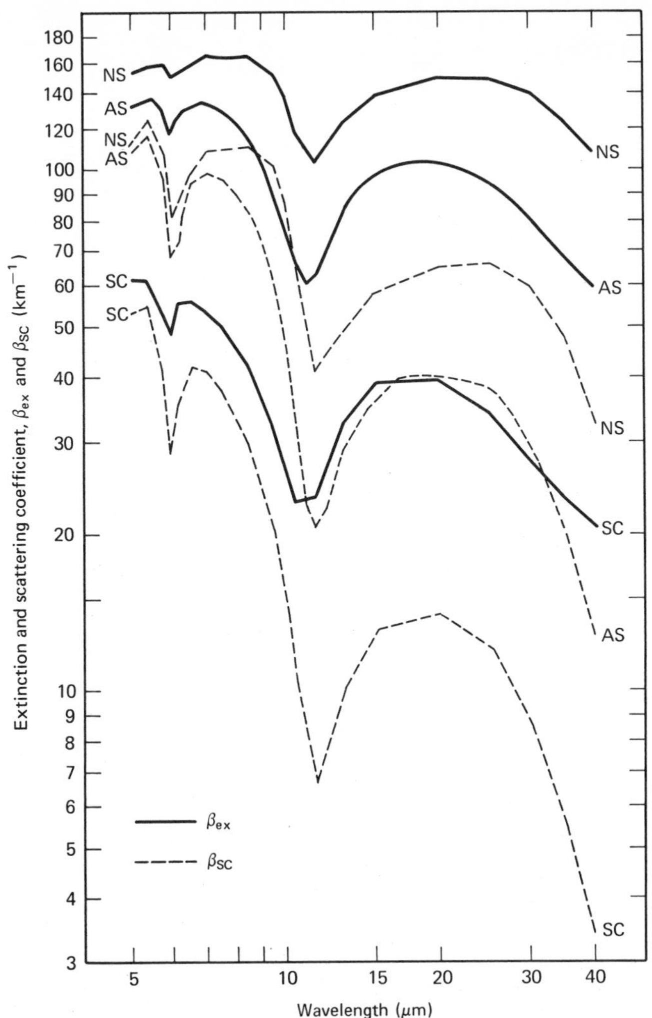

Figure 6.18 Extinction and scattering coefficients versus wavelength for the three cloud classes of Figure 3.23. Data from Yamamoto et al. (1971).

aircraft structure. Values of β_{ex} were computed from the droplet data and compared with "measured" values derived from the transmittance data. These investigators found that the measured values exceeded the computed values in the majority of cases. This may indicate the difficulty in accurately measuring droplet sizes and concentrations by airborne sampling techniques. Essentially, these give data for separated regions of an inhomogeneous medium.

TABLE 6.8 EXTINCTION COEFFICIENTS FOR THE MAJOR CLOUD MODELS OF FIGURE 3.22

	β_{ex} (m^{-1})				
	Wavelength (μm)				
Cloud type	0.488	0.694	1.06	4.0	10.6
Nimbostratus	1.28×10^{-1}	1.30×10^{-1}	1.32×10^{-1}	1.47×10^{-1}	1.36×10^{-1}
Altostratus	1.08×10^{-1}	1.09×10^{-1}	1.12×10^{-1}	1.30×10^{-1}	8.39×10^{-2}
Stratus II	1.00×10^{-1}	1.01×10^{-1}	1.03×10^{-1}	1.14×10^{-1}	1.04×10^{-1}
Cumulus congestus	6.92×10^{-2}	6.98×10^{-2}	7.13×10^{-2}	8.10×10^{-2}	6.76×10^{-2}
Stratus I	6.69×10^{-2}	6.79×10^{-2}	6.97×10^{-2}	9.01×10^{-2}	4.28×10^{-2}
Cumulonimbus	4.35×10^{-2}	4.38×10^{-2}	4.44×10^{-2}	4.82×10^{-2}	5.09×10^{-2}
Stratocumulus	4.53×10^{-2}	4.60×10^{-2}	4.71×10^{-2}	5.96×10^{-2}	2.48×10^{-2}
Fair-weather cumulus	2.10×10^{-2}	2.13×10^{-2}	2.19×10^{-2}	2.76×10^{-2}	1.17×10^{-2}

Source: Carrier et al. (1967).

6.4.3 Lidar and Cloud Reflectance

Although the principles and techniques of lidar are not covered in this book, it is instructive at this point to develop a basic lidar equation involving the backscattering and extinction coefficients. A more detailed, sophisticated treatment is given by Anderson and Browell (1972), which takes into account the factors of beam geometry and second-order scattering. Consider now a cloud irradiated by a pulsed lidar beam. It is necessary to distinguish two cases: (1) a lidar pulse that is very long in relation to the depth of cloud penetration, and (2) a lidar pulse whose length is approximately equal to or less than the depth of penetration. In the extreme the first case becomes equivalent to one of continuous irradiation, as by a searchlight beam, and this first case is considered here.

We assume that the lidar instrument is located outside the cloud at a distance large compared to the depth of cloud penetration, and that the transmitter beam is collimated. The flux backscattered at 180 deg by a unit volume of cloud at depth x in the cloud is proportional to $\beta(\pi)$ and to the irradiance of the incident beam. However, the flux in the beam and the backscattered flux are each attenuated in the cloud according to (1-35). From that equation and the meaning of $\beta(\pi)$, it is clear that the infinitesimal intensity $dI(\pi)$ backscattered from a cloud lamina of thickness dx at depth x in the cloud, as observed at the cloud face, is

$$dI(\pi) = E_0\beta(\pi)\exp(-2\beta_{ex}x)\,dx$$

Here E_0 is the lidar beam irradiance at the cloud face. Integration between the limits 0 and x gives

$$I(\pi, x) = \frac{E_0\beta(\pi)}{2\beta_{ex}}[1 - \exp(-2\beta_{ex}x)] \tag{6-51}$$

for the backscattered intensity, assuming single scattering only. At the near cloud face where $x = 0$, the backscattering is zero. When $x = \infty$ (effectively), as in a thick cloud, (6-51) becomes

$$I(\pi, \infty) = \frac{E_0\beta(\pi)}{2\beta_{ex}} \tag{6-52}$$

Thus the ratio $I(\pi, x)/E_0$ asymptotically approaches the value $\beta(\pi)/2\beta_{ex}$ as x becomes large. This latter ratio can be considered the theoretical reflectance of a thick cloud when it is irradiated by a collimated beam. This ratio is independent of droplet concentration and is dimensionless, although the unit sr^{-1} must be associated with it. Thus cloud reflectances can be predicted, within the limitations of the above assumptions, from values of coefficients such as are given in Tables 6.5 and 6.8. However, $I(\pi, \infty)/E_0$ from (6-52) expresses two measurable quantities determined in the calibration of a lidar transmitter and receiver. In principle the value of this ratio is not affected by changes in E_0 but depends only on the cloud parameters $\beta(\pi)$ and β_{ex}. It expresses the measured reflectance, which can be compared directly with the predicted value.

Figure 6.19 shows the ratio $I(\pi, x)/E_0$, computed from (6-51) and the values in Tables 6.5 and 6.8 for three of the cloud models, at the wavelength 0.694 μm. The asymptotic value of the ratio is indicated for each curve. For a nimbostratus cloud, whose droplets run to larger sizes as typically shown by Figure 3.20, most of the backscattered flux appearing at the cloud face originates in about the first 10 m of cloud penetration. In contrast, the flux from a cumulus cloud, whose droplets are numerous but smaller, originates in about the first 40 m. Despite the great differences between the size distributions, however, the asymptotic values do not differ much among the three models shown, nor among all the models listed in the cited tables. This suggests that a lidar employing long pulses can grossly detect the presence of clouds of different classes with nearly equal ease.

The situation is otherwise, however, when very short pulses are employed, which is necessary if data on cloud distance and structure are to be obtained. A basic problem here is the phenomenon called *pulse*

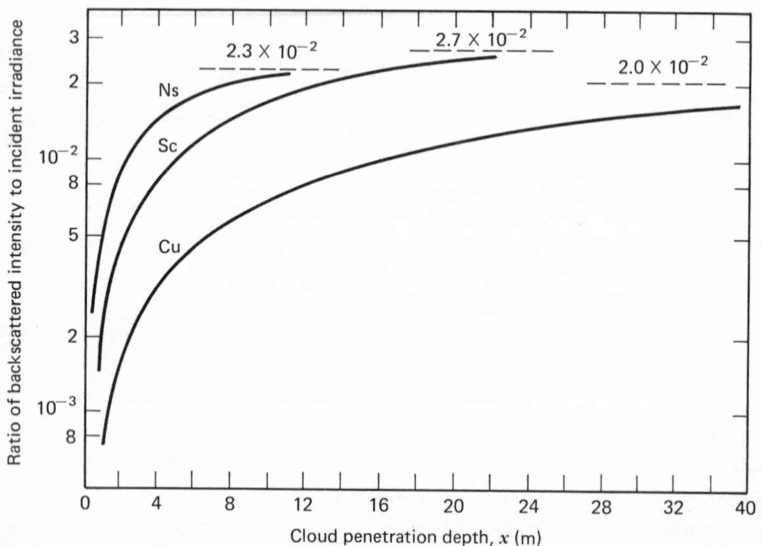

Figure 6.19 Ratio of backscattered intensity to incident irradiance for three cloud models of Figure 3.22 as function of penetration depth in the cloud at $\lambda = 0.694$ μm. Data from Carrier et al. (1967).

stretching, whereby the return from a short transmitted pulse becomes lengthened, with an exponential-like tail. This is due to the fact that backscattering comes from an extended volume along the penetration path. The received pulse thus consists of time-displaced components from all elements of the path. The reader can demonstrate this effect by postulating a narrow, rectangular pulse, for example, one having a duration of 3.3 nsec, hence a length of 1 m. Now let the pulse travel into any of the cloud models noted in Tables 6.5 and 6.8. When the backscattered flux is computed by (6-51) as a function of penetration depth (hence of time), the "stretching" is seen. The effect is evidently greater for short pulses and for clouds of lesser opacity.

Significant discrepancies have been found between measured and predicted cloud reflectances. Milton et al. (1972) reported that the values obtained by their pulsed lidar techniques exceeded by factors of 2 to 3 the theoretical values from single-scattering theory. As long as this theory is employed, such large differences are puzzling. They cannot be entirely attributed to differences between the assumed and actual size distributions because, as shown above for single scattering, the ratio is not very sensitive to the form of the distribution. In later computations, Anderson et al. (1973) took second-order scattering, pulse length, and droplet concentration into account. They found that the ratio of second-order to first-order scattering could vary from 0.10 for a 10.5-m pulse, to 0.30 for an infinitely long pulse. Not unexpectedly, higher ratios of the two orders occurred at greater concentrations of droplets. The comments by Brinkworth (1973), particularly on the effects of pulse length, should be studied. The reader can demonstrate by means of a simple time-distance diagram that the scattered flux reaching the receiver at any instant emanates from a region of cloud equivalent to one-half the pulse length, when this length is less than the penetration depth.

Lidar investigation of clouds—their distances, structures, and compositions—is an important and challenging task. It involves all the parameters of single scattering and of lidar operation, and relationships between geometric factors, pulse shape and length, and higher orders of scattering. In addition to the studies referenced above, the following should be noted. Liou and Schotland (1971) developed radiative transfer relationships for multiple backscattering and investigated depolarization of the backscattered flux as a quantity for distinguishing between water and ice clouds. Liou (1971) formulated a general theory of multiple backscattering by water clouds. He shows that orders of scattering higher than the second can be neglected when the lidar field of view is less than 10^{-2} rad and the cloud distance greater than 1 km. Liou (1972b) studied further the depolarization of multiple backscattering, looking for singularities that could be ascribed to differences between observed and model size distributions of droplets. He concluded that the use of depolarization data to determine droplet size distributions and concentrations would be difficult. Depolarization data do provide, however, a means for distinguishing between water and ice clouds, as found by Liou (1972a). Lidar measurements of cloud properties have been reported by Pal and Carswell (1973) and Cohen (1975). Also important in lidar investigations of clouds are the multiple-scattering studies referenced near the end of Section 6.4.1. The study by Plass and Kattawar (1971d) of the fractional flux returned to the receiver as a function of photon path length is particularly applicable.

6.4.4 Published Tabulations of Functions for Clouds

Although the tabulated cloud-scattering functions available in the literature have been noted in preceding sections, they are listed here to provide a single reference point.

1. Deirmendjian (1969) gives values of the four normalized phase functions $P_j(\theta)/4\pi$, the coefficient β_{ex}, and the albedo $\bar{\omega}$ for the cloud models of Table 3.2 and Figure 3.21. The intervals of θ used for the phase functions are sufficiently small to display the variations in the functions. For model C.1 the tabulations are given at 15 wavelengths in the range 0.45 to 16.6 μm, and for models C.2 and C.3 at 8 and 11 wavelengths, respectively, in this range.

2. Yamamoto et al. (1971) have tabulated similar data for the three cloud classes of Figure 3.23. They provide values of the average phase function $P(\theta)$, the three coefficients β_{sc}, β_{ab}, and β_{ex}, and the albedo. The angular intervals for $P(\theta)$ are $0\text{ deg} \leq \theta\ (1\text{ deg}) \leq 20\text{ deg}$ and $20\text{ deg} \leq \theta\ (2\text{ deg}) \leq 180\text{ deg}$. The data are tabulated at 38 wavelengths in the range 5.00 to 40.0 μm, making this the most detailed set of values for actual cloud classes that has yet appeared.

3. Carrier et al. (1967) give values of $\beta(\pi)$ and β at wavelengths of 0.0488, 0.694, 1.06, 4.0, and 10.6 μm for the cloud models of Figure 3.22. These tabulations are reproduced in Tables 6.5 and 6.8.

4. Zabrodskii and Morachevskii have tabulated β_{ex} at the wavelength 0.52 μm (effectively) for high fog and stratocumulus and nimbostratus clouds. Also listed are the corresponding values of altitude, visual range, liquid water content, and effective droplet radius.

Appendix A

Guide to Measurements of Atmospheric Scattering

Although numerous literature references to atmospheric scattering measurements have been given in Chaps. 1 through 6, the effort there was to emphasize the principles of scattering. Hence the referenced material was selected primarily to illustrate and exemplify the principles, rather than to form a source of scattering measurements data. This chapter, however, is intended to guide the reader to such data. It would not be feasible even to approach any ideal of completeness in this effort; the amount of relevant literature is simply too vast.

Instead, we present an organized listing of measurements found helpful by this author, and generally deemed representative. The organization of the listing follows the one used previously in discussing the characteristics and distributions of atmospheric particles (Chapter 3) and the scattering by polydispersions (Chapter 6). All the measurements cited here were made in the free atmosphere, so that the effects of molecular scattering are implicitly present. The references listed immediately below are general and inclusive, thus are variously applicable to all the meteorological categories of this appendix.

Ångström, A. (1951)
Bullrich, K. (1964)
Fritz, S. (1951)
Henderson, S. T. (1970)
Howard, J. N. and Garing, J. S. (1967)
Ivanov, A. I. (1969)
Johnson, J. C. (1954)
Middleton, W. E. K. (1951, 1952)
Neuberger, H. (1951)
Sekera, Z. (1951)
Zuev, V. E. (1974)

A.1 ANGULAR SCATTERING BY HAZE

The great manifestations of angular scattering by haze are the radiance and polarization of the sky. Thus the measurement of these two effects is in itself an implicit measurement of haze angular scattering. Because of the importance of sky radiance in particular, most workers have measured it directly and allowed the data to signify indirectly the angular scattering characteristic. This situation is analogous to that wherein the extinction by haze is signified by its most outstanding consequence—the value of the visual range. To some extent the angular scattering characteristic is revealed by measurements of diffuse transmission, as cited in Section A.1.5, but only in narrow angles centered about the forward direction. Even there the relationships are difficult to untangle in a practical sense. There is a surprising paucity of *direct* measurements of the angular scattering coefficients of haze, as indicated by the few entries in Section A.1.6.

A.1.1 Day Sky Radiance

Bell, E. E. et al. (1960)
Bennett, H. F. et al. (1960)
Brinkman, R. T. et al. (1967)
Bullrich, K. (1964)
Bullrich, K. et al. (1967)
Bullrich, K. et al. (1968)
Clark, W. M. (1969)
Clark, W. M. and Muldoon, H. A. (1964)
Dorian, M. and Harshbarger, F. (1967)
Gordon, J. I. and Church, P. V. (1966a)
Hopkinson, R. G. (1954)
Hulburt, E. O. (1941)
Kimball, H. H. and Hand, I. F. (1921)
Kimball, H. H. and Hand, I. F. (1922)
Knestrick, G. L. and Curcio, J. A. (1966)
Knestrick, G. L. and Curcio, J. A. (1967)
Knestrick, G. L. and Curcio, J. A. (1970)
Lloyd, J. W. et al. (1965)
MacQueen, R. M. (1968)
Nayatani, Y. and Wyszecki, G. (1963)
Packer, D. M. and Lock, C. (1951)
Sastri, V. D. and Das, S. R. (1968)
Tousey, R. and Hulburt, E. O. (1947)
Volz, F. E. (1970c)

Volz, F. E. (1971)
Volz, F. E. and Sheehan, L. (1971)
Winch, G. T. et al. (1966)

A.1.2 Sky Radiance during an Eclipse

Dandekar, B. S. (1968)
Dandekar, B. S. and Turtle, J. P. (1971)
Hall, W. N. (1971)
Lloyd, J. W. and Silverman, S. M. (1971)
Sharp, W. E. et al. (1966)
Sharp, W. E. et al. (1971)
Velasquez, D. A. (1971)

A.1.3 Twilight Sky Radiance

Kondratyev, K. Ya. et al. (1971)
Koomen, M. J. et al. (1952)
Volz, F. E. (1969b)
Volz, F. E. (1970a)
Volz, F. E. and Goody, R. M. (1962)

A.1.4 Skylight Polarization

Bullrich, K. (1964)
Foitzik, L. and Lenz, K. (1960, 1962)
Foitzik, L. and Zschacek, H. (1961)
Gehrels, T. (1962)
Holzworth, G. C. and Rao, C. R. (1965)
Sekera, Z. (1956)
Volz, F. E. and Bullrich, K. (1961)

A.1.5 Diffuse Transmission

Cantor, I. and Petriw, A. (1968)
Eldridge, R. G. and Johnson, J. C. (1958)
Eldridge, R. G. and Johnson, J. C. (1962)
Gibbons, M. G. (1958)
Gibbons, M. G. (1959)
Gibbons, M. G. et al. (1961)
Green, A. E. et al. (1971)
Herman, B. M. et al. (1971)
Newkirk, G. A., Jr. (1956)
Newkirk, G. A., Jr., and Eddy, J. A. (1964)
Stewart, H. S. and Curcio, J. A. (1952)

A.1.6 Angular Nephelometer Measurements

Pritchard, B. S. and Elliott, W. G. (1960)
Waldram, J. M. (1945a)
Waldram, J. M. (1945b)

A.2 TOTAL SCATTERING AND EXTINCTION BY HAZE

Measurements of the total scattering and extinction by haze fall into one or the other of two categories, depending on whether the solar or an artificial source is used. With the solar source, the optical path extends completely through the atmosphere and is always a slant path, practically. The spectral attentuation of the path thus is governed by the aerosol vertical profile. When an artificial source is used, the entire measurement path can be located near the ground, as desired, to encompass a wide variety of surface meteorological conditions. With either type of source, analysis of the spectral attenuation data provides much information on the aerosol characteristics and on the presence of gases if operations are conducted in the absorption bands. Measurements of visibility and contrast attentuation give further insight into the aerosol, and frequently the haze condition is described merely by stating the measured (even the estimated) value of the visual range.

A.2.1 Direct Transmission: Solar Source

Abbott, C. G. (1963)
Abbott, C. G. and Fowle, F. E. (1913)
Ångström, A. (1929, 1930, 1974a,b)
Arvesen, J. C. et al. (1969)
Dunkelman, L. and Scolnick, R. (1959)
Ellis, H. T. and Pueschel, R. F. (1971)
Fowle, F. E. (1914)
Guttman, A. (1968)
Guzzi, R. et al. (1972)
Irvine, W. M. and Peterson, F. W. (1970)
Kondratyev, K. Ya. et al. (1967)
Machta, L. (1972)
Penndorf, R. (1954)
Quenzel, H. (1970)
Rangajaan, S. (1972)
Roosen, R. G. et al. (1973)
Volz, F. E. (1970b)

A.2.2 Direct Transmission: Artificial Source

Arnulf, A. et al. (1957)
Baum, W. A. and Dunkelman, L. (1955)
Curcio, J. A. (1961)
Curcio, J. A. et al. (1958)
Curcio, J. A. and Durbin, K. A. (1959)
Curcio, J. A. et al. (1961)
Eldridge, R. G. (1967)
Gebbie, H. A. et al. (1951)
Hiser, H. W. et al. (1965)
Knestrick, G. L. et al. (1962)
U.S. (1961)
Yates, H. W. and Taylor, J. H. (1960)

A.2.3 Contrast and Visibility

Barber, D. R. (1950)
Beggs, S. S. and Waldram, J. M. (1943)
Blackwell, H. R. (1946)
Carman, D. D. and Carruthers, R. A. (1951)
Coleman, H. S. et al. (1949)
Coleman, H. S. and Rosenberger, H. E. (1950)
Duntley, S. Q. (1947)
Duntley, S. Q. (1948a)
Duntley, S. Q. (1948b)
Fry, G. A. et al. (1947)
Knoll, H. A., Tousey, R., and Hulburt, E. O. (1946)
Kochsmeider, H. (1930)
MacDonald, D. E. (1953)
Nelson, C. N. and Hamsher, D. H. (1950)
Pearson, C. A. et al. (1952)
Pueschel, R. F. and Noll, K. E. (1967)
Shallenberger, G. D. and Little, E. M. (1940)
Smith, A. G. (1955)
Stewart, H. S. et al. (1949)
Tousey, R. and Hulburt, E. O. (1948)
Vogt, H. (1968)
Winstanley, J. V. and Adams, M. J. (1975)

A.2.4 Integrating Nephelometer Measurements

Ahlquist and Charlson (1969)
Butcher and Charlson (1972)

Charlson, R. J. et al. (1967)
Charlson, R. J. et al. (1969)
Charlson, R. J. et al. (1972)
Ensor, D. S. et al. (1972)
Thielke, J. F. et al. (1972)

A.3 SCATTERING AND EXTINCTION BY FOG

In the early years of its technology, hopes were current that infrared would be attenuated far less than visual light by fog, similarly to the case of haze. The measurements by Houghton, Granath, Hulburt, and Sanderson cited below showed that the hopes were ill-founded, but the hopes did find expression in the use of yellow "fog lamps." Their use persists to this time, even though the work by Luckiesch and Holladay (1941) showed that "the advantage of yellow light in foggy and misty atmospheres is greatly overrated, and from a practical viewpoint is inappreciable if not entirely nonexistent." The measurements of direct transmission by Arnulf et al. (1957) are authoritative and are widely employed. For a given fog condition, the diffuse transmission is significantly greater than the direct, as indicated by the work of Eldridge and Johnson (1958, 1962).

Arnulf, A. et al. (1957)
Eldridge, R. G. and Johnson, J. C. (1958)
Eldridge, R. G. and Johnson, J. C. (1962)
Granath, L. P. and Hulburt, E. O. (1929)
Houghton, H. G. (1931)
Houghton, H. G. (1939)
Hulburt, E. O. (1934)
Kumai, M. (1973)
Kurnick, S. W. et al. (1960)
Luckiesch, M. and Holladay, L. L. (1941)
Pilie, R. J. et al. (1975b)
Sanderson, J. A. (1940)
Smith, P. N. and Hayes, H. V. (1940)
Spencer, D. E. (1960)
Zabrodskii, G. M. and Morachevskii, V. G. (1959)

A.4 SCATTERING AND EXTINCTION BY CLOUDS AND PRECIPITATION

Cloud cover to a varying extent is a condition existing about one-half of the time, on the average, in most of the habitable regions of the world. Such cover is a strong factor of course in determining the amount of incoming solar radiation (insolation) that reaches the lowest layers of the atmosphere and the ground. Hence the prevalence of cloud cover and its properties of direct and diffuse transmission are important influences on local climatology. Despite their importance, the various scattering properties of clouds have not been measured very extensively. This is due largely to the difficulty and expense of *in situ* experiments. Only a few measurements of the optical attentuation due to precipitation have been made; this contrasts strongly with the situation in microwave and radar technology. The works cited below are representative, and that by Chu and Hogg is widely quoted.

A.4.1 Clouds: Angular Scattering and Extinction

Bastin, J. A. et al. (1964)
Blau, H. H., Jr., et al. (1966)
Gates, D. M. and Shaw, C. C. (1960)
Gordon, J. I. and Church, P. V. (1966b)
Haurwitz, B. (1945)
Haurwitz, B. (1946)
Haurwitz, B. (1948)
Hovis, W. A., Jr., and Tobin, M. (1967)
Hovis, W. A., Jr., et al. (1970)
List R. J. (1966)
McDonald, R. K. and Deltenre, R. W. (1963)
Middleton, W. E. K. (1954)
Moon, P. and Spencer, D. E. (1942)
Ruff, I. et al. (1968)

A.4.2 Extinction by Rain and Snow

Chu, T. S. and Hogg, D. C. (1968)
Hogg, D. C. (1964)
Rensch, D. B. and Long, R. K. (1970)
Wilson, R. W. and Penzias, A. A. (1966)

A.5 OPTICAL PROBING OF ATMOSPHERIC AEROSOLS

The investigation of certain atmospheric properties by remote optical sensing is attractive for at least two reasons: (1) The troposphere and lower stratosphere can be examined from the ground with relative convenience and (2) the material and processes within the volume of space being examined are not appreciably disturbed by the operation, which is not true when physical sampling or probing is employed. Several techniques of optically probing the atmosphere have been developed and wisely used, such as measuring the angular scattering of searchlight beams and the backscattering of laser beams (lidar) by the atmospheric aerosol, and spectral and radiometric measurement of emission by the atmospheric gases. In this section are listed most of the accounts of searchlight probing available and several representative papers dealing with lidar probes. Although the latter technique is relatively new, it is coming into widespread use, as indicated by the number of papers listed.

A.5.1 Searchlight Beam

Elterman, L. (1966a,b, 1975)
Hulburt, E. O. (1937)
Johnson, E. A. et al. (1939)
Mikhailin, L. M. and Khvostikov, I. A. (1946)
Mironov, A. V. et al. (1946)
Romantzov, I. I. and Khvostikov, I. A. (1946)
Smirnov, I. P. (1946)

A.5.2 Lidar Beam

Barrett, E. W. and Ben-Dov, O. (1967)
Beneditti-Michelangeli, G. et al. (1972)
Cohen, A. and Graber, M. (1975)
Collis, R. T. (1969)
Collis, R. T. (1970)
Cook, C. S. et al. (1972)
Evans, W. E. and Collis, R. T. (1970)
Fiocco, G. and DeWolf, J. B. (1968)
Grams, G. and Fiocco, G. (1967)
Grams, G. et al. (1972)
Hall, F. F., Jr. (1970)
Hamilton, P. M. (1966)
Hamilton, P. M. (1969)
Horman, M. H. (1961)
Kent, G. S. et al. (1967)

McCormick, M. P. (1967)
McNeil, W. R. and Carswell, A. I. (1975)
Schuster, B. G. (1970)
Uthe, E. E. (1972)
Vogt, H. (1968)
Waggoner, A. P. et al. (1972)

Appendix B

Important Physical Constants

Velocity of light in vacuum	$c = 2.998 \times 10^{8}$ m sec^{-1}
Charge of electron	$e = 1.60 \times 10^{-19}$ C
Mass of electron	$m_e = 9.11 \times 10^{-31}$ kg
Permittivity of free space	$\epsilon_0 = 8.854 \times 10^{-12}$ F m^{-1}
Permeability of free space	$\mu_0 = 4\pi \times 10^{-7}$ H m^{-1}
Planck's constant	$h = 6.63 \times 10^{-34}$ J sec
Boltzmann's constant	$\kappa = 1.38 \times 10^{-23}$ J K^{-1}
Avogadro's number	$N_A = 6.023 \times 10^{23}$ mole^{-1}
Loschmidt's number (at STP)	$N_L = 2.687 \times 10^{19}$ cm^{-3}
Gas constant (universal)	$R* = 8.3143$ J mole^{-1} K^{-1}
Gas constant (dry air at STP)	$R = 2.8706 \times 10^{2}$ J kg^{-1} K^{-1}
Gas constant (water vapor)	$R_V = 4.615 \times 10^{2}$ J kg^{-1} K^{-1}
Acceleration of gravity (at sea level)	$g = 9.806$ m sec^{-2}
Density of air (STP)	$\rho = 1.292$ kg m^{-3}
Mean radius of earth	$= 6.3712 \times 10^{3}$ km
One arc-minute of longitude at the equator	= 1 nmi = 6080 ft
Radius of the moon	$= 1.74 \times 10^{3}$ km
Earth-moon distance	$= 3.84 \times 10^{5}$ km
Radius of the sun	$= 6.96 \times 10^{5}$ km
Earth-sun distance (1 astronomical unit)	$= 1.50 \times 10^{8}$ km
Molecular weight of dry air (effective)	= 28.964 (dimensionless)
Standard atmosphere (sea-level pressure)	= 1013.250 mb
	$= 1.0133 \times 10^{5}$ N m^{-2}
	$= 1.0133 \times 10^{6}$ dyne cm^{-2}
	= 760 mm Hg = 29.213 in Hg

Appendix C

Conversion Factors for Lengths and Areas

TABLE C.1 LENGTHS

Multiply number of → by to obtain ↓	Centimeters	Meters	Kilometers	Feet	Yards	Miles	Nautical miles
Centimeters	1	100	10^5	30.48	91.44	1.609×10^5	1.853×10^5
Meters	0.01	1	1000	0.3048	0.9144	1609	1853
Kilometers	10^{-5}	0.01	1	3.048×10^{-4}	9.144×10^{-4}	1.609	1.853
Feet	3.281×10^{-2}	3.281	3281	1	3	5280	6080
Yards	1.094×10^{-2}	1.094	1094	0.3333	1	1760	2027
Miles	6.214×10^{-6}	6.214×10^{-4}	0.6214	1.894×10^{-4}	5.682×10^{-4}	1	1.152
Nautical miles	5.396×10^{-6}	5.396×10^{-4}	0.5396	1.645×10^{-4}	4.934×10^{-4}	0.8684	1

TABLE C.2 Areas

Multiply number of → by to obtain ↓	Square centimeters	Square meters	Square kilometers	Square feet	Square yards	Square miles
Square centimeters	1	10^4	10^{10}	929.0	8361	2.590×10^{10}
Square meters	10^{-4}	1	10^6	9.290×10^{-2}	0.8361	2.590×10^6
Square kilometers	10^{-10}	10^{-6}	1	9.290×10^{-8}	8.361×10^{-7}	2.590
Square feet	1.076×10^{-3}	10.76	1.076×10^7	1	9	2.788×10^7
Square yards	1.196×10^{-4}	1.196	1.196×10^6	0.1111	1	3.098×10^6
Square miles	3.861×10^{-11}	3.861×10^{-7}	0.3861	3.587×10^{-8}	3.228×10^{-7}	1

Appendix D

Phototopic Spectral Luminous Efficiency of Radiant Flux

λ (nm)	Standard values	Values interpolated at intervals of 1 nm								
		1	2	3	4	5	6	7	8	9
380	0.00004	0.000045	0.000049	0.000054	0.000059	0.000064	0.000071	0.000080	0.000090	0.000104
390	0.00012	0.000138	0.000155	0.000173	0.000193	0.000215	0.000241	0.000272	0.000308	0.000350
400	0.0004	0.00045	0.00049	0.00054	0.00059	0.00064	0.00071	0.00080	0.00090	0.00104
410	0.0012	0.00138	0.00156	0.00174	0.00195	0.00218	0.00244	0.00274	0.00310	0.00352
420	0.0040	0.00455	0.00515	0.00581	0.00651	0.00726	0.00806	0.00889	0.00976	0.01066
430	0.0116	0.01257	0.01358	0.01463	0.01571	0.01684	0.01800	0.01920	0.02043	0.02170
440	0.023	0.0243	0.0257	0.0270	0.0284	0.0298	0.0313	0.0329	0.0345	0.0362
450	0.038	0.0399	0.0418	0.0438	0.0459	0.0480	0.0502	0.0525	0.0549	0.0574
460	0.060	0.0627	0.0654	0.0681	0.0709	0.0739	0.0769	0.0802	0.0836	0.0872
470	0.091	0.0950	0.0992	0.1035	0.1080	0.1126	0.1175	0.1225	0.1278	0.1333
480	0.139	0.1448	0.1507	0.1567	0.1629	0.1693	0.1761	0.1833	0.1909	0.1991
490	0.208	0.2173	0.2270	0.2371	0.2476	0.2586	0.2701	0.2823	0.2951	0.3087
500	0.323	0.3382	0.3544	0.3714	0.3890	0.4073	0.4259	0.4450	0.4642	0.4836
510	0.503	0.5229	0.5436	0.5648	0.5865	0.6082	0.6299	0.6511	0.6717	0.6914
520	0.710	0.7277	0.7449	0.7615	0.7776	0.7932	0.8082	0.8225	0.8363	0.8495

λ (nm)	Standard values	Values interpolated at intervals of 1 nm								
		1	2	3	4	5	6	7	8	9
530	0.862	0.8739	0.8851	0.8956	0.9056	0.9149	0.9238	0.9320	0.9398	0.9471
540	0.954	0.9604	0.9661	0.9713	0.9760	0.9803	0.9840	0.9873	0.9902	0.9928
550	0.995	0.9969	0.9983	0.9994	1.0000	1.0002	1.0001	0.9995	0.9984	0.9969
560	0.995	0.9926	0.9898	0.9865	0.9828	0.9786	0.9741	0.9691	0.9638	0.9581
570	0.952	0.9455	0.9386	0.9312	0.9235	0.9154	0.9069	0.8981	0.8890	0.8796
580	0.870	0.8600	0.8496	0.8388	0.8277	0.8163	0.8046	0.7928	0.7809	0.7690
590	0.757	0.7449	0.7327	0.7202	0.7076	0.6949	0.6822	0.6694	0.6565	0.6437
600	0.631	0.6182	0.6054	0.5926	0.5797	0.5668	0.5539	0.5410	0.5282	0.5156
610	0.503	0.4905	0.4781	0.4658	0.4535	0.4412	0.4291	0.4170	0.4049	0.3929
620	0.381	0.3690	0.3570	0.3449	0.3329	0.3210	0.3092	0.2977	0.2864	0.2755
630	0.265	0.2548	0.2450	0.2354	0.2261	0.2170	0.2082	0.1996	0.1912	0.1830
640	0.175	0.1672	0.1596	0.1523	0.1452	0.1382	0.1316	0.1251	0.1188	0.1128
650	0.107	0.1014	0.0961	0.0910	0.0862	0.0816	0.0771	0.0729	0.0688	0.0648
660	0.061	0.0574	0.0539	0.0506	0.0475	0.0446	0.0418	0.0391	0.0366	0.0343
670	0.032	0.0299	0.0280	0.0263	0.0247	0.0232	0.0219	0.0206	0.0194	0.0182
680	0.017	0.01585	0.01477	0.01376	0.01281	0.01192	0.01108	0.01030	0.00956	0.00886
690	0.0082	0.00759	0.00705	0.00656	0.00612	0.00572	0.00536	0.00503	0.00471	0.00440
700	0.0041	0.00381	0.00355	0.00332	0.00310	0.00291	0.00273	0.00256	0.00241	0.00225
710	0.0021	0.001954	0.001821	0.001699	0.001587	0.001483	0.001387	0.001297	0.001212	0.001130
720	0.00105	0.000975	0.000907	0.000845	0.000788	0.000736	0.000688	0.000644	0.000601	0.000560
730	0.00052	0.000482	0.000447	0.000415	0.000387	0.000360	0.000335	0.000313	0.000291	0.000270
740	0.00025	0.000231	0.000214	0.000198	0.000185	0.000172	0.000160	0.000149	0.000139	0.000130
750	0.00012	0.000111	0.000103	0.000096	0.000090	0.000084	0.000078	0.000074	0.000069	0.000064
760	0.00006	0.000056	0.000052	0.000048	0.000045	0.000042	0.000039	0.000037	0.000035	0.000032

Note: Unity at wavelength of maximum luminous efficacy.

Source: ANSI (1967). (Courtesy of Illuminating Engineering Society.)

Appendix E

USSA-1962: Temperature, Pressure, and Density versus Altitude

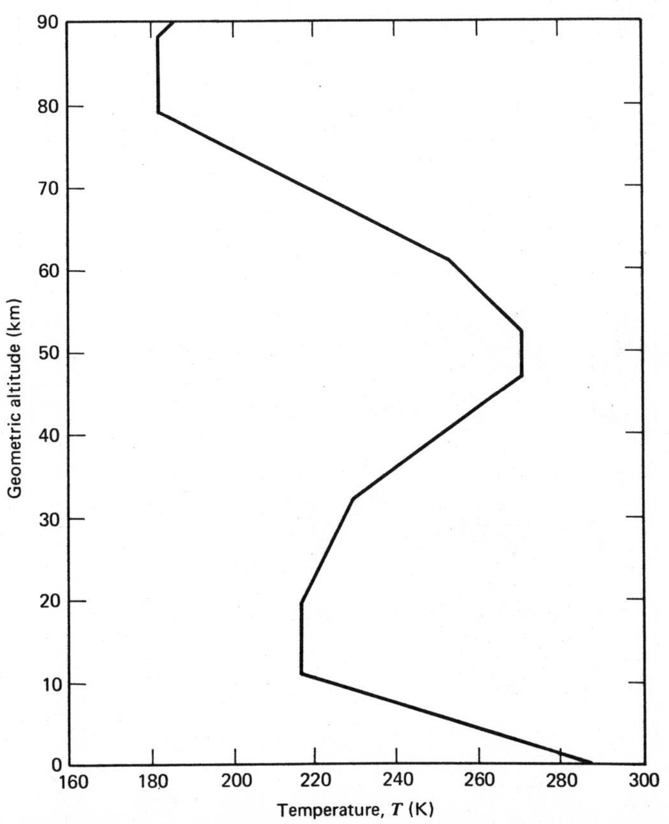

Figure E.1 Temperature T versus geometric altitude.

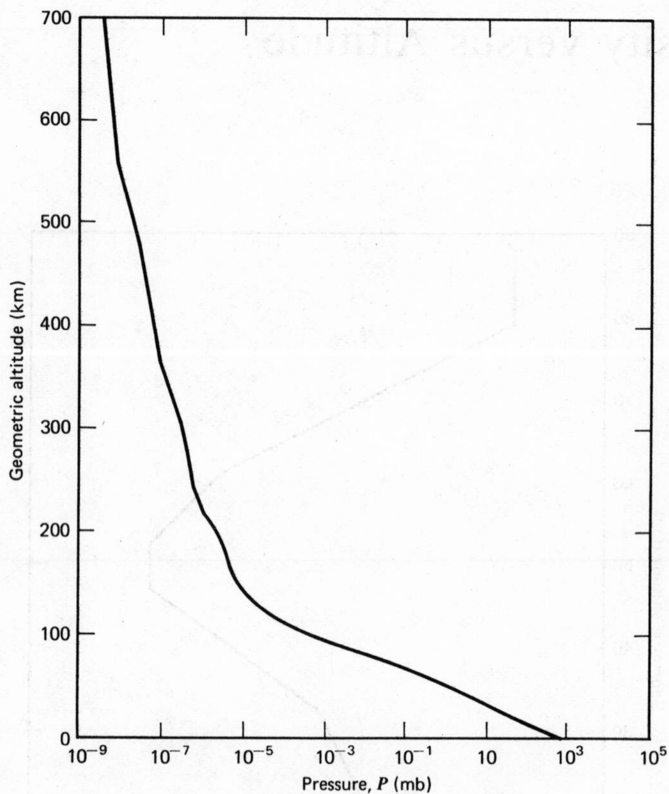

Figure E.2 Pressure P versus geometric altitude.

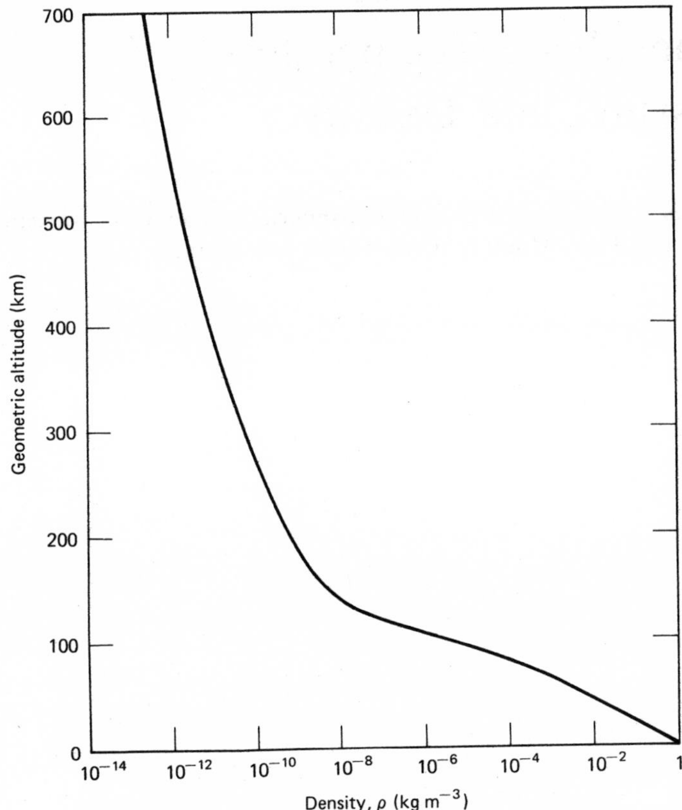

Figure E.3 Density ρ versus geometric altitude.

Appendix F

Monthly Profiles of Atmospheric Temperature, Pressure, and Density

Tables F.1 through F.3 list atmospheric temperature, pressure, and density to 79 km. Monthly mean values are given.

TABLE F.1 LATITUDE 30 deg NORTH

Altitude (m)	Jan.	Feb.	March	Apr.	May	June	July	Aug.	Sept.	Oct.	Nov.	Dec.
						Temperature (K)						
0	287.15	286.55	289.15	292.15	295.15	298.65	301.15	298.65	296.65	293.65	289.15	286.15
5,000	261.65	264.15	262.65	267.15	271.15	272.65	271.65	272.15	272.15	268.65	266.15	261.25
10,000	229.15	231.65	230.15	232.15	233.65	235.15	238.15	239.95	234.65	233.65	231.15	229.75
15,000	208.35	209.95	210.55	210.95	208.15	203.65	203.15	206.95	207.65	207.65	208.75	209.05
20,000	208.15	208.15	210.15	211.15	211.55	212.05	211.95	210.65	211.75	211.15	209.15	209.65
25,000	219.15	218.15	220.15	221.65	223.05	222.55	222.15	220.65	222.25	221.25	219.65	219.85
30,000	229.15	228.15	231.05	231.65	234.55	233.05	232.15	230.65	230.75	229.75	227.15	226.85
35,000	240.35	239.65	242.55	242.85	246.05	245.15	243.35	242.15	241.65	240.65	237.95	237.45
40,000	252.35	252.15	254.05	254.85	257.55	257.65	255.35	254.65	254.15	253.15	250.95	250.45
45,000	264.35	264.65	265.55	266.85	269.05	270.15	267.35	267.15	266.65	265.65	263.95	263.45
50,000	269.15	269.65	270.15	271.65	273.65	272.65	272.15	272.15	271.65	270.65	269.15	268.65
55,000	261.15	263.65	262.15	261.55	263.65	262.65	264.15	266.15	263.65	264.65	263.15	262.65
60,000	250.05	254.35	250.95	248.45	250.35	249.35	252.35	256.55	252.35	255.25	253.85	253.45
65,000	234.55	237.85	234.95	232.45	233.85	232.85	233.35	238.55	235.85	238.25	237.35	237.45
70,000	219.05	221.35	218.95	216.45	217.35	216.35	214.35	220.55	219.35	212.25	220.85	221.45
75,000	203.55	204.85	202.95	200.45	200.85	199.85	195.35	202.55	202.85	204.25	204.35	205.45
79,000	191.15	191.65	190.15	187.65	187.65	186.65	180.15	188.15	189.65	190.65	191.15	192.65
						Pressure (mb)						
0	1.0210	1.0190	1.0180	1.0170	1.0155	1.0140	1.0135	1.0135	1.0150	1.0170	1.0190	1.0200 +3*
5,000	5.5020	5.4909	5.4860	5.5191	5.5530	5.5692	5.5700	5.5612	5.5556	5.5372	5.5208	5.4866 +2
10,000	2.7402	2.7540	2.7400	2.7812	2.8188	2.8385	2.8515	2.8540	2.8278	2.8017	2.7743	2.7334
15,000	1.2437	1.2601	1.2499	1.2732	1.2888	1.2958	1.3127	1.3269	1.2953	1.2826	1.2638	1.2444
20,000	5.4096	5.5011	5.4942	5.6030	5.6479	5.6768	5.7429	5.7963	5.6629	5.6032	5.5031	5.4289 +1
25,000	2.4342	2.4680	2.4834	2.5460	2.5729	2.5861	2.6146	2.6247	2.5826	2.5469	2.4887	2.4589
30,000	1.1359	1.1478	1.1638	1.1981	1.2193	1.2216	1.2325	1.2310	1.2148	1.1940	1.1584	1.1444
35,000	5.4820	5.5235	5.5654	5.8269	5.9888	5.9757	6.0030	5.9704	5.8846	5.7660	5.5450	5.4696 +0

TABLE F.1 (continued)

Altitude (m)	Jan.	Feb.	March	Apr.	May	June	July	Aug.	Sept.	Oct.	Nov.	Dec.
40,000	2.7400	2.7572	2.8426	2.9327	3.0386	3.0286	3.0255	3.0012	2.9539	2.8863	2.7565	2.7151
45,000	1.4143	1.4234	1.4727	1.5234	1.5881	1.5852	1.5736	1.5591	1.5327	1.4938	1.4195	1.3964
50,000	7.4804	7.5365	7.8087	8.1051	8.4892	8.4673	8.3819	8.3041	8.1536	7.9284	7.5066	7.3756 − 1
55,000	3.9351	3.9771	4.1177	4.2810	4.5052	4.4830	4.4410	4.4083	4.3150	4.1939	3.9566	3.8829
60,000	2.0192	2.0602	2.1183	2.1925	2.3193	2.3019	2.2961	2.2977	2.2284	2.1779	2.0470	2.0063
65,000	0.9975	1.0289	1.0484	1.0772	1.1450	1.1331	1.1359	1.1521	1.1065	1.0896	1.0208	1.0001
70,000	4.6956	4.8878	4.9376	5.0310	5.3683	5.2945	5.2936	5.4721	5.2224	5.1789	4.8417	4.7491 − 2
75,000	2.0914	2.1919	2.1962	2.2161	2.3707	2.3289	2.2980	2.4393	2.3242	2.3193	2.1671	2.1326
79,000	1.0462	1.0999	1.0955	1.0956	1.1728	1.1480	1.1094	1.2115	1.1581	1.1606	1.0856	1.0731
						Density (kg m^{-3})						
0	1.2387	1.2384	1.2265	1.2127	1.1986	1.1828	1.1724	1.1822	1.1920	1.2065	1.2277	1.2418 + 0*
5,000	7.3255	7.2416	7.2764	7.1970	7.1344	7.1159	7.1431	7.1186	7.1116	7.1802	7.2263	7.3162 − 1
10,000	4.1657	4.1416	4.1474	4.1734	4.2028	4.2051	4.1712	4.1435	4.1982	4.1773	4.1812	4.1447
15,000	2.0796	2.0908	2.0680	2.1026	2.1570	2.2166	2.2510	2.2337	2.1731	2.1517	2.1091	2.0738
20,000	9.0537	9.2069	9.1078	9.2441	9.3007	9.3262	9.4393	9.5858	9.3165	9.2444	9.1662	9.0209 − 2
25,000	3.8694	3.9412	3.9297	4.0016	4.0184	4.0481	4.1002	4.1440	4.0481	4.0101	3.9472	3.8963
30,000	1.7269	1.7525	1.7547	1.8018	1.8110	1.8261	1.8494	1.8593	1.8339	1.8104	1.7766	1.7574
35,000	7.9457	8.0292	8.1242	8.3587	8.4792	8.4917	8.5937	8.5893	8.4833	8.3469	8.1181	8.0246 − 3
40,000	3.7825	3.8093	3.8980	4.0089	4.1101	4.0950	4.1277	4.1057	4.0490	3.9720	3.8266	3.7766
45,000	1.8638	1.8736	1.9320	1.9888	2.0563	2.0442	2.0505	2.0332	2.0024	1.9590	1.8735	1.8465
50,000	0.9682	0.9737	1.0070	1.0394	1.0807	1.0819	1.0729	1.0630	1.0456	1.0205	0.9716	0.9564
55,000	5.2494	5.2551	5.4719	5.6999	5.9528	5.9460	5.8570	5.7701	5.7016	5.5206	5.2379	5.1501 − 4
60,000	2.8131	2.8217	2.9406	3.0743	3.2274	3.2160	3.1697	3.1201	3.0762	2.9725	2.8091	2.7577
65,000	1.4815	1.5069	1.5545	1.6144	1.7058	1.6952	1.6958	1.6825	1.6344	1.5932	1.4983	1.4673
70,000	7.4676	7.6926	7.8562	8.0971	8.6043	8.5252	8.6033	8.6434	8.2942	8.1544	7.6372	7.4709 − 5
75,000	3.5794	3.7276	3.7699	3.8514	4.1119	4.0596	4.0980	4.1954	3.9915	3.9558	3.6944	3.6160
79,000	1.9067	1.9993	2.0071	2.0339	2.1773	2.1426	2.1453	2.2432	2.1273	2.1207	1.9784	1.9405

*Power of 10 by which preceeding numbers should be multiplied.
Source: Kantor and Cole (1965).

TABLE F.2 LATITUDE 45 deg NORTH

Altitude (m)	Jan.	Feb.	March	Apr.	May	June	July	Aug.	Sept.	Oct.	Nov.	Dec.
						Temperature (K)						
0	272.15	273.15	274.15	279.15	284.65	288.15	294.15	292.15	288.15	284.15	278.15	273.15
5,000	249.65	249.65	253.15	257.15	260.15	263.40	267.15	265.40	262.40	257.15	254.15	251.15
10,000	219.65	217.15	223.15	224.65	225.15	231.90	235.15	231.90	230.90	227.15	224.15	218.65
15,000	217.15	217.15	217.15	218.15	218.15	216.15	215.65	215.15	215.15	215.15	215.35	216.15
20,000	215.15	217.15	217.15	218.15	218.15	219.45	219.25	218.75	217.65	215.15	214.65	216.15
25,000	215.15	219.65	220.15	221.65	223.15	224.95	225.25	224.75	222.65	220.15	218.65	216.15
30,000	219.90	224.65	225.15	228.65	231.65	233.15	233.65	233.15	227.65	225.15	222.65	216.15
35,000	229.90	231.65	233.40	239.15	244.90	245.40	245.40	244.40	239.15	234.15	229.65	227.90
40,000	240.65	242.15	246.65	252.40	258.40	259.15	258.40	257.40	251.90	246.90	241.90	241.40
45,000	255.60	255.90	260.65	264.65	268.90	269.65	268.15	268.40	268.65	258.90	255.15	257.15
50,000	264.40	263.40	266.65	269.40	272.40	273.40	271.65	271.40	269.40	265.40	262.65	265.40
55,000	259.90	258.15	258.65	262.15	265.90	267.65	268.40	265.15	262.40	262.40	259.15	264.15
60,000	247.65	247.90	244.15	251.15	253.15	254.65	256.65	253.15	248.40	251.90	249.65	255.65
65,000	238.65	237.40	233.15	238.15	235.90	238.65	235.40	233.65	232.15	238.40	239.15	241.90
70,000	232.15	227.40	225.40	222.90	216.90	216.40	212.15	213.15	213.65	223.65	225.15	226.65
75,000	218.65	218.15	216.90	207.15	198.40	191.40	190.15	194.65	196.15	207.90	211.15	212.65
79,000	206.65	210.95	209.30	193.55	182.80	173.40	171.85	177.85	183.35	195.50	201.55	203.85
						Pressure (mb)						
0	1.0180	1.0165	1.0160	1.0150	1.0145	1.0130	1.0135	1.0150	1.0165	1.0175	1.0180	1.0180 +3
5,000	5.3072	5.3097	5.3368	5.3928	5.4428	5.4709	5.5224	5.5176	5.4805	5.4178	5.3714	5.3316 +2
10,000	2.5602	2.5510	2.6024	2.6509	2.6888	2.7422	2.7963	2.7730	2.7393	2.6734	2.6271	2.5736
15,000	1.1711	1.1617	1.1876	1.2143	1.2319	1.2614	1.2924	1.2721	1.2556	1.2189	1.1983	1.1730
20,000	5.3098	5.2900	5.4080	5.5497	5.6300	5.7439	5.8760	5.7735	5.6891	5.5103	5.4085	5.3222 +1
25,000	2.4004	2.4144	2.4707	2.5442	2.5959	2.6627	2.7244	2.6722	2.6185	2.5137	2.4584	2.4148
30,000	1.0898	1.1191	1.1471	1.1914	1.2211	1.2608	1.2919	1.2651	1.2262	1.1671	1.1335	1.0956
35,000	5.0963	5.2839	5.4308	5.7231	5.9557	6.1632	6.3226	6.1799	5.8972	5.5287	5.3123	5.0765 +0

TABLE F.2 (continued)

Altitude (m)	Jan.	Feb.	March	Apr.	May	June	July	Aug.	Sept.	Oct.	Nov.	Dec.
40,000	2.4649	2.5660	2.6616	2.8573	3.0225	3.1325	3.2098	3.1268	2.9442	2.7177	2.5693	2.4471
45,000	1.2347	1.2905	1.3580	1.4758	1.5830	1.6441	1.6804	1.6351	1.5173	1.3834	1.2931	1.2328
50,000	6.4418	6.7176	7.1345	7.8173	8.4457	8.7858	8.9418	8.7112	8.0229	7.2327	6.7035	6.4418 −1
55,000	3.3637	3.4996	3.7390	4.1224	4.4903	4.6867	4.7590	4.6185	4.2366	3.7965	3.4972	3.3829
60,000	1.7211	1.7818	1.8943	2.1173	2.3265	2.4394	2.4905	2.3919	2.1708	1.9574	1.7862	1.7593
65,000	0.8475	0.8812	0.9238	1.0554	1.1593	1.2210	1.2470	1.1899	1.0674	0.9750	0.8888	0.8851
70,000	4.1199	4.2233	4.3841	5.0260	5.4425	5.7846	5.7972	5.5146	4.9601	4.6582	4.2643	4.2737 −2
75,000	1.9325	1.9611	2.0266	2.2728	2.3919	2.4938	2.4803	2.3908	2.1506	2.1094	1.9449	1.9579
79,000	1.0162	1.0372	1.0672	1.1488	1.1673	1.1782	1.1648	1.1473	1.0463	1.0711	1.0028	1.0157
						Density(kg m^{-3})						
0	1.3031	1.2964	1.2910	1.2667	1.2416	1.2247	1.2003	1.2103	1.2289	1.2475	1.2750	1.2983 +0
5,000	7.4058	7.4093	7.3442	7.3058	7.2885	7.2357	7.2013	7.2425	7.2760	7.3396	7.3626	7.3954 −1
10,000	4.0606	4.0925	4.0627	4.1108	4.1604	4.1194	4.1426	4.1657	4.1329	4.1001	4.0830	4.1004
15,000	1.8788	1.8636	1.9052	1.9392	1.9672	2.0330	2.0877	2.0598	2.0331	1.9737	1.9385	1.8905
20,000	8.5976	8.4866	8.6760	8.8624	8.9907	9.1182	9.3364	9.1945	9.1059	8.9221	8.7778	8.5777 −2
25,000	3.8866	3.8292	3.9096	3.9988	4.0526	4.1236	4.2135	4.1420	4.0971	3.9777	3.9169	3.8919
30,000	1.7265	1.7353	1.7749	1.8151	1.8364	1.8838	1.9262	1.8903	1.8764	1.8058	1.7736	1.7659
35,000	7.7224	7.9462	8.1059	8.3368	8.4719	8.7492	8.9755	8.8089	8.5724	8.2256	8.0585	7.7599 −3
40,000	3.5683	3.6916	3.7593	3.9437	4.0748	4.2109	4.3274	4.2318	4.0718	3.8346	3.7002	3.5314
45,000	1.6825	1.7568	1.8150	1.9426	2.0508	2.1241	2.1830	2.1223	2.0048	1.8615	1.7655	1.6700
50,000	0.8488	0.8884	0.9321	1.0109	1.0801	1.1195	1.1467	1.1182	1.0375	0.9494	0.8891	0.8456
55,000	4.5087	4.7226	5.0360	5.4783	5.8829	6.1002	6.1769	6.0680	5.6246	5.0403	4.7011	4.4615 −4
60,000	2.4210	2.5039	2.7029	2.9369	3.2016	3.3372	3.3806	3.2916	3.0444	2.7071	2.4925	2.3973
65,000	1.2371	1.2931	1.3803	1.5438	1.7120	1.7824	1.8454	1.7741	1.6017	1.4248	1.2947	1.2746
70,000	6.1810	6.4699	6.7759	7.8551	8.7414	9.3123	9.5195	9.0130	8.0878	7.2559	6.5981	6.5688
75,000	3.0790	3.1318	3.2550	3.8222	4.1998	4.5390	4.5441	4.2789	3.8195	3.5346	3.2088	3.2075
79,000	1.7130	1.7129	1.7763	2.0676	2.2246	2.3671	2.3626	2.2474	1.9881	1.9086	1.7334	1.7358

*Power of 10 by which preceding numbers should be multiplied.
Source: Kantor and Cole (1965).

TABLE F.3 LATITUDE 60 deg NORTH

Altitude (m)	Jan.	Feb.	March	Apr.	May	June	July	Aug.	Sept.	Oct.	Nov.	Dec.
						Temperature (K)						
0	257.15	256.65	261.65	269.15	276.65	282.65	287.15	284.15	281.15	275.15	266.15	259.15
5,000	240.95	244.90	245.65	250.15	253.25	259.65	260.15	262.15	256.15	252.65	248.65	243.90
10,000	217.15	218.65	219.65	222.15	223.65	224.65	225.15	224.15	221.15	220.15	218.65	217.65
15,000	217.15	218.65	219.65	222.15	223.65	224.65	225.15	224.15	221.15	220.15	218.65	217.65
20,000	214.15	218.65	219.65	222.15	223.65	224.65	225.15	224.15	221.15	218.35	216.65	214.15
25,000	211.15	219.55	215.15	222.15	223.65	228.65	228.15	227.75	222.15	218.65	214.65	210.65
30,000	216.15	222.55	217.65	225.65	231.15	233.65	235.65	232.25	227.15	221.15	219.65	212.65
35,000	222.65	225.55	225.35	238.15	244.65	246.65	247.65	245.15	236.65	229.65	224.65	219.65
40,000	235.15	236.15	240.85	250.65	259.65	261.65	262.65	260.15	249.15	242.15	235.65	233.15
45,000	247.65	248.65	256.35	263.15	270.85	273.05	273.85	271.15	261.65	254.65	248.15	246.65
50,000	260.15	261.15	265.65	270.65	274.15	276.65	277.15	275.15	271.65	267.15	260.65	260.15
55,000	258.35	259.35	260.65	266.65	269.15	272.65	273.15	269.15	268.75	264.95	259.15	258.65
60,000	250.65	251.15	248.15	256.65	256.65	260.05	260.45	252.85	253.35	253.65	251.65	251.15
65,000	248.15	246.15	240.15	242.15	236.65	237.05	236.95	231.35	234.35	241.15	243.25	243.65
70,000	244.75	240.35	234.15	224.65	216.65	214.05	213.45	209.85	215.35	228.65	231.25	236.15
75,000	237.75	231.35	224.15	207.15	196.65	191.05	189.95	188.35	196.35	216.15	219.25	228.65
79,000	232.15	224.15	216.15	193.15	180.65	172.65	171.15	171.15	181.15	206.15	209.65	222.65
						Pressure (mb)						
0	1.0135	1.0140	1.0140	1.0130	1.0125	1.0105	1.0100	1.0105	1.0120	1.0110	1.0212	1.0125 +3*
5,000	5.1582	5.1916	5.2147	5.2668	5.3293	5.3952	5.4076	5.4051	5.3689	5.3034	5.2373	5.1866 +2
10,000	2.4160	2.4504	2.4793	2.5302	2.5771	2.6614	2.6715	2.6735	2.6211	2.5719	2.4940	2.4399
15,000	1.1002	1.1219	1.1392	1.1728	1.2007	1.2442	1.2510	1.2477	1.2107	1.1838	1.1419	1.1131
20,000	4.9826	5.1364	5.2341	5.4359	5.5940	5.8166	5.8583	5.8232	5.5922	5.4384	5.2092	5.0456 +1
25,000	2.2315	2.3528	2.3856	2.5196	2.6063	2.7338	2.7489	2.7309	2.5839	2.4827	2.3592	2.2576
30,000	1.0031	1.0864	1.0834	1.1749	1.2296	1.3056	1.3159	1.2994	1.2079	1.1417	1.0743	1.0072
35,000	4.5949	5.0684	4.9826	5.6237	5.9849	6.3943	6.4717	6.3355	5.7657	5.3288	4.9793	4.5697 +0

TABLE F.3 (continued)

Altitude (m)	Jan.	Feb.	March	Apr.	May	June	July	Aug.	Sept.	Oct.	Nov.	Dec.
40,000	2.1782	2.4120	2.3938	2.7952	3.0392	3.2645	3.3127	3.2216	2.8535	2.5828	2.3655	2.1484
45,000	1.0733	1.1920	1.2039	1.4375	1.6004	1.7274	1.7570	1.6983	1.4617	1.2983	1.1673	1.0539
50,000	5.4764	6.0979	6.2852	7.6063	8.5635	9.2938	9.4652	9.0955	7.7204	6.7449	5.9638	5.3701 − 1
55,000	2.8388	3.1690	3.2961	4.0389	4.5820	5.0033	5.1013	4.8755	4.1139	3.5568	3.0956	2.7839
60,000	1.4488	1.6214	1.6840	2.1023	2.3923	2.6410	2.6959	2.5367	2.1399	1.8410	1.5858	1.4243
65,000	0.7304	0.8157	0.8343	1.0623	1.1964	1.3277	1.3558	1.2521	1.0617	0.9228	0.7959	0.7140
70,000	3.6558	4.0453	4.0647	5.1078	5.6282	6.2216	6.3457	5.7689	4.9647	4.4591	3.8736	3.5032 − 2
75,000	1.8008	1.9605	1.9286	2.3144	2.4610	2.6746	2.7181	2.4442	2.1640	2.0682	1.8143	1.6797
79,000	1.0066	1.0759	1.0367	1.1690	1.1921	1.2607	1.2743	1.1421	1.0487	1.0826	0.9592	0.9167
Density ($kg\ m^{-3}$)												
0	1.3730	1.3764	1.3501	1.3112	1.2750	1.2454	1.2253	1.2389	1.2540	1.2800	1.3246	1.3611 + 0*
5,000	7.4578	7.3850	7.3952	7.3347	7.3310	7.2386	7.2414	7.1827	7.3018	7.3126	7.3377	7.4082 − 1
10,000	3.8759	3.9041	3.9322	3.9677	4.0142	4.1271	4.1335	4.1551	4.1288	4.0697	3.9737	3.9052
15,000	1.7650	1.7874	1.8067	1.8391	1.8792	1.9294	1.9357	1.9392	1.9071	1.8732	1.8193	1.7816
20,000	8.1054	8.1837	8.3014	8.5244	8.7136	9.0199	9.0644	9.0503	8.8091	8.6767	8.3763	8.2079 − 2
25,000	3.6818	3.7333	3.8627	3.9512	4.0597	4.1652	4.1973	4.1772	4.0520	3.9556	3.8289	3.7335
30,000	1.6167	1.7006	1.7340	1.8138	1.8532	1.9467	1.9454	1.9491	1.8526	1.7985	1.7038	1.6501
35,000	7.1894	7.8282	7.7025	8.2264	8.5222	9.0313	9.1037	9.0030	8.4876	8.0836	7.7215	7.2477 − 3
40,000	3.2270	3.5582	3.4624	3.8850	4.0777	4.3464	4.3939	4.3140	3.9898	3.7157	3.4970	3.2102
45,000	1.5098	1.6700	1.6360	1.9030	2.0585	2.2039	2.2351	2.1820	1.9461	1.7761	1.6387	1.4886
50,000	0.7333	0.8134	0.8242	0.9790	1.0882	1.1703	1.1897	1.1516	0.9901	0.8795	0.7971	0.7191
55,000	3.8279	4.2567	4.4053	5.2766	5.9306	6.3928	6.5061	6.3105	5.3327	4.6766	4.1613	3.7496 − 4
60,000	2.0137	2.2490	2.3641	2.8536	3.2473	3.5380	3.6060	3.4950	2.9425	2.5284	2.1953	1.9756
65,000	1.0254	1.1544	1.2103	1.5282	1.7612	1.9512	1.9933	1.8854	1.5783	1.3332	1.1399	1.0209
70,000	0.5204	0.5863	0.6048	0.7921	0.9050	1.0126	1.0357	0.9577	0.8031	0.6794	0.5835	0.5168
75,000	2.6387	2.9521	2.9974	3.8922	4.3596	4.8770	4.9850	4.5207	3.8394	3.3333	2.8827	2.5592 − 5
79,000	1.5105	1.6721	1.6708	2.1084	2.2989	2.5439	2.5937	2.3247	2.0168	1.8295	1.5939	1.4342

*Power of 10 by which preceding numbers should be multiplied.
Source: Kantor and Cole (1965).

Appendix G

Vertical Profile Quantities of Mean Annual Moisture at Midlatitude

Altitude (km)	Frost point (°C)	Vapor pressure (mb)	Vapor density (g m^{-3})	Relative humidity (%)	Mixing ratio (ppm)	Precipitable water (cm)
		+0	+0		+3	−1
0	3	7.575	5.947	75	4.686	8.893
2	−7	3.618	2.946	51	2.843	4.020
4	−20	1.254	1.074	47	1.268	1.452
		−1	−1		+2	−2
6	−32	4.205	3.779	48	5.548	4.951
8	−44	1.239	1.172	48	2.165	1.355
		−2	−2		+1	−3
10	−56	1.838	1.834	47	4.324	2.205
		−3	−3			−4
12	−68	3.511	3.708	20	1.130	4.549
		−4	−4		+0	
14	−78	7.577	8.413	4	3.342	1.455
16	−80	5.472	6.138	3	3.308	1.059
						−5
18	−82	3.925	4.449	2	3.253	8.898
20	−82	3.925	4.449	2	4.459	9.679
						−4
22	−81	4.638	5.230	2	7.212	1.137
					+1	
24	−80	5.472	6.138	2	1.162	1.333
26	−79	6.444	7.191	2	1.862	1.242
						−5
28	−81	4.638	5.230	1	1.819	9.008
30	−83	3.316	3.778	1	1.760	6.488

Altitude (km)	Frost point (°C)	Vapor pressure (mb)	Vapor density (g m^{-3})	Relative humidity (%)	Mixing ratio (ppm)	Precipitable water (cm)
32	−85	2.353	2.710	—	1.686	6.501
35	−88	1.388	1.624	—	1.545	5.726
		−5	−5			
40	−93	5.542	6.665	—	1.242	2.454
45	−97	2.561	3.150	—	1.113	1.082
		−6			+0	−6
50	−102	9.287	1.176	—	7.606	3.974

Notes: A one-digit number, preceded by a plus or minus sign, between lines indicates the power of 10 by which each succeeding entry of that column is to be multiplied. The value of precipitable water is that contained in the next height interval.

Source: Adapted from Sissenwine et al. (1968).

Appendix H

Rayleigh Angular Scattering Coefficient and Phase Function

Below are given the Rayleigh angular scattering coefficient and phase function for air at 288.15 K and 1013.25 mb at $\lambda = 0.55\ \mu m$.

θ (deg)	$P(\theta)$	$\beta_m(\theta)$ (km^{-1})	θ (deg)
0	1.4379	1.363×10^{-3}	180
1	1.4737	1.363	179
2	1.4730	1.362	178
3	1.4719	1.361	177
4	1.4704	1.360	176
5	1.4685	1.358	175
6	1.4661	1.356	174
7	1.4634	1.353	173
8	1.4601	1.350	172
9	1.4565	1.347	171
10	1.4525	1.343	170
11	1.4481	1.339	169
12	1.4432	1.334	168
13	1.4379	1.330	167
14	1.4323	1.324	166
15	1.4263	1.319	165
16	1.4199	1.313	164
17	1.4131	1.307	163
18	1.4060	1.301	162
19	1.3985	1.293	161
20	1.3908	1.286	160
21	1.3826	1.278	159
22	1.3741	1.270	158
23	1.3654	1.263	157
24	1.3563	1.254	156
25	1.3469	1.245	155

θ (deg)	$P(\theta)$	$\beta_m(\theta)$ (km^{-1})	θ (deg)
26	1.3373	1.237	154
27	1.3274	1.227	153
28	1.3172	1.218	152
29	1.3068	1.208	151
30	1.2962	1.198	150
31	1.2853	1.188	149
32	1.2743	1.178	148
33	1.2630	1.168	147
34	1.2516	1.157	146
35	1.2400	1.146	145
36	1.2283	1.136	144
37	1.2164	1.125	143
38	1.2044	1.114	142
39	1.1923	1.102	141
40	1.1801	1.091	140
41	1.1679	1.080	139
42	1.1556	1.069	138
43	1.1432	1.057	137
44	1.1308	1.046	136
45	1.1184	1.034	135
46	1.1060	1.023	134
47	1.0936	1.011	133
48	1.0812	0.9997	132
49	1.0689	0.9883	131
50	1.0567	0.9771	130
51	1.0445	0.9659	129
52	1.0324	0.9547	128
53	1.0204	0.9434	127
54	1.0086	0.9325	126
55	0.9968	0.9218	125
56	0.9852	0.9111	124
57	0.9738	0.9004	123
58	0.9626	0.8900	122
59	0.9515	0.8798	121
60	0.9407	0.8698	120
61	0.9300	0.8599	119
62	0.9196	0.8502	118
63	0.9095	0.8408	117
64	0.8996	0.8316	116
65	0.8899	0.8227	115
66	0.8805	0.8141	114
67	0.8715	0.8059	113
68	0.8627	0.7978	112

θ (deg)	$P(\theta)$	$\beta_m(\theta)$ (km^{-1})	θ (deg)
69	0.8542	0.7899	111
70	0.8461	0.7822	110
71	0.8383	0.7751	109
72	0.8308	0.7682	108
73	0.8237	0.7616	107
74	0.8169	0.7555	106
75	0.8105	0.7494	105
76	0.8045	0.7438	104
77	0.7989	0.7387	103
78	0.7936	0.7338	102
79	0.7888	0.7292	101
80	0.7843	0.7252	100
81	0.7803	0.7216	99
82	0.7767	0.7180	98
83	0.7735	0.7152	97
84	0.7707	0.7127	96
85	0.7683	0.7104	95
86	0.7664	0.7086	94
87	0.7649	0.7073	93
88	0.7638	0.7063	92
89	0.7631	0.7055	91
90	0.7629	0.7053	90

Source: Adapted from Penndorf (1957a).

Appendix I

Rayleigh Total Scattering Coefficient and Refractive Index Term

Below are given the Rayleigh scattering coefficient and refractive index term for air at 288.15 K and 1013 mb at $\lambda = 0.30$ to 20 μm.

λ (μm)	β_m (km^{-1})	$\beta_m/4\pi$ (km^{-1})	$(n^2-1)^2$
0.30	144.6×10^{-3}	11.50×10^{-3}	3.4012×10^{-7}
0.31	125.7	10.00	3.3702
0.32	109.8	8.735	3.3425
0.33	96.31	7.664	3.3177
0.34	84.92	6.759	3.2955
0.35	75.16	5.981	3.2754
0.36	66.78	5.314	3.2572
0.37	59.55	4.739	3.2406
0.38	53.27	4.239	3.2255
0.39	47.80	3.804	3.2116
0.40	43.04	3.425	3.1989
0.41	38.84	3.091	3.1873
0.42	35.15	2.797	3.1765
0.43	31.90	2.538	3.1665
0.44	29.01	2.308	3.1572
0.45	26.44	2.103	3.1486
0.46	24.15	1.922	3.1407
0.47	22.11	1.760	3.1332
0.48	20.29	1.614	3.1263
0.49	18.64	1.483	3.1198
0.50	17.16	1.365	3.1137
0.51	15.82×10^{-3}	1.259×10^{-3}	3.1080
0.52	14.62	1.163	3.1026
0.53	13.52	1.076	3.0976
0.54	12.52	9.963×10^{-4}	3.0928
0.55	11.62×10^{-3}	9.247×10^{-4}	3.0884

λ (μm)	β_m (km^{-1})	$\beta_m/4\pi$ (km^{-1})	$(n^2-1)^2$
0.56	10.80	8.591	3.0841
0.57	10.05	7.995	3.0801
0.58	9.361	7.449	3.0763
0.59	8.732	6.948	3.0728
0.60	8.156	6.489	3.0694
0.61	7.626	6.066	3.0661
0.62	7.139	5.680	3.0631
0.63	6.689	5.323	3.0602
0.64	6.275	4.993	3.0574
0.65	5.893	4.689	3.0547
0.66	5.540	4.408	3.0522
0.67	5.212	4.147	3.0498
0.68	4.908	3.905	3.0475
0.69	4.627	3.6810	3.0453
0.70	4.365	3.473	3.0432
0.71	4.121	3.279	3.0412
0.72	3.895	3.099	3.0393
0.73	3.684	2.931	3.0375
0.74	3.486	2.774	3.0357
0.75	3.302	2.628	3.0340
0.76	3.130×10^{-3}	2.490×10^{-4}	3.0324
0.77	2.969	2.362	3.0309
0.78	2.818	2.242	3.0293
0.79	2.677	2.130	3.0279
0.80	2.544	2.024	3.0265
0.90	1.583	1.260	3.0151
1.00	1.035	8.236×10^{-5}	3.0071
1.10	7.059×10^{-4}	5.617	3.0012
1.20	4.976	3.960	2.9967
1.30	3.609	2.871×10^{-5}	2.9932
1.40	2.681	2.133	2.9904
1.50	2.032	1.617	2.9882
1.60	1.569	1.248	2.9864
1.70	1.230	9.789×10^{-6}	2.9849
1.80	9.792×10^{-5}	7.791	2.9836
1.90	7.881	6.270	2.9826
2.0	6.417	5.106	2.9816
2.5	2.626	2.089	2.9786
3.0	1.265	1.006	2.9770
3.5	6.829×10^{-6}	5.433×10^{-7}	2.9759
4.0	4.002	3.184	2.9753
4.5	2.499	1.988	2.9749
5.0	1.639	1.304	2.9747

λ (μm)	β_m (km^{-1})	$\beta_m/4\pi$ (km^{-1})	$(n^2-1)^2$
5.5	1.119	9.398×10^{-8}	2.9743
6.0	7.903×10^{-7}	6.288	2.9741
6.5	5.738×10^{-7}	4.565×10^{-8}	2.9740
7.0	4.030	3.207	2.9739
7.5	3.236	2.575	2.9738
8.0	2.500	1.989	2.9737
8.5	1.961	1.560	2.9737
9.0	1.561	1.242	2.9736
9.5	1.257	1.000	2.9736
10.0	1.024	8.146×10^{-9}	2.9736
11.0	6.994×10^{-8}	5.565	2.9735
12.0	4.938	3.929	2.9734
13.0	3.585	2.852	2.9734
14.0	2.665	2.121	2.9734
15.0	2.023	1.610	2.9734
16.0	1.562	1.243	2.9734
17.0	1.226	9.753×10^{-10}	2.9734
18.0	9.753×10^{-9}	7.762	2.9733
19.0	7.857	6.253	2.9733
20.0	6.399	5.092	2.9733

Adapted from Penndorf (1957a).

Appendix J

Scattering Efficiency Factors

	Refractive index			Refractive index	
α	1.33	1.50	α	1.33	1.50
0.1	0.000011090	0.000023084	3.0	1.7534	3.4181
0.2	0.00017704	0.00037007	3.1	1.8769	3.4581
0.3	0.00089246	0.0018787	3.2	1.9900	3.5317
0.4	0.0028019	0.0059551	3.3	2.0909	3.6733
0.5	0.0067731	0.014567	3.4	2.1833	3.8788
0.6	0.013847	0.030183	3.5	2.2747	4.0785
0.7	0.025151	0.055609	3.6	2.3726	4.1849
0.8	0.041764	0.093635	3.7	2.4811	4.1842
0.9	0.064544	0.14648	3.8	2.5979	4.1264
1.0	0.093924	0.21510	3.9	2.7142	4.0671
1.1	0.12977	0.29869	4.0	2.8197	4.0525
1.2	0.17138	0.39496	4.1	2.9089	4.1187
1.3	0.21771	0.50159	4.2	2.9845	4.2555
1.4	0.26786	0.61897	4.3	3.0545	4.3593
1.5	0.32171	0.75282	4.4	3.1283	4.3326
1.6	0.38045	0.91435	4.5	3.2125	4.2025
1.7	0.44654	1.1142	4.6	3.3070	4.0413
1.8	0.52306	1.3489	4.7	3.4034	3.8997
1.9	0.61213	1.5901	4.8	3.4878	3.8191
2.0	0.71295	1.7984	4.9	3.5502	3.8381
2.1	0.82096	1.9553	5.0	3.5910	3.9278
2.2	0.92912	2.0753	5.1	3.6200	3.9325
2.3	1.0312	2.1916	5.2	3.6505	3.7832
2.4	1.1252	2.3383	5.3	3.6935	3.5683
2.5	1.2135	2.5395	5.4	3.7535	3.3567
2.6	1.3023	2.7924	5.5	3.8238	3.1816
2.7	1.3985	3.0512	5.6	3.8868	3.0835
2.8	1.5068	3.2510	5.7	3.9241	3.1101
2.9	1.6270	3.3648	5.8	3.9295	3.1905

α	Refractive index 1.33	Refractive index 1.50	α	Refractive index 1.33	Refractive index 1.50
5.9	3.9122	3.1117	9.8	2.3130	2.6454
6.0	3.8892	2.9039	9.9	2.2866	2.8567
6.1	3.8777	2.6852	10.0	2.2065	2.8820
6.2	3.8889	2.4861	10.1	2.1000	2.7829
6.3	3.9226	2.3300	10.2	1.9939	2.7575
6.4	3.9625	2.2713	10.3	1.9119	2.9621
6.5	3.9824	2.3673	10.4	1.8772	2.9680
6.6	3.9647	2.4480	10.5	1.9026	2.8014
6.7	3.9127	2.3145	10.6	1.9603	2.9159
6.8	3.8442	2.1462	10.7	1.9812	3.1116
6.9	3.7806	1.9949	10.8	1.9395	2.9812
7.0	3.7396	1.8483	10.9	1.8616	2.8732
7.1	3.7299	1.7456	11.0	1.7741	2.8763
7.2	3.7428	1.7720	11.1	1.7002	3.1493
7.3	3.7498	1.9722	11.2	1.6657	2.7988
7.4	3.7214	2.0010	11.3	1.6943	2.7509
7.5	3.6511	1.8848	11.4	1.7793	2.9338
7.6	3.5547	1.8318	11.5	1.8487	2.9538
7.7	3.4554	1.7642	11.6	1.8501	2.7597
7.8	3.3745	1.6723	11.7	1.8082	2.6751
7.9	3.3275	1.6367	11.8	1.7488	2.8112
8.0	3.3158	1.7777	11.9	1.6909	2.6360
8.1	3.3156	2.0283	12.0	1.6604	2.4760
8.2	3.2881	1.9519	12.1	1.6869	2.4949
8.3	3.2145	1.9488	12.2	1.7847	2.7365
8.4	3.1070	2.0127	12.3	1.8951	2.5066
8.5	2.9899	1.9849	12.4	1.9343	2.3791
8.6	2.8866	1.9263	12.5	1.9242	2.3395
8.7	2.8166	1.9569	12.6	1.8930	2.4582
8.8	2.7904	2.2264	12.7	1.8519	2.1643
8.9	2.7939	2.3240	12.8	1.8255	2.1101
9.0	2.7831	2.2384	12.9	1.8459	2.2405
9.1	2.7231	2.3708	13.0	1.9413	2.2814
9.2	2.6201	2.4753	13.1	2.0807	2.1088
9.3	2.5004	2.4231	13.2	2.1512	2.0463
9.4	2.3892	2.3817	13.3	2.1619	1.9359
9.5	2.3087	2.4791	13.4	2.1533	2.2768
9.6	2.2766	2.7985	13.5	2.1238	1.8688
9.7	2.2918	2.6198	13.6	2.0949	1.8887
			13.7	2.1018	2.2079

α	Refractive index 1.33	Refractive index 1.50	α	Refractive index 1.33	Refractive index 1.50
13.8	2.1797	1.9725	18.0	2.5836	2.4603
13.9	2.3299	1.9411	18.1	2.6015	2.5650
14.0	2.4271	1.9518	18.2	2.5202	2.4032
14.1	2.4462	1.9022	18.3	2.5463	2.4588
14.2	2.4538	1.8285	18.4	2.5243	2.7143
14.3	2.4323	1.8280	18.5	2.3995	2.4218
14.4	2.3933	1.9817	18.6	2.3011	2.3471
14.5	2.3780	2.0205	18.7	2.2574	2.2767
14.6	2.4248	2.0204	18.8	2.3202	2.2247
14.7	2.5611	2.0364	18.9	2.3812	2.1771
14.8	2.6795	2.1235	19.0	2.2745	2.1819
14.9	2.6939	1.9897	19.1	2.2947	2.2687
15.0	2.7095	1.9847	19.2	2.3203	2.2299
15.1	2.6976	2.0342	19.3	2.1986	2.1603
15.2	2.6461	2.2960	19.4	2.0930	2.1026
15.3	2.6038	2.2137	19.5	2.0390	2.0423
15.4	2.6126	2.2763	19.6	2.0718	1.9971
15.5	2.7131	2.5288	19.7	2.1793	1.9757
15.6	2.8442	2.2679	19.8	2.0565	2.0333
15.7	2.8439	2.2544	19.9	2.0662	2.1018
15.8	2.8527	2.2675	20.0	2.1401	2.0358
15.9	2.8537	2.4850	20.1	2.0397	2.0127
16.0	2.7926	2.4199	20.2	1.9349	1.9769
16.1	2.7240	2.5164	20.3	1.8823	1.9415
16.2	2.6967	2.5646	20.4	1.8970	1.9167
16.3	2.7517	2.5317	20.5	2.0431	1.9353
16.4	2.8844	2.5026	20.6	1.9178	2.0574
16.5	2.8712	2.4884	20.7	1.9192	2.0418
16.6	2.8561	2.5121	20.8	2.0239	2.0241
16.7	2.8704	2.5584	20.9	1.9645	2.0443
16.8	2.8068	2.6350	21.0	1.8625	2.0823
16.9	2.7140	2.6973	21.1	1.8179	2.0076
17.0	2.6565	2.6762	21.2	1.8258	2.0154
17.1	2.6690	2.6221	21.3	1.9922	2.0826
17.2	2.7903	2.5811	21.4	1.8825	2.1960
17.3	2.7811	2.5536	21.5	1.8821	2.1321
17.4	2.7324	2.6128	21.6	1.9937	2.1817
17.5	2.7561	2.5835	21.7	1.9938	2.2126
17.6	2.7033	2.6652	21.8	1.8920	2.1920
17.7	2.5901	2.6702	21.9	1.8555	2.1986
17.8	2.5086	2.5812	22.0	1.8614	2.2241
17.9	2.4876	2.5164	22.1	2.0217	2.3546

α	Refractive index 1.33	1.50	α	Refractive index 1.33	1.50
22.2	1.9452	2.3216	26.2	2.5101	2.1094
22.3	1.9475	2.3129	26.3	2.4787	2.0087
22.4	2.0495	2.3961	26.4	2.5066	2.0299
22.5	2.1178	2.3861	26.5	2.6199	2.0270
22.6	2.0120	2.3771	26.6	2.5497	2.1395
22.7	1.9815	2.3898	26.7	2.4987	1.9592
22.8	1.9856	2.4268	26.8	2.4640	1.9347
22.9	2.1163	2.5320	26.9	2.4441	1.9910
23.0	2.0799	2.4169	27.0	2.4447	2.0201
23.1	2.0833	2.4623	27.1	2.4019	2.0284
23.2	2.1696	2.5185	27.2	2.4182	2.0713
23.3	2.2984	2.4711	27.3	2.4953	2.0770
23.4	2.1849	2.4698	27.4	2.4574	2.0550
23.5	2.1576	2.4659	27.5	2.4037	2.0140
23.6	2.1585	2.5597	27.6	2.3641	2.0234
23.7	2.2507	2.4631	27.7	2.3321	2.1915
23.8	2.2473	2.4165	27.8	2.3211	2.1022
23.9	2.2467	2.4974	27.9	2.2713	2.1757
24.0	2.3164	2.4649	28.0	2.2781	2.2048
24.1	2.4836	2.4224	28.1	2.3310	2.2893
24.2	2.3614	2.4111	28.2	2.3176	2.1677
24.3	2.3329	2.4046	28.3	2.2661	2.1603
24.4	2.3286	2.4586	28.4	2.2264	2.2067
24.5	2.3852	2.3079	28.5	2.1901	2.2342
24.6	2.4005	2.3075	28.6	2.1716	2.2799
24.7	2.3916	2.3881	28.7	2.1217	2.3475
24.8	2.4454	2.2908	28.8	2.1232	2.3687
24.9	2.6206	2.2730	28.9	2.1650	2.3233
25.0	2.4983	2.2498	29.0	2.1713	2.2993
25.1	2.4634	2.3196	29.1	2.1273	2.2995
25.2	2.4500	2.1704	29.2	2.0918	2.4394
25.3	2.4762	2.1161	29.3	2.0566	2.3459
25.4	2.4960	2.1758	29.4	2.0347	2.4209
25.5	2.4765	2.1595	29.5	1.9902	2.4390
25.6	2.5167	2.1149	29.6	1.9903	2.4177
25.7	2.6690	2.1103	29.7	2.0301	2.3738
25.8	2.5658	2.0911	29.8	2.0555	2.3481
25.9	2.5217	2.0898	29.9	2.0254	2.3663
26.0	2.4968	1.9861	30.0	1.9984	2.3528
26.1	2.4970	1.9792			

Source: From Penndorf (1956).

Appendix K

Glossary of Principal Symbols

The principal symbols of physical quantities used in this book are listed below. Some symbols formed from principal symbols by adding subscripts, and others that are used in only one place, are not listed.

Symbol	First used in Section	Definition
a_n	5.3.1	Amplitude function in Mie theory
A	1.2.3	Area
b_n	5.3.1	Amplitude function in Mie theory
B	5.5.4	Parameter in anomalous diffraction theory
c	1.2.3	Speed of light
c	3.2.3	Parameter of power-law size distribution
C	1.5.2	Object-background contrast
C	2.1.1	Speed of a molecule
e	4.2.1	Charge of an electron
e_s	2.2.5	Water vapor pressure
E	4.2.2	Electric field strength
E	1.2.2	Flux irradiance
E	2.1.1	Molecular kinetic energy
g	2.1.4	Acceleration of gravity
h	1.2.3	Planck's constant
H_P	2.3.3	Pressure scale height of molecular atmosphere
H_ρ	2.3.2	Density scale height of molecular atmosphere
H_p	3.3.2	Scale height of haze atmosphere
i_1	5.3.1	Intensity distribution function in Mie theory
i_2	5.3.1	Intensity distribution function in Mie theory
I	4.6.3	Stokes parameter
I	1.2.2	Flux intensity

Symbol	First used in Section	Definition
$I(\theta)$	1.4.1	Flux scattered intensity
K	1.2.6	Luminous efficacy of flux
$\bar{l}$	2.1.1	Molecular mean free path
L	1.4.1	Dimensional unit of length
L	1.2.3	Flux radiance
m	2.1.1	Mass of a molecule
m	4.2.1	Mass of an electron
m	5.2.3	Complex refractive index
M	1.2.3	Flux radiant exitance
M_A	2.1.1	Molar mass
n	4.2.3	Refractive index, general
n	5.2.3	Real component of complex refractive index
n_i	5.2.3	Imaginary component of complex refractive index
$n(r)$	5.2.1	Number of particles per unit radius and volume
N	2.1.1	Number of molecules per unit volume
N	3.2.1	Number of particles per unit volume
N_A	2.1.1	Avogadro's number
N_L	2.1.1	Loschmidt's number
p	4.2.2	Electric dipole moment
P	4.4.2	Degree of polarization
P	2.1.1	Gas pressure
$P(\theta)$	4.3.4	Phase function for scattering
Q	4.6.3	A Stokes parameter
Q	1.2.3	Radiant energy
Q_{ab}	5.5.3	Efficiency factor of absorption
Q_{ex}	5.5.3	Efficiency factor of extinction
Q_{pr}	5.5.3	Efficiency factor of light pressure
Q_{sc}	4.3.2	Efficiency factor of scattering
r	3.1.4	Particle radius
r_e	3.1.4	Effective mean particle radius
R	2.1.2	Gas constant (unit mass)
R_*	2.1.2	Universal gas constant (molar mass)
R	1.5.1	Range or distance along an optical path
R_m	1.5.4	Meteorological range
R_v	1.5.4	Visual range

Symbol	First used in Section	Definition
$\bar{R}$	2.5.3	Equivalent path distance
$\bar{R}_{sl}$	2.5.3	Equivalent sea-level path distance
$\bar{S}$	4.2.2	Average value of Poynting vector
t	1.2.3	Time
T	4.6.1	Atmospheric turbidity
T	2.1.2	Absolute (Kelvin) temperature
U	4.6.3	Stokes parameter
U	2.2.5	Relative humidity
v	3.2.3	Parameter of the power-law size distribution
V	4.6.3	Stokes parameter
V	1.2.6	Luminous efficiency of flux
V	2.1.2	Volume
V_A	2.1.1	Molar volume
W	1.2.3	Radiant energy density
W_v	2.2.5	Mixing ratio of water vapor
x	1.4.2	Distance along an x axis
XYZ	1.4.1	Set of orthogonal coordinate axes
y	4.2.1	Distance along a Y axis
Z	2.1.4	Altitude above earth's surface
Z'	2.4.3	Reduced height of atmosphere from any reference altitude
Z'_0	2.4.3	Reduced height of atmosphere from sea level
α	5.2.2	Particle size parameter in Mie theory
β	1.4.2	Volume extinction coefficient, general (scattering or extinction)
β_{ab}	1.4.2	Volume absorption coefficient
β_{ex}	1.4.2	Volume extinction coefficient
β_m	1.4.2	Volume total scattering coefficient (molecular)
β_p	1.4.2	Volume total scattering coefficient (particle)
β_{sc}	1.4.2	Volume total scattering coefficient (general)
$\beta(\phi)$	1.4.1	Volume angular scattering coefficient, general (molecular or particle)
$\beta_m(\theta)$	1.4.1	Volume angular scattering coefficient (molecular, unpolarized light)

Symbol	First used in Section	Definition
$\beta_p(\theta)$	1.4.1	Volume angular scattering coefficient (particle, unpolarized light)
$\beta_m(\phi)$	4.4.1	Volume angular scattering coefficient (molecular, polarized light)
$\beta_p(\phi)$	4.4.1	Volume angular scattering coefficient (particle, polarized light)
γ	2.2.1	Environmental lapse rate of temperature
δ	2.5.2	Relative azimuth or bearing
ϵ	1.5.3	Threshold contrast of visual perception
ϵ_0	4.2.1	Permittivity, or absolute dielectric constant, of free space
ζ	2.5.1	Zenith angle
θ	1.2.2	Direction-defining angle
θ	1.4.1	Observation angle of scattering
κ	2.1.2	Boltzmann's constant
λ	1.2.3	Electromagnetic wavelength
ν	1.2.3	Electromagnetic frequency
π	1.2.3	Ratio of circle circumference to diameter
π_n	5.3.1	Angular function in Mie theory
ρ	2.1.1	Mass density
ρ	5.5.4	Normalized size parameter in anomalous diffraction theory
σ_m	4.3.2	Total scattering cross section of a molecule
$\sigma_m(\phi)$	4.3.1	Angular scattering cross section of a molecule, polarized light
σ_p	4.3.2	Total scattering cross section of a particle
$\sigma_p(\phi)$	4.3.1	Angular scattering cross section of a particle in Rayleigh theory, polarized light
$\sigma_p(\theta)$	5.4.1	Angular scattering cross section of a particle in Mie theory
τ	1.4.4	Optical transmittance
τ_n	5.3.1	Angular function in Mie theory
T	1.4.4	Optical thickness
ϕ	4.2.2	Angle between dipole axis and direction of observation
Φ	1.2.2	Radiant flux or radiant power

Symbol	First used in Section	Definition
χ	2.5.4	Parameter of the Chapman function
ψ	5.2.1	Orientation angle of the plane of polarization
ω	1.2.3	Solid angle
ω	4.2.1	Electromagnetic circular or angular frequency

References

Abbott, C. G. (1963). "Solar variation and weather," *Smithsonian Misc. Coll.*, **146**, No. 3 (Publ. No. 4545).

Abbott, C. G. and Fowle, F. E. (1913). "Volcanoes and climate." *Ann. Astrophys. Obs. Smithsonian Inst.*, **3**, 211–231. Also *Smithsonian Misc. Coll.*, **60**, No. 29.

Adams, G. W. (1970). "N_2, O_2, and O Densites in the 80–120 Km Region," ESSA Tech. Rept. ERL-184-SDL-17. GPO, Washington, D.C.

Ahlquist, N. C. and Charlson, R. J. (1969). "Measurement of the wavelength dependence of atmospheric extinction due to scattering," *J. Atmos. Environ.*, **3**, 551–564.

AIP (undated). *Polarized Light: Selected Reprints.* American Institute of Physics, New York.

Aitken, J. (1888). "On the number of dust particles in the atmosphere." In *Collected Scientific Papers*, C. G. Knott, ed. Cambridge University Press, Cambridge, England (1923).

Allard, E. (1876). *Mémoire sur l'intensité et la portée des phares.* Dunod, Paris.

Altshuler, T. L. (1961). "Infrared Transmission and Background Radiation by Clear Atmospheres," Doc. No. 61SD199. Missile and Space Vehicle Department, General Electric Company, Valley Forge, Pa.

Amelin, A. G. (1967). *Theory of Fog Condensation.* Israel Program for Scientific Translations, Jerusalem.

Anderson, R. C. and Browell, E. V. (1972). "First and second-order backscattering from clouds illuminated by finite beams," *Appl. Opt.*, **11**, 1345–1351.

Anderson, R. C. et al. (1973). "Reply to: Pulsed lidar reflectance of clouds," *Appl. Opt.*, **12**, 428.

Ångström, A. (1929). "On the atmospheric transmission of sun radiation and on dust in the air," *Geogr. Ann., Stockholm*, **11**, 156–166.

Ångström, A. (1930). "On the atmospheric transmission of sun radiation, II," *Geogr. Ann., Stockholm*, **12**, 130–159.

Ångström, A. (1951). "Actinometric measurements." In *Compendium of Meteorology*, T. F. Malone, ed. AMS, Boston.

Ångström, A. (1974a). "Circumsolar sky radiation and turbidity of the atmosphere," *Appl. Opt.*, **13**, 474–477.

Ångström, A. (1974b). "Circumsolar radiation as a measure of the turbidity of the atmosphere," *Appl. Opt.*, **13**, 1477–1480.

ANSI (1967). *American National Standard: Nomenclature and Definitions for Illuminating Engineering, Z7.1-1967.* Illuminating Engineering Society, New York.

AO (1964). *Appl. Opt.*, **3**, No. 10 (Rayleigh issue).

Arnulf, A. et al. (1957). "Transmission by haze and fog in the spectral region 0.35 to 10 microns," *J. Opt. Soc. Am.*, **47**, 491–498.

Arvesen, J. C. et al. (1969). "Determination of extraterrestrial solar spectral irradiance from a research aircraft," *Appl. Opt.*, **8**, 2215–2232.

Asano, S. and Yamamoto, G. (1975). "Light scattering by a spheroidal particle," *Appl. Opt.*, **14**, 29–49.

Auer, A. H., Jr., and Veal, D. L. (1970). "The dimensions of ice crystals in natural clouds," *J. Atmos. Sci.*, **27**, 919–926.

aufm Kampe, H. J. (1950). "Visibility and liquid water content in clouds in the free atmosphere," *J. Meteorol.*, **7**, 54–57.

aufm Kampe, H. J. and Weickman, H. K. (1952). "Trabert's formula and the determination of the water content in clouds," *J. Meteorol.*, **9**, 167–171.

aufm Kampe, H. J. and Weickman, H. K. (1957). "Physics of clouds," *Meteorol. Monogr.* **3**, No. 18, pp. 182–225. AMS, Boston.

Averitt, J. M. and Rushkin, R. E. (1967). "Cloud particle replication in stormfury tropical cumulus," *J. Appl. Meteorol.*, **6**, 88–94.

AW (1965). "Aviation Weather," Federal Aviation Agency and Department of Commerce, Cat. No. FAA 5.8/2:W37. GPO, Washington, D.C.

Baker, D. J. (1974). "Rayleigh, the unit for light radiance," *Appl. Opt.*, **13**, 2160–2163.

Barber, D. R. (1950). "Note on the brightness profile and photometric contrast of a test object having small angular dimensions and silhouetted against the horizon sky," *Proc. Roy. Soc. London*, **B63**, 364–369.

Barber, P. and Yeh, C. (1975). "Scattering of electromagnetic waves by arbitrarily shaped dielectric bodies," *Appl. Opt.*, **14**, 2864–2872.

Bar-Isaac, C. and Hardy, A. (1975). "Simple derivation of the factor two in the Mie theory," *Am. J. Phys.*, **43**, 275–276.

Barnhardt, E. A. and Streete, J. L. (1970). "A method for predicting atmospheric aerosol scattering coefficients in the infrared," *Appl. Opt.*, **9**, 1337–1344.

Barrett, E. W. and Ben-Dov, O. (1967). "Application of the lidar to air pollution measurements," *J. Appl. Meteorol.*, **6**, 500–515.

Bastin, J. A. et al. (1964). "Spectroscopy at extreme infrared wavelengths. III. Astrophysical and atmospheric measurements," *Proc. Roy. Soc. London*, **A278**, 543–573.

Bates, D. R. (1953). "The Physics of the Upper Atmosphere." In *The Earth as a Planet*, G. Kuiper, ed. University of Chicago Press, Chicago.

Battan, L. (1974). *Weather.* Prentice-Hall, Englewood Cliffs, N.J.

Bauer, E. (1964). "The scattering of infrared radiation from clouds," *Appl. Opt.*, **3**, 197–202.

Baum, W. A. and Dunkelman, L. (1955). "Horizontal attenuation of ultraviolet light by the lower atmosphere," *J. Opt. Soc. Am.*, **45**, 166–175.

Beard, K. V. (1974). "Experimental and numerical collision efficiencies for submicron particles scavenged by small raindrops," *J. Atmos. Sci.*, **31**, 1595–1603.

Beggs, S. S. and Waldram, J. M. (1943). "A method of estimating visibility from the air near the ground at night," Rept. 8303. General Electric Research Laboratory, Wembley, England.

Beiser, A. (1962). *The Earth.* Time-Life Books, New York.

Bell, E. E. (1959). "Radiometric quantities, symbols, and units," *Proc. IRE*, **47**, 1432–1434.

Bell, E. E. et al. (1960). "Spectral radiance of sky and terrain at wavelengths between 1 and 20 microns. II. Sky measurements," *J. Opt. Soc. Am.*, **50**, 1313–1320.

Bemporad, A. (1907). "Search for a new empirical formula for the representation of the variation of the intensity of solar radiation with zenith angle," *Meteorol. Z.*, **24**, 306–313. NASA Technical Translation, Publ. No. NASA TT F-302. NTIS, Springfield, Va.

Beneditti-Michelangeli, G. et al. (1972). "Measurement of aerosol motion and wind velocity in the lower troposphere by Doppler optical radar," *J. Atmos. Sci.*, **29**, 906–910.

Bennett, H. F. et al. (1960). "Distribution of infrared radiance over a clear sky," *J. Opt. Soc. Am.*, **50**, 100–106.

Bernoulli, D. (1738). "Hydrodynamica." In *A Source Book of Physics*. McGraw-Hill, New York (1935). Also in *The World of the Atom*," H. A. Boorse and L. Motz, eds. Basic Books, New York (1966).

Beuttell, R. G. and Brewer, A. W. (1949). "Instruments for the measurement of the visual range," *J. Sci. Instr.*, **26**, 357–359.

Bigg, E. K. et al. (1970). "Aerosols at altitudes between 20 and 37 km," *Tellus*, **22**, 550–563.

Blackwell, H. R. (1946). "Contrast thresholds of the human eye," *J. Opt. Soc. Am.*, **36**, 624–643.

Blanchard, D. C. (1953). "Raindrop size-distribution in Hawaiian rains," *J. Meteorol.*, **10**, 457–473.

Blanchard, D. C. (1967). *From Raindrops to Volcanos*, Doubleday, Garden City, N.Y.

Blanchard, D. C. and Spencer, A. T. (1970). "Experiments on the generation of raindrop size-distributions by drop breakup," *J. Atmos. Sci.*, **27**, 101–108.

Blättner, W. G. et al. (1974). "Monte Carlo studies of the sky radiation at twilight," *Appl. Opt.*, **13**, 534–547.

Blau, H. H., Jr., et al. (1966). "Near infrared scattering by sunlit terrestrial clouds," *Appl. Opt.*, **5**, 555–564.

Blau, H. H., Jr., et al. (1970). "Scattering by individual transparent spheres," *Appl. Opt.*, **9**, 2522–2528.

Blifford, I. H., Jr. (1970). "Tropospheric aerosols," *J. Geophys. Res.*, **75**, 3099–3103.

Blifford, I. H., Jr., ed. (1971). "Particulate Models: Their Validity and Application," Publ. No. NCAR-TN/PROC-68. Boulder, Colo.

Blifford, I. H., Jr., and Ringer, L. D. (1969). "The size and number distribution of aerosols in the continental troposphere," *J. Atmos. Sci.*, **26**, 716–726.

Bowditch, N. (1958). "American Practical Navigator," 70th ed., HO Publ. No. 9. GPO, Washington, D.C.

Bragg, W. (1940). *The Universe of Light*, Bell, London. Also Dover, New York (1959).

Breeding, R. J. et al. (1975). "The urban plume as seen at 80 and 120 km by five different sensors," *J. Appl. Meteorol.*, **14**, 204–216.

Brillouin, L. (1949). "The scattering cross section of spheres for electromagnetic waves," *J. Appl. Phys.*, **20**, 1110–1125.

Brinkman, R. T. et al. (1967). "Atmospheric scattering of the solar flux in the middle ultraviolet," *Appl. Opt.*, **6**, 373–383.

Brinkworth, B. J. (1973). "Pulsed-lidar reflectance of clouds," *Appl. Opt.*, **12**, 427.

Brumberger, H. et al. (1968). "Light scattering," *Sci. Technol.*, November, pp. 34–60.

Bull. Am. Meteorol. Soc. (1971). "High atmospheric dust pollution," *Bull. Am. Meteorol. Soc.*, **52**, 1124.

Bull. Am. Meteorol. Soc. (1972). "Atmospheric visibility decrease," *Bull. Am. Meteorol. Soc.*, **53**, 462.

Bullrich, K. (1964). "Scattered radiation in the atmosphere," In *Advances in Geophysics*, Vol. 10, Academic Press, New York.

Bullrich, K. et al. (1967). "Research on Atmospheric Optical Radiation Transmission," Rept. AFCRL-67-0207. AFCRL, Bedford, Mass. AD 653 757. NTIS, Springfield, Va.

Bullrich, K. et al. (1968). "Research on Optical Radiation Transmission," Rept. AFCRL-68-0186. AFCRL, Bedford, Mass. AD 670 210, NTIS, Springfield, Va.

Burroughs, W. J. et al. (1966). "Transmission of submillimeter waves in fog," *Nature*, **212**, 337–338.

Butcher, S. S. and Charlson, R. J. (1972). *Introduction to Air Chemistry*. Academic Press, New York.

Byers, H. R. (1965). *Elements of Cloud Physics*. University of Chicago Press, Chicago.

Byers, H. R. (1974). *General Meteorology*, 4th ed. McGraw-Hill, New York.

Cadle, R. D. (1965). *Particle Size*. Van Nostrand Reinhold, New York.

Cadle, R. D. (1966). *Particles in the Atmosphere and Space*. Van Nostrand Reinhold, New York.

Cadle, R. D. (1975). "The Measurement of Airborne Particles," Wiley, New York.

Cadle, R. D. et al. (1970). "Some aspects of atmospheric chemical reactions of atomic oxygen," *Tellus*, **18**, 176–185.

Cadle, R. D. et al. (1975). "A comparison of the Langer, Rosen, Nolan-Pollack, and SANDS condensation nucleus counters," *J. Appl. Meteorol.*, **14**, 1566–1571.

Campen, C. F., Jr., et al., eds. (1960). *Handbook of Geophysics*. Macmillan, New York.

Cantor, I. and Petriw, A. (1968). "Atmospheric light transmission in a Wisconsin area," *Appl. Opt.*, **7**, 1365–1381.

Carman, D. D. and Carruthers, R. A. (1951). "Brightness of fine detail in air photography," *J. Opt. Soc. Am.*, **41**, 305–310.

Carpenter, R. O'B. et al. (1957). "Predicting Infrared Molecular Absorption for Long Slant Paths in the Upper Atmosphere." AFCRC Rept. TN-58-253. AFCRL, Bedford, Mass.

Carrier, L. W. et al. (1967). "The backscattering and extinction of visible and infrared radiation by selected major cloud models," *Appl. Opt.*, **6**, 1209–1216.

Centeno, M. (1941). "The refractive index of liquid water in the near infrared spectrum," *J. Opt. Soc. Am.*, **31**, 244–247.

Chagnon, C. W. and Junge, C. E. (1961). "The vertical distribution of sub-micron particles in the stratosphere," *J. Meteorol.*, **18**, 746–752.

Chamberlain, J. W. (1961). *Physics of the Aurora and Airglow*. Academic Press, New York.

Champion, K. S. (1965). "Mean Atmospheric Properties in the Range 30 to 300 Km," Rept. AFCRL-65-443. AFCRL, Bedford, Mass.

Chandrasekhar, S. (1950). *Radiative Transfer*, Oxford University Press, London. Also Dover, New York (1960).

Chandrasekhar, S. and Elbert, D. D. (1954). "The illumination and polarization of the sunlit sky on Rayleigh scattering, *Trans. Am. Phil. Soc.*, **44**, 643–728.

Chapman, S. (1930). "A theory of upper-atmospheric ozone," *Mem. Roy. Meteorol. Soc.*, **3**, 103–125.

Chapman, S. (1931). "The absorption and dissociative or ionizing effect of monochromatic radiation in an atmosphere on a rotating earth. II, Grazing incidence, *Proc. Phys. Soc.*, **43**, 483–501.

Charlson, R. J. (1972). "Multiwavelength nephelometer measurements in Los Angeles smog aerosol." In *Aerosols and Atmospheric Chemistry*, G. M. Hidy, ed. Academic Press, New York.

Charlson, R. J. et al. (1967). "The direct measurement of atmospheric light scattering coefficient for studies of visibility and air pollution," *J. Atmos. Environ.*, **1**, 469–478.

Charlson, R. J. et al. (1969). "Atmospheric visibility related to aerosol mass concentration," *Environ. Sci. Technol.*, **3**, 913–918.

Charlson, R. J. et al. (1972). "Multiwavelength nephelometer measurements in Los Angeles smog aerosol. III. Comparison to light extinction by CO_2." In *Aerosols and Atmospheric Chemistry*, G. M. Hidy, ed. Academic Press, New York.

Chen, C. S. (1974). "Evaluation of the vapor pressure over an aerosol particle," *J. Atmos. Sci.*, **31**, 847–849.

Christie, A. D. (1969). "The genesis and distribution of noctilucent cloud," *J. Atmos. Sci.*, **26**, 168–176.

Christy, R. W. and Pytte, A. (1965). *The Structure of Matter: An Introduction to Modern Physics.* W. A. Benjamin, New York.

Chu, B. (1967). *Molecular Forces* (based on the Baker Lectures of Debye). Wiley, New York.

Chu, T. S. and Hogg, D. C. (1968). "Effects of precipitation on propagation at 0.63, 3.5, and 10.6 microns," *Bell Sys. Tech. J.*, **47**, 723–759.

Clark, W. M. (1969). "High altitude daytime sky radiance," *Soc. Photo. Instr. Eng.*, **7**, 40–45.

Clark, W. M. and Muldoon, H. A. (1964). "High altitude sky luminance measurement," *Infrared Phys.*, **4**, 127–136.

Clark, W. E. and Whitby, K. T. (1967). "Concentration and size distribution measurements of atmospheric aerosols and a test of the theory of self-preserving size distributions," *J. Atmos. Sci.*, **24**, 677–687.

Clemeshaw, B. R. et al. (1967). "Laser radar for atmospheric studies," *J. Appl. Meteorol.*, **6**, 386–395.

Cohen, A. (1975). "Cloud-base water content measurement using single wavelength laser-radar data," *Appl. Opt.*, **14**, 2873–2877.

Cohen, A. and Graber, M. (1975). "Laser-radar polarization measurements of the lower stratospheric aerosol layer over Jerusalem," *J. Appl. Meteorol.*, **14**, 400–406.

Cole, F. W. (1975). *Introduction to Meteorology*, Wiley, New York.

Cole, A. E. et al. (1969). "Precipitation and Clouds," a revision of Chap. 5 in *Handbook of Geophysics and Space Environment*, Rept. AFCRL-69-0487. AFCRL, Bedford, Mass.

Cole, A. E. and Kantor, A. J. (1964). "Horizontal and Vertical Distributions of Atmospheric Density, up to 90 Km." Rept. AFCRL-64-483. AFCRL, Bedford, Mass.

Coleman, H. S. et al. (1949). "A photoelectric method of measuring the atmospheric attenuation of brightness contrast along a horizontal path," *J. Opt. Soc. Am.*, **39**, 515–521.

Coleman, H. S. and Rosenberger, H. E. (1950). "A comparison of visual and photoelectric measurements of the attenuation of brightness contrast by the atmosphere," *J. Opt. Soc. Am.*, **40**, 371–372.

Collins, D. G. (1968). "Study of Polarization of Atmospheric Scattered Light Using Monte Carlo Methods," Rept. RRA-T86. Radiation Research Associates. Rept. AFCRL-68-0310. AFCRL, Bedford, Mass.

Collins, D. G. et al. (1972). "Backward Monte Carlo calculations of the polarization characteristics of the radiation emerging from spherical-shell atmospheres," *Appl. Opt.*, **11**, 2684–2696.

Collis, R. T. (1969). "Lidar," in Advances in Geophysics, Vol. 13, Academic Press, New York.

Collis, R. T. (1970). "Lidar," *Appl. Opt.*, **9**, 1782–1788.

Cook, C. S. et al. (1972). "Remote measurement of smoke plume transmittance using lidar," *Appl. Opt.*, **11**, 1742–1748.

Cooke, D. D. and Kerker, M. (1975). "Response calculations for light-scattering aerosol particle counters," *Appl. Opt.*, **14**, 734–739.

Corby, G. A., ed. (1970). *The Global Circulation of the Atmosphere*, Royal Meteorological Society, London.

Coulson, K. L. (1968). "Effect of surface reflection on the angular and spectral distribution of skylight," *J. Atmos. Sci.*, **35**, 759–770.

Coulson, K. L. et al. (1960). *Tables Related to Radiation Emerging from a Planetary Atmosphere with Rayleigh Scattering*." University of California Press, Berkeley.

Craig, R. A. (1965). *The Upper Atmosphere: Meteorology and Physics*. Academic Press, New York.

Crutzen, P. J. (1970). "The influence of nitrogen oxides on the atmospheric ozone content," *Quart. J. Roy. Meteorol. Soc.*, **96**, 320–325.

Crutzen, P. J. (1972). "SST's, a threat to the earth's ozone shield," *Ambio*, **1**, 41–51.

Cunnold, D. et al. (1975). "A three-dimensional dynamical-chemical model of atmospheric ozone," *J. Atmos. Sci.*, **31**, 170–194.

Curcio, J. A. (1961). "Evaluation of atmospheric particle size from scattering measurements in the visible and infrared," *J. Opt. Soc. Am.*, **51**, 548–551.

Curcio, J. A. and Durbin, K. A. (1959). "Atmospheric Transmission in the Visible," NRL Rept. 5368. NRL, Washington, D.C.

Curcio, J. A. et al. (1958). NRL Rept. 5143. NRL, Washington, D.C.

Curcio, J. A. et al. (1961). "Atmospheric Scattering in the Visible and Infrared," NRL Rept. 5567. NRL, Washington, D.C. AD 250 945. NTIS, Springfield, Va.

Dandekar, B. S. (1968). "Measurements of the zenith sky brightness and color during the total solar eclipse of 12 November 1966 at Quehua, Bolivia," *Appl. Opt.* **7**, 705–710.

Dandekar, B. S. and Turtle, J. P. (1971). "Day sky brightness and polarization during the total solar eclipse of 7 March 1970," *Appl. Opt.*, **10**, 1220–1224.

Danielson, R. E. et al. (1969). "The transfer of visible radiation through clouds," *J. Atmos. Sci.*, **26**, 1078–1087.

Dave, J. V. (1965). "Multiple scattering in a non-homogeneous Rayleigh atmosphere," *J. Atmos. Sci.*, **22**, 273–279.

Dave, J. V. (1970). "Intensity and polarization of the radiation emerging from a plane-parallel atmosphere containing monodispersed aerosols," *Appl. Opt.* **9**, 2673–2684.

Dave, J. V. and Furukawa, P. M. (1966a). "Scattered radiation in the ozone absorption bands at selected levels of a terrestrial, Rayleigh atmosphere," *Meteorol. Monogr*, **7**, No. 29 AMS, Boston.

Dave, J. V. and Furukawa, P. M. (1966b). "Intensity and polarization of the radiation emerging from an optically thick Rayleigh atmosphere," *J. Opt. Soc., Am.*, **56**, 394–400.

Dave, J. V. and Gazdag, J. (1970). "A modified Fourier transform method for multiple scattering calculations in a plane parallel Mie atmosphere," *Appl. Opt.*, **9**, 1457–1465.

Davies, C. N., ed. (1966). *Aerosol Science.* Academic Press, New York.

de Bary, E. (1964). "Influence of multiple scattering of the intensity and polarization of diffuse sky radiation," *Appl. Opt.*, **3**, 1293–1303.

de Bary, E. (1972). "Verification of an approximation method for calculating multiple scattering of sky radiation," *Appl. Opt.*, **11**, 2717–2718.

de Bary, E. et al. (1965). "Tables Related to Light Scattering in a Turbid Atmosphere," 3 vols., Rept. AFCRL-65-710. AFCRL, Bedford, Mass. AD 628 874, AD 629 123, AD 629 127. NTIS, Springfield, Va.

Debye, P. (1909). "The light pressure upon a sphere of arbitrary material," *Ann. Phys.*, **30**, No. 4, 57. In German.

Deirmendjian, D. (1960). "Atmospheric extinction of infrared radiation," *Quart. J. Roy. Meteorol. Soc.*, **86**, 371–381.

Deirmendjian, D. (1963a). "Scattering and polarization properties of polydispersed suspensions with partial absorption." In *Electromagnetic Scattering*, M. Kerker, ed. Macmillan, New York.

Deirmendjian, D. (1963b). "Tables of Mie Scattering Cross Sections and Amplitudes," Rept. R-407-PR. Rand Corporation, Santa Monica, Calif.

Deirmendjian, D. (1964). "Scattering and polarization properties of water clouds and hazes in the visible and near infrared," *Appl. Opt.*, **3**, 187–196.

Deirmendjian, D. (1969). *Electromagnetic Scattering on Spherical Polydispersions*, American Elsevier, New York.

Deirmendjian, D. (1975). "Far-infrared and submillimeter wave attenuation by clouds and rain," *J. Appl. Meteorol.*, **14**, 1584–1593.

Deirmendjian, D. and Clasen, R. J. (1962). "Light Scattering on Partially Absorbing Homogeneous Spheres of Finite Size," Rept. R-393-PR. Rand Corporation, Santa Monica, Calif.

Deirmendjian, D. and Sekera, Z. (1953). "Quantitative evaluation of multiply scattered and diffusely reflected light in the direction of a stellar source in a Rayleigh atmosphere," *J. Opt. Soc. Am.*, **43**, 1158–1165.

Deirmendjian, D. et al. (1961). "Mie scattering with complex index of refraction," *J. Opt. Soc. Am.*, **51**, 620–633.

Denman, H. H. et al. (1966). *Angular Scattering Functions for Spheres.* Wayne State University Press, Detroit.

Dessens, H. (1949). "The use of spiders' threads in the study of condensation nuclei," *Quart. J. Roy. Meteorol. Soc.*, **75**, 23–26.

Diem, M. (1942). "Messung der Grosse von Wolkenelementen," *Ann. Hydrog. Meteorol.*, **32**, 142–150.

Die, M. (1948). "Messung der Grosse von Wolkenelementen II," *Meteorol. Rundsch.*, No. 9/10, 261–273.

Dobson, G. M. (1930). "Observations of the amount of ozone in the earth's atmosphere and its relation to other geophysical conditions. Part IV," *Proc. Roy. Soc.*, **A129**, 411–433.

Dobson, G. M. (1968). "Forty years research on atmospheric ozone at Oxford: A history," *Appl. Opt.*, **7**, 387–405.

Donahue, T. M. et al. (1972). "Noctilucent clouds in daytime," *J. Atmos. Sci.*, **29**, 1205–1209.

Dorian, M. and Harshbarger, F. (1967). "Measurement of the atmospheric spectral radiance at 35 km in the near ultraviolet," *Appl. Opt.*, **6**, 1487–1491.

Dowling, J., Jr., and Green, A. E. (1966). "Second-order scattering contributions to the earth's radiance in the middle ultraviolet." In *The Middle Ultraviolet*, A. E. Green, ed. Wiley, New York.

Draegert, D. A. et al. (1966). "Far-infrared spectrum of liquid water," *J. Opt. Soc. Am.*, **56**, 64–69.

Drummond, A. J., ed. (1970). "Precision Radiometry," Vol. 14 of *Advances in Geophysics*, **14**, Academic Press, New York.

Dubin, M. and McCracken, C. W. (1962). "Measurements of distributions of interplanetary dust," *Astron. J.*, **67**, 248–256.

Dunkelman, L. (1952). "Horizontal Attenuation of Ultraviolet and Visible Light by the Lower Atmosphere," NRL Rept. 4031. NRL, Washington, D.C.

Dunkelman, L. and Scolnick, R. (1959). "Solar spectral irradiance and vertical atmospheric attenuation in the visible and ultraviolet," *J. Opt. Soc. Am.*, **49**, 356–367.

Dunlap, G. D. and Shufeldt, H. H. (1969). *Dutton's Navigation and Piloting*, 12th ed. United States Naval Institute, Annapolis, Md.

Duntley, S. Q. (1947). "The visibility of objects seen through the atmosphere," *J Opt. Soc. Am.*, **37**, 635–641.

Duntley, S. Q. (1948a). "The reduction of apparent contrast by the atmosphere," *J. Opt. Soc. Am.*, **38**, 179–191.

Duntley, S. Q. (1948b). "The visibility of distant objects," *J. Opt. Soc. Am.*, **38**, 237–249.

Ehhalt, D. H. et al. (1975). "Concentrations of CH_4, CO, CO_2, H_2, H_2O, and N_2O in the upper stratosphere," *J. Atmos. Sci.*, **32**, 163–169.

Eiden, R. (1966). "The elliptical polarization of light scattered by a volume of atmospheric air," *Appl. Opt.*, **5**, 569–575.

Eiden, R. (1971). "Determination of the complex index of refraction of spherical aerosol particles," *Appl. Opt.*, **10**, 749–754.

EK (1965). "How to Understand and Use Photometric Quantities," Kodak Tech Bits, No. 1. Eastman Kodak Company, Rochester, N.Y.

Eldridge, R. G. (1961). "A few fog drop-size distributions," *J. Meteorol.*, **18**, 671–676.

Eldridge, R. G. (1966). "Haze and fog aerosol distributions," *J. Atmos. Sci.*, **23**, 605–613.

Eldridge, R. G. (1967). "A comparison of computed and experimental spectral transmissions through haze," *Appl. Opt.*, **6**, 929–933.

Eldridge, R. G. (1969). "Mist—The transition from haze to fog," *Bull. Am. Meteorol. Soc.*, **50**, 422–426.

Eldridge, R. G. (1971). "The relationship between visibility and liquid water content in fog," *J. Atmos. Sci.*, **28**, 1183–1186.

Eldridge, R. G. and Johnson, J. C. (1958). "Diffuse transmission through real atmospheres," *J. Opt. Soc. Am.*, **48**, 463–468.

Eldridge, R. G. and Johnson, J. C. (1962). "Distribution of irradiance in haze and fog," *J. Opt. Soc. Am.*, **52**, 787–796.

Ellis, H. T. and Pueschel, R. F. (1971). "Solar radiation, absence of air pollution at Mauna Loa," *Science*, **172**, 845–846.

Elterman, L. (1964). "Parameters for attenuation in the atmospheric windows for fifteen wavelengths," *Appl. Opt.*, **3**, 745–749.

Elterman, L. (1966a). "Aerosol measurements in the troposphere and stratosphere, *Appl. Opt.*, **5**, 1769–1776.

Elterman, L. (1966b). "An Atlas of Aerosol Attenuation and Extinction Profiles for the Troposphere and Stratosphere," Rept. AFCRL-66-828. AFCRL, Bedford, Mass.

Elterman, L. (1968). "UV, Visible, and IR Attenuation for Altitudes to 50 Km, 1968," Rept. AFCRL-68-0153. AFCRL, Bedford, Mass.

Elterman, L. (1970a). "Vertical Attenuation Model with Eight Surface Meteorological Ranges 2 to 13 Km," Rept. AFCRL-70-0200. AFCRL, Bedford, Mass. AD 707 488, NTIS, Springfield, Va.

Elterman, L. (1970b). "Relationships between vertical attenuation and surface meteorological range," *Appl. Opt.*, **9**, 1804–1810.

Elterman, L. (1975). "Stratospheric aerosol parameters for the Fuego volcanic incursion," *Appl. Opt.*, **14**, 1262–1263.

Elterman, L. et al. (1969). "Features of tropospheric and stratospheric dust," *Appl. Opt.*, **8**, 893–903.

Elterman, L. et al. (1973). "Stratospheric aerosol measurements with implications for global climate," *Appl. Opt.*, **12**, 330–337.

Engelman, R. J. and Slinn, W. G., eds. (1970). "Precipitation Scavenging," *Symposium Proceedings*, Rept. CONF-700601. NTIS, Springfield, Va.

Engstrom, R. W., ed. (1974). *Electro-Optics Handbook.* RCA Commerical Engineering, Harrison, N.J.

Ensor, D. S. et al. (1972). "Multiwavelength nephelometer measurements in Los Angeles smog aerosol. I. Comparison of calculated and measured light scattering." In *Aerosols and Atmospheric Chemistry*, G. M. Hidy, ed. Academic Press, New York.

Erkovich, S. P. et al. (1965). "Influence of fog on the range of ground communications using an optical carrier," *Telecommun. Radio Eng.*, December, pp. 12–16.

Eshelman, V. R. and Gallagher, P. B. (1962). "Radar studies of 15th-magnitude meteors," *Astron. J.*, **67**, 245–248.

Evans, W. E. and Collis, R. T. (1970). "Meteorological applications of lidar," *Soc. Photogr. Instr. Eng.*, **8**, 38–45.

Farlow, N. H. et al. (1970). "Examination of surfaces exposed to a noctilucent cloud, August 1, 1968," *J. Geophys. Res.*, **75**, 6736–6750.

Feather, N. (1959). *An Introduction to the Physics of Mass, Length, and Time.* Edinburgh University Press, Edinburgh.

Fenn, R. W. and Oser, H. (1965). "Scattering properties of concentric soot-water spheres for visible and infrared light," *Appl. Opt.*, **4**, 1504–1509.

Ferarra, R. et al. (1970). "Evolution of the fog droplet size distribution observed by laser scattering," *Appl. Opt.*, **9**, 2517–2522.

Feynman, R. P. et al. (1963). *Lectures on Physics*, Vol. 1. Addison-Wesley, Reading, Mass.

Fiocco, G. and Columbo, G. (1964). "Optical radar results and meteoritic fragmentation," *J. Geophys. Res.*, **69**, 1795–1803.

Fiocco, G. and De Wolf, J. B. (1968). "Frequency spectrum of laser echoes from atmospheric constituents and determination of the aerosol content of air," *J. Atmos. Sci.*, **25**, 488–496.

Fiocco, G. and Smullin, L. D. (1963). "Determination of scattering layers in the upper atmosphere (60–140 km) by optical radar," *Nature*, **199**, 1275–1276.

Fischer, K. (1975). "Mass absorption indices of various types of natural aerosol particles in the infrared," *Appl. Opt.*, **14**, 2851–2856.

Fischer, K. and Hänel, G. (1972). "Bestimmung physikalischer Eigenschaften atmospharischer Aerosolteilchen uber dem Atlantik," *Meteor. Forschungs. Ergebnisse*, **B8**, 59–62.

Fleagle, R. G. and Businger, J. A. (1963). *An Introduction to Atmospheric Physics.* Academic Press, New York.

Foitzik, L. (1938). "Über die Lichtdurchlassigkeit der stark getrubten Atmosphare im sichtbaren Spektralbereich," *Wiss. Abh. Reichsamt Wetterdienst*, Berlin, **4**, No. 5.

Foitzik, L. and Lenz, K. (1960). *Monatsber. Deutsch. Akad. Wiss., Berlin*, **2**, 682–684.

Foitzik, L. and Lenz, K. (1962). *Optik und Spectroscopie aller Wellenlangen* (Jena meeting 1960). Akademie Verlag, Berlin.

Foitzik, L. and Zschacek, H. (1961). "Messungen der spektralen Strahldichte in der Polarisation des wolkenlosen Himmels," *Gerl. Beitr. Geophys.*, **70**, 350.

Forsythe, W. E., ed. (1937). *Measurement of Radiant Energy.* McGraw-Hill, New York.

Fowle, F. E. (1914). "Avogadro's constant and atmospheric transparency," *Astrophys. J.*, **40**, 435–442.

Frank, N. H. (1950). *Introduction to Electricity and Optics*, 2d ed. McGraw-Hill, New York.

Fraser, R. S. and Walker, W. H. (1968). "Effect of specular reflection at the ground on light scattered from a Rayleigh atmosphere," *J. Opt. Soc. Am.*, **58**, 636–644.

Friedlander, S. K. (1961). "Theoretical considerations for the particle size spectrum of the stratospheric aerosol," *J. Meteorol.*, **18**, 753–759.

Friedlander, S. K. (1965). "Theory of Self-Preserving Size Distributions in a Coagulation Dispersion," Rept. Radioactive Fallout from Nuclear Weapons Tests, 253–259. U.S. Atomic Energy Commission, Washington, D.C.

Friend, J. P. et al. (1973). "On the formation of stratospheric aerosols," *J. Atmos. Sci.*, **30**, 465–479.

Fritz, S. (1951). "Solar radiant energy and its modification by the earth and its atmosphere." In *Compendium of Meteorology*, T. F. Malone, ed. AMS, Boston.

Fry, G. A. et al. (1947). "Effect of atmospheric scattering upon the appearance of a dark object against a sky background," *J. Opt. Soc. Am.*, **37**, 635–641.

Fuchs, N. (1964). *The Mechanics of Aerosols*, Macmillan, New York.

Gaertner, H. (1947). "The Transmission of Infrared in Cloudy Atmosphere," Naval Ordnance Rept. 429, U.S. Navy Bureau of Ordnance. GPO, Washington, D.C.

Gates, D. M. and Shaw, C. C. (1960). "Infrared transmission of clouds," *J. Opt. Soc. Am.*, **50**, 876–882.

Gebbie, H. A. et al. (1951). "Atmospheric transmission in the 1 to 14 μ region," *Proc. Roy. Soc.*, **A206**, 87–107.

Gehrels, T. (1962). "Wavelength dependence of the polarization of the sunlit sky," *J. Opt. Soc. Am.*, **52**, 1164–1173.

George, J. J. (1951). "Fog." In *Compendium of Meteorology*, T. F. Malone, ed. AMS, Boston.

Germogenova, O. A. et al. (1970). "Atmospheric Haze: A review," Bolt, Beranek, and Newman, Inc., Rept. 1821. National Air Pollution Control Administration, Cincinnati, Ohio.

Gerson, N. C. (1952). "Unsolved problems in physics of the upper atmosphere." In *Advances in Geophysics*, Vol. 1. Academic Press; New York.

Gibbons, M. G. (1958). "Radiation received by an uncollimated receiver from a 4π source," *J. Opt. Soc. Am.*, **48**, 550–555.

Gibbons, M. G. (1959). "Experimental study of the effect of field of view on transmission measurements," *J. Opt. Soc. Am.*, **49**, 702–709.

Gibbons, M. G. et al. (1961). "Transmission and scattering properties of a Nevada desert atmosphere," *J. Opt. Soc. Am.*, **51**, 633–640.

Giese, R. H. (1961). "Streuung elektromagnetischer Wellen and absorbierenden und dielektrischen kugelformigen Einzelteilchen und an Gemischen solcher Teilchen," *Z. Astrophys.*, **51**, 119.

Giese, R. H. et al. (1961). "Tables related to scattering functions and scattering cross sections of particles according to the Mie theory," (In German.) *Abh. Deut. Akad. Wiss. Berlin, Kl. Math. Phys. Tech.*, No. 6.

Gillette, D. A. and Blifford, I. H., Jr. (1971). "Composition of tropospheric aerosols as a function of altitude," *J. Atmos. Sci.*, **28**, 1197–1210.

Glueckauf, E. (1951). "The Composition of Atmospheric Air." In *Compendium of Meteorology*, T. F. Malone, ed. AMS, Boston.

Goody, R. M. (1964). *Atmospheric Radiation: I, Theoretical Basis.* Oxford University Press, London.

Goody, R. M. and Walker, J. C. (1972). *Atmospheres.* Prentice-Hall, Englewood Cliffs, N.J.

Gordon, J. I. and Church, P. V. (1966a). "Sky luminances and the directional luminous reflectances of objects and backgrounds for a moderately high sun," *Appl. Opt.*, **5**, 793–801.

Gordon, J. I. and Church, P. V. (1966b). "Overcast sky luminances and directional luminous reflectances of objects and backgrounds under overcast skies," *Appl. Opt.*, **5**, 919–923.

Gordon, J. I., Edgerton, C. F., and Duntley, S. Q. (1975). "Signal light nomogram," *J. Opt. Soc. Am.*, **65**, 111–118.

Götz, F. W. (1951). "Ozone in the atmosphere." In *Compendium of Meteorology*, T. F. Malone, ed. AMS, Boston.

Götz, F. W. et al. (1934). "The vertical distribution of ozone in the atmosphere," *Proc. Roy. Soc.* **A145**, 416–446.

Goyer, G. G. and Watson, R. D. (1968). "Laser techniques for observing the upper atmosphere," *Bull. Am. Meteorol. Soc.*, **49**, 890–895.

Grams, G. and Fiocco, G. (1967). "Stratospheric aerosol layer during 1964 and 1965," *J. Geophys. Res.*, **72**, 3523–3542.

Grams, G. et al. (1972). "Complex index of refraction of airborne fly ash determined by laser radar and collection of particles at 13 km," *J. Atmos. Sci.*, **29**, 900–905.

Granath, L. P. and Hulburt, E. O. (1929). "The absorption of light by fog," *Phys. Rev.*, **34**, 140–144.

Green, H. N. (1932). "The Atmospheric Transmission of Coloured Light." RAE Rept. E. and I: 720. Royal Aircraft Establishment, Farnborough, England.

Green, H. L. and Lane, W. R. (1957). *Particulate Clouds: Dusts, Smokes, and Mists.*" Van Nostrand Reinhold, New York.

Green, A. E. and Griggs, M. (1963). "Infrared transmission through the atmosphere," *Appl. Opt.*, **2**, 561–570.

Green, A. E. and Martin, J. D. (1966). "A generalized Chapman function." In *The Middle Ultraviolet*, A. E. Green, ed. Wiley, New York.

Green, A. E. et al. (1971). "Interpretation of the sun's aureole based on atmospheric aerosol models," *Appl. Opt.*, **10**, 1263–1269.

Green, A. E. et al. (1972). "Light scattering and the size-altitude distribution of atmospheric aerosols," *J. Colloid Interface Sci.*, **39**, 520–535.

Greenfield, S. M. (1957). "Rain scavenging of particulate matter from the atmosphere," *J. Meteorol.*, **14**, 115–125.

Griggs, M. (1966). "Atmospheric ozone." In *The Middle Ultraviolet*, A. E. Green, ed. Wiley, New York.

Gringorten, I. I. et al. (1966). "Atmospheric Humidity Atlas—Northern Hemisphere." Rept. AFCRL-66-621, Air Force Surveys in Geophysics No. 186. AFCRL, Bedford, Mass.

Groves, G. V. (1971). "Atmospheric Structure and Its Variations in the Region from 25 to 120 Km." Rept. AFCRL-71-0140. AFCRL, Bedford, Mass.

Gucker, F. T. and Egan, J. J. (1961). "Measurement of the angular variation of light scattered from single aerosol droplets," *J. Colloid Sci.*, **16**, 68–84.

Gumprecht, R. O. and Sliepceich, C. M. (1951a). *Tables of Ricatti-Bessel Functions for Large Arguments and Orders.* Engineering Research Institute, University of Michigan Press, Ann Arbor.

Gumprecht, R. O. and Sliepcevich, C. M. (1951b). *Tables of Scattering Functions for Spherical Particles.* Engineering Research Institute, University of Michigan Press, Ann Arbor.

Gumprecht, R. O. and Sliepcevich, C. M. (1951c). *Tables of Functions of First and Second Partial Derivatives of Legendre Polynomials.* Engineering Research Institute, University of Michigan Press, Ann Arbor.

Gumprecht, R. O. et al. (1952). "Angular distribution of intensity of light scattered by large droplets of water," *J. Opt. Soc. Am.*, **42**, 226–231.

Gunn, K. L. and Marshall, J. S. (1958). "The distribution with size of aggregate snowflakes," *J. Meteorol.*, **15**, 452–461.

Gutnick, M. (1961). "How dry the sky?" *J. Geophys. Res.*, **66**, 2867–2871.

Gutnick, M. (1962a). "Mean Annual Mid-Latitude Moisture Profiles," Rept. AFCRL-62-681, Air Force Surveys in Geophysics, No. 147. AFCRL, Bedford, Mass.

Gutnick, M. (1962b). "Mean atmospheric moisture profiles to 31 km for middle latitudes," *Appl. Opt.*, **1**, 670–62.

Guttman, A. (1968). "Extinction coefficient measurements on clear atmospheres and thin cirrus clouds," *Appl. Opt.*, **7**, 2377–2381.

Guzzi, R. et al. (1972). "Evidence of particulate extinction in the near infrared spectrum of the sun," *J. Atmos. Sci.*, **29**, 517–523.

Hale, G. M. and Querry, M. R. (1973). "Optical constants of water in the 200-nm to 200-μm wavelength region," *Appl. Opt.*, **12**, 555–563.

Hall, F. F., Jr. (1968). "A physical model of cirrus 8–13 μ radiance," *Appl. Opt.*, **7**, 2264–2269.

Hall, F. F., Jr. (1970). "Laser measurements of turbidity in the atmosphere," *Opt. Spectra*, July/Aug., pp. 67–70.

Hall, W. N. (1971). "Spectral color changes in the zenith skylight during total solar eclipses," *Appl. Opt.*, **10**, 1225–1231.

Hamilton, P. M. (1966). "The use of lidar in air pollution studies," *Air Water Pollut. Int. J.*, **10**, 427–434.

Hamilton, P. M. (1969). "The application of a pulsed light rangefinder (lidar) to the study of chimney plumes," *Phil. Trans. Roy. Soc. London*, **A265**, 153–172.

Hampl, V. et al. (1971). "Scavenging of aerosol particles by a falling water droplet," *J. Atmos. Sci.*, **28**, 1211–1221.

Hänel, G. (1976). "Properties of Atmospheric Particles as Functions of Relative Humidity, at Thermodynamic Equilibrium with the Surrounding Air." In *Advances in Geophysics*, Vol. 19, Academic Press, New York.

Hansen, J. E. (1969). "Exact and approximate solutions for multiple scattering by cloudy and hazy planetary atmospheres," *J. Atmos. Sci.*, **26**, 478–487.

Hansen, J. E. (1971a). "Multiple scattering of polarized light in planetary atmospheres. Part I. The doubling method," *J. Atmos. Sci.*, **28**, 120–125.

Hansen, J. E. (1971b). "Multiple scattering of polarized light in planetary atmospheres. Part II. Sunlight reflected by terrestrial water clouds," *J. Atmos. Sci.*, **28**, 1400–1426.

Hansen, J. E. and Pollack, J. B. (1970). "Near infrared light scattering by terrestrial clouds," *J. Atmos. Sci.*, **27**, 265–281.

Hardy, A. C. (1967). "How large is a point source?" *J. Opt. Soc. Am.*, **57**, 44–47.

Hardy, A. C. and Perrin, F. H. (1932). *Principles of Optics.* McGraw-Hill, New York.

Harris, F. S., Jr. (1969). "Changes in polarization and angular distribution of scattered radiation during cloud formation," *Appl. Opt.*, **8**, 143–145.

Harris, F. S., Jr. (1971). "Water and ice cloud discrimination by laser beam scattering," *Appl. Opt.*, **10**, 732–737.

Harris, F. S., Jr. (1972). "Calculated Mie scattering properties in the visible and infrared of measured Los Angeles aerosol size distributions," *Appl. Opt.*, **11**, 2697–2705.

Harrison, H. et al. (1972). "Mie theory computations of lidar and nephelometric scattering parameters for power law aerosols," *Appl. Opt.*, **11**, 2880–2885.

Haurwitz, B. (1945). "Insolation in relation to cloudiness and cloud density," *J. Meteorol.*, **2**, 154–166.

Haurwitz, B. (1946). "Insolation in relation to cloud type," *J. Meteorol.*, **3**, 123–124.

Haurwitz, B. (1948). "Insolation in relation to cloud type," *J. Meteorol.*, **5**, 110–113.

Havard, J. B. (1960). "On the radiational characteristics of water clouds at infrared wavelengths," Ph.D. Thesis, University of Washington, Seattle.

Hawkins, G. S. (1962). "Radar determination of meteor orbits," *Astron. J.*, **67**, 241–244.

Hawkins, G. S. (1964). "Interplanetary debris near the earth," *Ann. Rev. Astron. Astrophys.*, **2**, 149–164.

Hawkins, G. S., ed. (1967). "Meteor Orbits and Dust," Symposium Proceedings, Cambridge, Mass., Aug. 9–13, 1965, NASA Special Publ. SP-135. GPO, Washington, D.C.

Heggestad, H. M. (1971). "Multiple scattering model for light transmission through optically thick clouds," *J. Opt. Soc. Am.*, **61**, 1293–1300.

Heller, W. (1963). "Light scattering of colloidal spheres." In *Electromagnetic Scattering*, M. Kerker, ed. Macmillan, New York.

Hemenway, C. L. and Soberman, R. K. (1962). "Studies of micrometeorites obtained from a recoverable sounding rocket," *Astron. J.*, **67**, 256–266.

Hemenway, C. L. et al. (1965). "Investigations of Noctilucent Cloud Particles," Rept. AFCRL-65-122. AFCRL, Bedford, Mass.

Henderson, S. T. (1970). *Daylight and Its Spectrum.* American Elsevier, New York.

Hering, W. S. (1964). "Ozonesonde Observations over North America," Rept. AFCRL-64-30(I). AFCRL, Bedford, Mass.

Hering, W. S. and Borden, T. R., Jr. (1964). "Ozonesonde Observations over North America," Rept. AFCRL-64-30 (II and III). AFCRL, Bedford, Mass.

Hering, W. S. and Borden, T. R., Jr. (1965). "Mean Distributions of Ozone Density over North America, 1963–1964," Rept. AFCRL-65-913. AFCRL, Bedford, Mass.

Hering, W. S. and Dütsch, H. U. (1965). "A comparison of chemiluminescent and electrochemical observations," *J. Geophys. Res.*, **70**, 5483–5490.

Herman, B. M. et al. (1971). "The effect of atmospheric aerosols on scattered sunlight," *J. Atmos. Sci.*, **28**, 419–428.

Hess, S. L. (1959). *Introduction to Theoretical Meteorology.* Holt, New York.

Hidy, G. M., ed. (1972). *Aerosols and Atmospheric Chemistry.* Academic Press, New York.

Hiser, H. W. et al. (1965). "Study of Light Attenuation under Subtropical Climatic Conditions," Rept. No. 1 on Contract DA-36-039-AMC-0352 E, University of Miami, AD 626 198. NTIS, Springfield, Va.

Hodges, J. A. (1972). "Aerosol extinction contribution to atmospheric attenuation in infrared wavelengths," *Appl. Opt.*, **11**, 2304–2310.

Hodkinson, J. R. (1963). "Light Scattering and Extinction by Irregular Particles Larger than the Wavelength." In *Electromagnetic Scattering*, M. Kerker, ed. Macmillan, New York.

Hodkinson, J. R. (1966a). "Particle sizing by means of the forward scattering lobe," *Appl. Opt.*, **5**, 839–844.

Hodkinson, J. R. (1966b). "The optical measurement of aerosols." In *Aerosol Science*, C. N. Davies, ed. Academic Press, New York.

Hodkinson, J. R. and Greenleaves, I. (1963). "Computations of light-scattering and extinction by spheres according to diffraction and geometrical optics, and some comparisons with the Mie theory," *J. Opt. Soc. Am.*, **53**, 577–588.

Hodkinson, J. R. and Greenfield, J. R. (1965). "Response calculations for light-scattering aerosol counters and photometers," *Appl. Opt.* **4**, 1463.

Hoffman, D. J. et al. (1975). "Stratospheric aerosol measurements. I: Time variations at northern midlatitudes," *J. Atmos. Sci.*, **32**, 1446–1456.

Hogan, A. et al. (1975). "A portable aerosol detector of high sensitivity," *J. Appl. Meteorol.*, **14**, 39–45.

Hogg, D. C. (1964). "Scattering and attenuation due to snow at optical wavelengths," *Nature*, **203**, 396.

Holl, H. (1948). *Optik*, **4**, 173.

Holland, A. C. and Draper, J. S. (1967). "Analytical and experimental investigations of light scattering from polydispersions of Mie particles," *Appl. Opt.*, **6**, 511–518.

Holland, A. C. and Gagne, G. (1970). "The scattering of polarized light by polydisperse systems of irregular particles," *Appl. Opt.*, **9**, 1113–1121.

Holland, A. C. and Gagne, G. (1971). "Comment on: The scattering of polarized light by polydisperse systems of irregular particles," *Appl. Opt.*, **10**, 1173–1174.

Holzworth, G. C. and Rao, C. R. (1965). "Studies of skylight polarization," *J. Opt. Soc. Am.*, **55**, 403–408.

Hopkinson, R. G. (1954). "Measurements of sky luminance distribution at Stockholm," *J. Opt. Soc. Am.*, **44**, 455–459.

Horman, M. H. (1961). "Measurement of atmospheric transmissivity using backscattered light from a pulsed light beam," *J. Opt. Soc. Am.*, **51**, 681–691.

Horman, M. H. (1967). "Visibility of light sources against a background of uniform luminance," *J. Opt. Soc. Am.*, **57**, 1516–1521.

Houghton, H. G. (1931). "The transmission of visible light through fog," *Phys. Rev.*, **38**, 152–158.

Houghton, H. G. (1934). "Research on fog at the Round Hill station of the MIT," *J. Aeronaut. Sci.*, **1**, 109.

Houghton, H. G. (1939). "On the relation between visibility and the constitution of cloud and fog," *J. Aeronaut. Sci.*, **6**, 408–411.

Houghton, H. G. (1951). "On the physics of clouds and precipitation." In *Compendium of Meteorology*, T. F. Malone, ed. AMS, Boston.

Houghton, H. G. and Chalker, W. R. (1949). "Scattering cross-sections of water drops in air for visible light," *J. Opt. Soc. Am.*, **39**, 955–957.

Houghton, H. F. and Radford, W. H. (1938). "On the measurement of drop-size and liquid water content in fogs and clouds," *Pap. Phys. Oceanog. Meteorol.*, **6**, No. 4. Mass. Inst. Tech., Cambridge, Mass.

Hovis, W. A., Jr., and Tobin, M. (1967). "Spectral measurements from 1.6μ to 5.4μ of natural surfaces and clouds," *Appl. Opt.*, **6**, 1399–1402.

Hovis, W. A., Jr., et al. (1970). "Infrared reflectance of high altitude clouds," *Appl. Opt.*, **9**, 561–563.

Howard, J. N. and Garing, J. S. (1967). "Atmospheric optics and radiation transfer" (bibliography), *Trans. Am. Geophys. Union*, **48**, 471–485.

Hudson, N. W. (1963). "Raindrop size distribution in high intensity storms," *Rhodesian J. Agri. Res.*, **1**, 6–11.

Hudson, R. D., Jr. (1969). *Infrared System Engineering.* Wiley, New York.

Hulburt, E. O. (1934). "Absorption of heat rays by fog," *Physics*, **5**, 101–102.

Hulburt, E. O. (1937). "Observations of a searchlight beam to an altitude of 28 kilometers," *J. Opt. Soc. Am.*, **27**, 377–382.

Hulburt, E. O. (1941). "Optics of atmospheric haze," *J. Opt. Soc. Am.*, **31**, 467–476.

Hulburt, E. O. (1946). "Optics of searchlight illumination," *J. Opt. Soc. Am.*, **36**, 483–491.

Hulburt, E. O. (1956). "Some recent papers in the *Journal of the Optical Society of America*," *J. Opt. Soc. Am.*, **46**, 5–9.

Hulburt, E. O. (1957). "Physics of the upper atmosphere," In *Meteorological Monographs*, Vol. 3, No. 17, pp. 160–181. AMS, Boston.

Humphreys, W. J. (1940). *Physics of the Air.* McGraw-Hill, New York. Also Dover, New York (1964).

Hunten, D. M. (1973). "The escape of light gases from planetary atmospheres," *J. Atmos. Sci.*, **30**, 1481–1494.

Huschke, R. E. (1959). *Glossary of Meteorology*. AMS, Boston.

Hutchinson, G. E. (1954). "The Biochemistry of the Terrestrial Atmosphere." In *The Earth as a Planet*, G. P. Kuiper, ed. University of Chicago Press, Chicago.

Hynek, J. A. (1951). "Photographing stars in the daytime," *Sky Telescope*, January, pp. 61–62.

Ingrao, H. C. (1973). "A Bibliography on Methods of Atmospheric Visibility Measurements Relevant to Air Traffic Control and Related Subjects," Rept. No. FAA-RD-73-128, November 1973. U.S. Department of Transportation, Transportation Systems Center, Cambridge, Mass.

Irvine, W. M. (1965a). "Light scattering by spherical particles: Radiation pressure, asymmetry factor, and extinction cross-section," *J. Opt. Soc. Am.*, **55**, 16–21.

Irvine, W. M. (1965b). "Multiple scattering by large particles," *Astrophys. J.*, **142**, 1563–1575.

Irvine, W. M. (1968). "Multiple scattering by large particles. II: Optically thick layers," *Astrophys. J.*, **152**, 823–834.

Irvine, W. M. and Peterson, F. W. (1970). "Observations of atmospheric extinction from 0.315 to 1.06 microns," *J. Atmos. Sci.*, **27**, 62–69.

Irvine, W. M. and Pollack, J. B. (1968). "Infrared optical properties of water and ice spheres," *Icarus*, **8**, 324–360.

Ivanov, A. I. et al. (1969). "Light Scattering in the Atmosphere, Part 2," NASA Technical Translation, Publ. No. NASA TT F-477. NTIS, Springfield, Va.

Jeans, J. H. (1925). *The Dynamical Theory of Gases*. Cambridge University Press, Cambridge, England.

Jeans, J. H. (1952). *An Introduction to the Kinetic Theory of Gases*. Cambridge University Press, Cambridge, England.

Johnson, E. A. et al. (1939). "The measurement of light scattered by the upper atmosphere from a search-light beam," *J. Opt. Soc. Am.*, **29**, 512–517.

Johnson, I. and La Mer, V. K. (1947). "The determination of the particle size of monodispersed systems by the scattering of light," *J. Am. Chem. Soc.*, **69**, 1184–1192.

Johnson, J. C. (1954). *Physical Meteorology*. Technology Press, MIT, Cambridge, Mass., and Wiley, New York.

Johnson, J. C., Eldridge, R. G., and Terrell, J. R. (1954). "An Improved Infrared Transmissometer for Cloud Drop Sizing," Scientific Rept. No. 4, Contract AF 19 (122)-45. Department of Meteorology, MIT, Cambridge, Mass.

Johnson, J. C. and Terrell, J. R. (1955). "Transmission cross sections for water spheres illuminated by infrared radiation," *J. Opt. Soc. Am.*, **45**, 451–454.

Johnston, H. S. (1971). "Reduction of stratospheric ozone by nitrogen oxide catalysts from SST exhaust," *Science*, **173**, 517–522.

Jones, O. C. and Preston, J. S. (1969). "Photometric Standards and the Unit of Light," Notes on Applied Science No. 24, National Physical Laboratory. Her Majesty's Stationery Office, London. Also, British Information Services, New York.

J. Opt. Soc. Am., (1956). "Bibliography of Edward O. Hulburt," *J. Opt. Soc. Am.*, **46**, 1–5.

J. Opt. Soc. Am. (1960). "Publications by W. E. Knowles Middleton," *J. Opt. Soc. Am.*, **50**, 95–96.

Junge, C. E. (1951). "Nuclei of atmospheric condensation." In *Compendium of Meteorology*, T. F. Malone, ed. AMS, Boston.

Junge, C. E. (1958). "Atmospheric Chemistry," In *Advances in Geophysics*, Vol. 4. Academic Press, New York.

Junge, C. E. (1960a). "Sulfur in the atmosphere," *J. Geophys. Res.*, **65**, 227–237.

Junge, C. E. (1960b). "Aerosols." In *Handbook of Geophysics*, C. F. Campen et al., eds. Macmillan, New York.

Junge, C. E. (1961). "Vertical profiles of condensation nuclei in the stratosphere," *J. Meteorol.* **18**, 501–509.

Junge, C. E. (1963). *Air Chemistry and Radioactivity.* Academic Press, New York.

Junge, C. E. (1969). "Comments on concentration and size distribution measurements of atmospheric aerosols, and a test of the theory of self-preserving size distributions," *J. Atmos. Sci.*, **26**, 603–608.

Junge, C. E. and Manson, J. E. (1961). "Stratospheric aerosol studies," *J. Geophys. Res.*, **66**, 2163–2182.

Junge, C. E. et al. (1961). "Stratospheric aerosols," *J. Meteorol.*, **18**, 81–108.

Kano, M. (1968). "Effect of a concentrated turbid layer on the polarization of skylight," *J. Opt. Soc. Am.*, **58**, 789–797.

Kantor, A. J. (1966). "Horizontal and Vertical Distributions of Atmospheric Pressure, 30 to 90 Kilometers," Rept. AFCRL-66-346. AFCRL, Bedford Mass.

Kantor, A. J. and Cole, A. E. (1965). "Monthly atmospheric structure, surface to 80 km," *J. Appl. Meteorol.*, **4**, 228–237. Also Rept. AFCRL-65-634, August 1965. AFCRL, Bedford, Mass.

Kattawar, G. W. and Plass, G. N. (1967). "Resonance scattering from absorbing spheres," *Appl. Opt.*, **6**, 1549–1554.

Kattawar, G. W. and Plass, G. N. (1971). "Radiance and polarization of light reflected from optically thick clouds," *Appl. Opt.*, **10**, 74–80.

Kattawar, G. W. and Plass, G. N. (1972). "Degree and direction of polarization of multiple scattered light. 1: Homogeneous cloud layers," *Appl. Opt.*, **11**, 2851–2865.

Katz, U. and Kocmond, W. C. (1973). "An investigation of the size-supersaturation relationship of soluble condensation nuclei," *J. Atmos. Sci.*, **30**, 160–165.

Keith, C. H. and Arons, A. B. (1954). "The growth of sea-salt particles by condensation of atmospheric water vapor," *J. Meteorol.*, **11**, 173–184.

Kennard, E. H. (1938). *Kinetic Theory of Gases.* McGraw-Hill, New York.

Kent, G. S. et al. (1967). "High altitude atmospheric scattering from a laser beam," *J. Atmos. Terr. Phys.*, **29**, 169–181.

Kerker, M. (1969). *The Scattering of Light and Other Electromagnetic Radiation.* Academic Press, New York.

Kerker, M. et al. (1966). "Color effects in the scattering of white light by micron and submicron spheres," *J. Opt. Soc. Am.*, **56**, 1248–1258.

Kerker, M. and Hampl, V. (1974). "Scavenging of aerosol particles by a falling water drop and calculation of washout coefficients," *J. Atmos. Sci.*, **31**, 1368–1376.

Khvostikov, I. A. (1965). "The Upper Layers of the Atmosphere," NASA Technical Translation, Publ. No. NASA TT F-315. NTIS, Springfield, Va.

Kimball, H. H. and Hand, I. F. (1921). "Sky brightness and daylight illumination measurements," *Mon. Weather Rev.*, **49**, 481–488.

Kimball, H. H. and Hand, I. F. (1922). "Daylight illumination on horizontal, vertical, and sloping surfaces, *Mon. Weather Rev.*, **50**, 615–628.

Knestrick, G. L. and Curcio, J. A. (1966). "Spectral radiance of the horizon sky," *J. Opt. Soc. Am.*, **56**, 1455.

Knestrick, G. L. and Curcio, J. A. (1967). "Measurements of spectral radiance of the horizon sky," *Appl. Opt.*, **6**, 2105–2109.

Knestrick, G. L. and Curcio, J. A. (1970). "Measurements of ultraviolet spectral radiance of the horizon sky," *Appl. Opt.*, **9**, 1574–1576.

Knestrick, G. L. et al. (1962). "Atmospheric scattering coefficients in the visible and infrared regions," *J. Opt. Soc. Am.*, **52**, 1010–1016.

Knoll, H. A., Tousey, R., and Hulburt, E. O. (1946). "Visual thresholds of steady point sources of light in fields of brightness from dark to light," *J. Opt. Soc. Am.*, **36**, 480–482.

Kochsmeider, H. (1924). "Theorie der horizontalen Sichtweite," *Beitr. Phys. Freien Atmos.*, **12**, 33–53, 171–181.

Kochsmeider, H. (1930). "Measurements of visibility at Danzig," *Mon. Weather Rev.*, **58**, 439–444.

Kondratyev, K. Ya. (1969). *Radiation in the Atmosphere*, Academic Press, New York.

Kondratyev, K. Ya. et al. (1967). "Direct solar radiation up to 30 km and stratification of attenuation components in the stratosphere," *Appl. Opt.*, **6**, 197–207.

Kondratyev, K. Ya. et al. (1971). "Visual observations and spectral investigations of the earth's twilight aureole from the Soyuz-5 spacecraft," *Appl. Opt.*, **10**, 2521–2533.

Koomen, M. J. (1959). "Visibility of stars at high altitude in daylight," *J. Opt. Soc. Am.*, **49**, 626–629.

Koomen, M. J. et al. (1952). "Measurements of the brightness of the twilight sky," *J. Opt. Soc. Am.*, **42**, 353–356.

Kruse, P. W. et al. (1963). *Elements of Infrared Technology*, Wiley, New York.

Kuiper, G. P., ed. (1954). *The Earth as a Planet*. University of Chicago Press, Chicago.

Kumai, M. (1973). "Arctic fog droplet size distribution and its effect on light attenuation," *J. Atmos. Sci.*, **30**, 635–643.

Kurnick, S. W. et al. (1960). "Attenuation of infrared radiation by fogs," *J. Opt. Soc. Am.*, **50**, 578–583.

La Mer, V. K. and Sinclair, D. (1943). "Progress Report on the Verification of Mie Theory. Calculations and Measurements of Light Scattering by Dielectric Spherical Particles," OSRD Rept. 1857, Sept. 29, 1943. Department of Commerce, Washington, D. C.

Latimer, P. (1972). "Dependence of extinction efficiency of spherical scatterers on photometer geometry," *J. Opt. Soc. Am.*, **62**, 208–211.

Laws, J. O. and Parsons, D. A. (1943). "The relation of raindrop size to intensity," *Trans. Am. Geophys. Union*, **24**, 452–460.

Lazrus, A. L. and Gandrud, B. W. (1974). "Distribution of stratospheric nitric acid vapor," *J. Atmos. Sci.*, **31**, 1102–1108.

Levi, L. (1968). *Applied Optics: A Guide to Modern Optical System Design*. Wiley, New York.

Levin, R. E. (1968). "Luminance—A tutorial paper," *J. Soc. Mot. Pict. Telev. Eng.*, **77**, 1005–1011.

Lewis, W. (1951). "Meteorological aspects of aircraft icing," In *Compendium of Meteorology*, T. F. Malone, ed. AMS, Boston.

Lin, Chin-I, et al. (1973). "Absorption coefficient of atmospheric aerosol: A method for measurement," *Appl. Opt.*, **12**, 1356–1363.

Lindberg, J. D. and Laude, L. S. (1974). "Measurement of the absorption coefficient of atmospheric dust," *Appl. Opt.*, **13**, 1923–1927.

Lindsay, R. B. (1970). *Lord Rayleigh, the Man and His Work.* Pergamon Press, London.

Link, F. and Sekera, Z. (1940). *Dioptric Tables of the Earth's Atmosphere*, and *Extension of the Dioptric Tables.* Prometheus Press, Prague. Translated by R. J. Stirton in Navweps Rept. 8504, June 1964, AD 444 993. U.S. Naval Ordnance Test Station, China Lake, Calif.

Link, F. and Neuzil, L. (1969). *Tables of Light Trajectories in the Terrestrial Atmosphere.* Hermann, Paris, France.

Liou, K. (1971). "Time-dependent multiple backscattering," *J. Atmos. Sci.*, **28**, 824–827.

Liou, K. (1972a). "Light scattering by ice clouds in the visible and infrared: A theoretical study," *J. Atmos. Sci.*, **29**, 524–536.

Liou, K. (1972b). "On depolarization of visible light from water clouds for a monostatic lidar," *J. Atmos. Sci.*, **29**, 1000–1003.

Liou, K. and Schotland, R. M. (1971). "Multiple backscattering and depolarization from water clouds for a pulsed lidar system," *J. Atmos. Sci.*, **28**, 772–784.

List, R. J. (1966). *Smithsonian Meteorological Tables*, 6th rev. ed. Smithsonian Institution, Washington, D.C.

Liu, B. Y. et al. (1975). "Calibration of the Pollak counter with monodisperse aerosols," *J. Appl. Meteorol.*, **14**, 46–51.

Lloyd, J. W. et al. (1965). "Day skylight intensity from 20 km to 90 km at 5500 A," *Appl. Opt.*, **4**, 1602–1606.

Lloyd, J. W. and Silverman, S. M. (1971). "Measurements of the zenith sky intensity and spectral distribution during the solar eclipse of 12 November 1966 at Bage, Brazil, and on an aircraft," *Appl. Opt.*, **10**, 1215–1219.

Loeb, L. B. (1934). *The Kinetic Theory of Gases.* McGraw-Hill, New York. Also Dover, New York (1961).

Logan, N. A. (1965). "Survey of some early studies of the scattering of plane waves by a sphere," *Proc. IEEEE*, **53**, 773–785.

London, J. (1962). "The distribution of total ozone over the northern hemisphere," *Sun Work*, **7**, 11–12.

Longley, R. W. (1970). *Elements of Meteorology.* Wiley, New York.

Lowan, A. N. (1949). "Tables of Scattering Functions for Spherical Particles," Applied Mathematics Series 4, National Bureau of Standards, Jan. 25, 1949.

Luckiesch, M. and Holladay, L. L. (1941). "Penetration of fog by light from sodium and tungsten lamps," *J. Opt. Soc. Am.*, **31**, 528–530.

Ludlum, D. M., ed. (1971). *Weatherwise*, **24**, No. 5, October (AMS, Boston).

MacAdam, D. L. (1967). "Editor's Page: Nomenclature and symbols for radiometry and photometry," *J. Opt. Soc. Am.*, **57**, 854.

MacDonald, D. E. (1953). "Air photography," *J. Opt. Soc. Am.*, **43**, 290–298.

Machta, L. (1972). "Mauna Loa and global trends in air quality," *Bull. Am. Meteorol. Soc.*, **53**, 402–420.

MacQueen, R. M. (1968). "A note on the stratospheric infrared sky radiance," *J. Atmos. Sci.*, **25**, 335–337.

Malone, T. F., ed. (1951). *Compendium of Meteorology.* AMS, Boston.

Manson, J. E. (1965). "Aerosols." In *Handbook of Geophysics and Space Environment*, S. L. Valley, ed. McGraw-Hill, New York.

Manson, J. E. et al. (1961). "The possible role of gas reactions in the formation of the stratospheric aerosol layer." In *Chemical Reactions in the Lower and Upper Atmosphere.* Interscience, New York.

Marshall, J. S. and Palmer, W. (1948). "The distribution of raindrops with size," *J. Meteorol.*, **5**, 165–166.

Mason, B. J. (1961). "Physical and chemical processes involved in the formation of atmospheric condensation nuclei." In *Chemical Reactions in the Lower and Upper Atmosphere.* Interscience, New York.

Mason, B. J. (1971). *The Physics of Clouds*, 2d ed. Oxford University Press, London.

Mason, B. J. (1975). *Clouds, Rain, and Rainmaking*, Cambridge University Press, Cambridge, England.

Mastenbrook, H. J. (1968). Water vapor distribution in the stratosphere and high troposphere," *J. Atmos. Sci.*, **25**, 299–311.

Mastenbrook, H. J. (1971). "The variability of water vapor in the stratosphere," *J. Atmos. Sci.*, **28**, 1495–1501.

McClatchey, R. A. et al. (1971). "Optical Properties of the Atmosphere" (revised), Rept. AFCRL-71-0279. AFCRL, Bedford, Mass. AD 726 116. NTIS, Springfield, Va.

McCormick, M. P. (1967). "Laser backscatterer measurements of the lower atmosphere," Ph.D. Thesis, College of William and Mary, Williamsburg, Va. University Microfilms No. 68-810, University Microfilms, Ann Arbor.

McCormick, M. P. and Fuller, W. H., Jr. (1975). "Lidar measurements of two intense stratospheric dust layers, *Appl. Opt.*, **14**, 4–5.

McCormick, M. P. et al. (1968). "Mie total and differential backscattering cross sections at laser wavelengths for Junge aerosol models," *Appl. Opt.*, **7**, 2424–2425.

McCormick, R. A. and Ludwig, J. H. (1967). "Climate modification by atmospheric aerosols," *Science*, **156**, 1358–1359.

McDonald, R. K. and Deltenre, R. W. (1963). "Cirrus infrared reflection measurements," *J. Opt. Soc. Am.*, **53**, 860–868.

McElroy, M. B. et al. (1974). "Atmospheric ozone: Possible impact of stratospheric aviation," *J. Atmos. Sci.*, **31**, 287–303.

McNally, D. (1975). *Positional Astronomy.* Wiley, New York.

McNeil, W. R. and Carswell, A. I. (1975). "Lidar polarization studies of the troposphere," *Appl. Opt.* **14**, 2158–2168.

Meyer-Arendt, J. R. (1968). "Radiometry and photometry: Units and conversion factors, *Appl. Opt.*, **7**, 2081–2084.

Middleton, W. E. K. (1949). "The effect of the angular aperture of a telephotometer on the telephotometry of collimated and non-collimated beams," *J. Opt. Soc. Am.*, **39**, 576–581.

Middleton, W. E. K. (1951). "Visibility in meteorology. "In *Compendium of Meteorology*, T. F. Malone, ed. AMS, Boston.

Middleton, W. E. K. (1952). *Vision through the atmosphere.* University of Toronto Press, Toronto.

Middleton, W. E. K. (1954). "The color of the overcast sky," *J. Opt. Soc. Am.*, **44**, 793–798.

Middleton, W. E. K. (1960). "Random reflections on the history of atmospheric optics." *J. Opt. Soc. Am.*, **50**, 97–100.

Mie, G. (1908). "A contribution to the optics of turbid media, especially colloidal metallic suspensions," *Ann. Phys.*, **25**, No. 4, 377–445. In German.

Mikhailin, L. M. and Khvostikov, I. A. (1946). "Investigation of the atmosphere by means of a searchlight beam to altitudes of up to 25 km," *Compt. Rend. Acad. Sci. USSR (Moscow)*, **54**, 223–226.

Millman, P. M. (1962). "The meteor radar echo—An observational survey," *Astron. J.*, **67**, 235–240.

Milton, J. E. et al. (1972). "Lidar reflectance of fair-weather cumulus clouds at 0.903 μ," *Appl. Opt.*, **11**, 697–698.

Minnaert, M. (1940). *The Nature of Light and Color in the Open Air.* Bell, London. Also Dover, New York (1954).

Mironov, A. V. et al. (1946). "Method for optical investigation of the atmosphere under daylight conditions by means of a searchlight beam," *Compt. Rend. Acad. Sci. USSR (Moscow)*, **54**, 483–486.

Mirtov, B. A. (1964). "Gaseous Composition of the Atmosphere and Its Analysis," NASA Technical Translation, Publ. No. NASA TT F-145. NTIS, Springfield, Va.

Mitra, S. K. (1952). *The Upper Atmosphere.* Asiatic Society, Calcutta.

Moon, P. (1936) *The Scientific Basis of Illuminating Engineering.* McGraw-Hill, New York. Also Dover, New York (1961).

Moon, P. and Spencer, D. E. (1942). "Illumination from a non-uniform sky," *Illum. Eng.*, December, pp. 707–726.

Mossop, S. C. (1970). "Concentrations of ice crystals in clouds," *Bull. Am. Meteorol. Soc.*, **51**, 474–479.

MTP (1945). *Mathematical Tables Project: Tables of Associated Legendre Functions*, Columbia University Press, New York.

Mueller, H. (1948). "The foundations of optics," *J. Opt. Soc. Am.*, **38**, 661.

Mueller, E. A. and Sims, A. L. (1966). "Investigation of the Quantitative Determination of Point and Areal Precipitation by Radar Echo Measurements," Tech. Rept. No. ECOM-00032-F, Contract DA-28-043AMC-00032(E). Illinois State Water Survey, Urbana, Ill.

Mugele, R. A. and Evans, H. D. (1951). "Droplet size distributions in sprays," *Ind. Chem. Eng.*, **43**, 1317–1324.

Muray, J. J. et al. (1971). "Proposed supplement to the SI nomenclature for radiometry and photometry," *Appl. Opt.*, **10**, 1465–1468.

Murcray, D. G. et al. (1969). "Presence of HNO_3 in the upper atmosphere," *J. Opt. Soc. Am.*, **59**, 1131–1134.

Murcray, D. G. et al. (1973). "Nitric acid distribution in the stratosphere," *J. Geophys. Res.*, **78**, 7033–7038.

Myers, J. N. (1968). "Fog," *Sci. Am.*, December, pp. 75–82.

NA (annual). "The Nautical Almanac." U.S. Naval Observatory. GPO, Washington, D.C.

Nassau, J. J. (1948). *Practical Astronomy.* McGraw-Hill, New York.

Nayatani, Y. and Wyszecki, G. (1963). "Color of daylight from North sky," *J. Opt. Soc. Am.*, **53**, 626–629.

Neiburger, M. and Wurtele, M. G. (1949). "On the nature and size of particles in haze, fog, and stratus of the Los Angeles region," *Chem. Rev.*, **44**, 321–335.

Neiburger, M. and Chien, C. W. (1960). "Computations of the growth of cloud drops by condensation using an electronic digital computer." In *Physics of Precipitation*, Monograph No. 5, American Geophysical Union. Available from University Microfilms, Ann Arbor.

Nelson, C. N. and Hamsher, D. H. (1950). "Photography of high altitude aerial objects," *J. Opt. Soc. Am.*, **40**, 863–877.

Nesti, A. J., Jr. (1970). "The condensation nuclei counter as an air pollution weapon," *Opt. Spectra*, July/Aug., pp. 76–79.

Neuberger, H. (1950). "Arago's neutral point: A neglected tool in meteorological research," *Bull. Am. Meteorol. Soc.*, **31**, 119–125.

Neuberger, H. (1951). "General meteorological optics." In *Compendium of Meteorology*, T. F. Malone, ed. AMS, Boston.

Neumann, J. (1973). "Radiation absorption by droplets of sulfuric acid water solutions and by ammonium sulfate particles," *J. Atmos. Sci.*, **30**, 95–100.

Newell, R. E. (1964). "The circulation of the upper atmosphere," *Sci. Am.*, March, pp. 62–74.

Newell, R. E. (1971). "The global circulation of atmospheric pollutants," *Sci. Am.*, January, pp. 32–42.

Newkirk, G. A., Jr. (1956). "Photometry of the solar aureole," *J. Opt. Soc. Am.*, **46**, 1028–1037.

Newkirk, G. A., Jr., and Eddy, J. A. (1964). "Light scattering by particles in the upper atmosphere," *J. Atmos. Sci.*, **21**, 35–60.

Nicodemus, F. E. (1967). "Radiometry." In *Applied Optics and Optical Engineering*, R. Kingslake, ed. Academic Press, New York.

Nicodemus, F. E. (1968). "Radiometry with spectrally sensitive detectors," *Appl. Opt.*, **7**, 1649–1652.

Nicodemus, F. E. (1969). "Optical resource letter on radiometry," *J. Opt. Soc. Am.*, **59**, 243–248. Also *Am. J. Phys.*, **38**, 43–50.

Nicodemus, F. E. (1970). "Reflectance nomenclature and directional reflectance and emissivity," *Appl. Opt.*, **9**, 1474–1475.

Nicolet, M. (1953). "Dynamic Effects in the High Atmosphere." In *The Earth as a Planet*, G. Kuiper, ed. University of Chicago Press, Chicago.

Northam, G. B. et al. (1974). "Dustsonde and lidar measurements of stratospheric aerosols: A comparison," *Appl. Opt.*, **13**, 2416–2420.

Ohtake, T. (1970). "Factors affecting the size distribution of raindrops and snowflakes," *J. Atmos. Sci.*, **27**, 804–813.

Ono, A. (1969). "The shape and riming properties of ice crystals in natural clouds," *J. Atmos. Sci.*, **26**, 138–147.

Ono, A. (1970). "Growth mode of ice crystals in natural clouds," *J. Atmos. Sci.*, **27**, 649–658.

Osgood, T. H. et al. (1964). *Atoms, Radiation, and Nuclei.* Wiley, New York.

Packer, D. M. and Lock, C. (1951). "The brightness and polarization of the daylight sky at altitudes of 18,000 to 38,000 feet above sea level," *J. Opt. Soc. Am.*, **41**, 473–478.

Pal, S. R. and Carswell, A. I. (1973). "Polarization properties of lidar backscattering from clouds," *Appl. Opt.*, **12**, 1530–1535.

Palmer, K. F. and Williams, D. (1975). "Optical constants of sulfuric acid; application to the clouds of Venus," *Appl. Opt.*, **14**, 208–219.

Pangonis, W. J. and Heller, W. (1960). *Angular Scattering Functions for Spherical Particles*, Wayne State University Press, Detroit.

Pangonis, W. J. et al. (1957). *Tables of Light Scattering Functions of Spherical Particles*. Wayne State University Press, Detroit.

Pasceri, R. E. and Friedlander, S. K. (1965). "Measurements of the particle size distribution of the atmospheric aerosol. II: Experimental results and discussion," *J. Atmos. Sci.*, **22**, 577–584.

Pearson, C. A. et al. (1952). "Visual measurements of atmospheric transmission of light at night," *Bull. Am. Meteorol. Soc.*, **33**, 117–121.

Penndorf, R. (1953). "On the Phenomenon of the Colored Sun, Especially the "Blue" Sun of September 1950." Geophysical Research Papers No. 20. AFCRL, Bedford, Mass.

Penndorf, R. (1954). "The Vertical Distribution of Mie Particles in the Troposphere." Rept. AFCRC-54-5, AFCRL, Bedford, Mass.

Penndorf, R. (1956). "New Tables of Mie Scattering Functions for Spherical Particles: Part 6. Total Mie Scattering Coefficients for Real Refractive Indices." Rept. AFCRC-TR-56-204(6). AFCRL, Bedford, Mass. AD 98 772. NTIS, Springfield, Va.

Penndorf, R. (1957a). "Tables of the refractive index for standard air and the Rayleigh scattering coefficient for the spectral region between 0.2 and 20.0μ, and their application to atmospheric optics," *J. Opt. Soc. Am.*, **47**, 176–182.

Penndorf, R. (1975b). "Total Mie scattering coefficients for spherical particles of refractive index $n \approx 1.0$," *J. Opt. Soc. Am.*, **47**, 603–605.

Penndorf, R. (1957c). "New tables of total Mie scattering coefficients for spherical particles of real refractive index ($1.33 \leq n \leq 1.50$)," *J. Opt. Soc. Am.*, **47**, 1010–1015.

Penndorf, R. (1962). "Angular Mie scattering," *J. Opt. Soc. Am.*, **52**, 402–408.

Penndorf, R. (1965). "Mie Scattering Coefficients for Selected Aerosol Size Distributions," Rept. RAD-TR-65-39 on Contract AF 19 (628)-4072 by Avco Corporation, Wilmington, Mass., AD 626 228. NTIS, Springfield, Va.

Penndorf, R. (1967). "Review of *Tables of light scattering*, Part I, by K. S. Shifrin and I. L. Zelmanovich, (1966)," *Appl. Opt.*, **6**, 2019.

Penndorf, R. (1968). "Review of *Tables of light scattering*, Part II, by K. S. Shifrin, and I. L. (1967)," *Appl. Opt.*, **7**, 1896.

Penndorf, R. (1969). "Review of *Tables of light scattering*, Part III, by I. L. Zelmanovich, and K. S. Shifrin, (1968)," *Appl. Opt.*, **8**, 892.

Penndorf, R. (1971). "Review of *Tables of Light Scattering*, Part 4, by I. L. Zelmanovich, and K. S. Shifrin, (1971)." *Appl. Opt.*, **10**, 2805.

Penndorf, R. and Goldberg, B. (1956). "New Tables of Mie Scattering Functions for Spherical Particles," (Parts 1 through 5: Amplitude Functions a_m and b_m), Rept. AFCRC-TR-56-204. AFCRL, Bedford, Mass. AD 98 767 through AD 98 771. NTIS, Springfield, Va.

Petterson, H. (1960). "Cosmic spherules and meteoritic dust," *Sci. Am.*, February, pp. 123–132.

Petterssen, S. (1969). *Introduction to Meteorology*, McGraw-Hill, New York.

Phillips, D. T. and Wyatt, P. J. (1972). "Single-particle light-scattering measurement: Photochemical aerosols and atmospheric particulates," *Appl. Opt.*, **11**, 2082–2087.

Pilat, M. J. (1967). "Optical efficiency factors for concentric spheres," *Appl. Opt.*, **6**, 1555–1558.

Pilié, R. J. et al. (1975a). "The life cycle of valley fog. Part I: Micrometeorological characteristics," *J. Appl. Meteorol.*, **14**, 347–363.

Pilié, R. J. et al. (1975b). "The life cycle of valley fog. Part II: Fog microphysics," *J. Appl. Meteorol.*, **14**, 364–374.

Pinnick, R. G. et al. (1976). "Stratospheric aerosol measurements III: Optical model calculations," *J. Atmos. Sci.*, **33**, 304–314.

Plass, G. N. (1966). "Mie scattering and absorption cross-sections for absorbing particles," *Appl. Opt.*, **5**, 279–285.

Plass, G. N. and Kattawar, G. W. (1971a). "Radiative transfer in water and ice clouds in the visible and infrared region," *Appl. Opt.*, **10**, 738–748.

Plass, G. N. and Kattawar, G. W. (1971b). "Comment on: The scattering of polarized light by polydisperse systems of irregular particles," *Appl. Opt.*, **10**, 1172–1173.

Plass, G. N. and Kattawar, G. W. (1971c). "Radiance and polarization of the earth's atmosphere with haze and clouds," *J. Atmos. Sci.*, **28**, 1187–1198.

Plass, G. N. and Kattawar, G. W. (1971d). "Reflection of light pulses from clouds," *Appl. Opt.*, **10**, 2304–2310.

Plass, G. N. and Kattawar, G. W. (1972a). "Effect of aerosol variation on radiance in the earth's atmosphere-ocean system," *Appl. Opt.*, **11**, 1598–1604.

Plass, G. N. and Kattawar, G. W. (1972b). "Degree and direction of polarization of multiple scattered light. 2: Earth's atmosphere with aerosols," *Appl. Opt.*, **11**, 2866–2879.

Plass, G. N. and Yates, H. (1965). "Atmospheric phenomena." In *Handbook of Military Infrared Technology*, W. L. Wolfe, ed. GPO, Washington, D.C.

Plass, G. N. et al. (1973). "Matrix operator theory of radiative transfer. I: Rayleigh scattering," *Appl. Opt.*, **12**, 314–329.

Platt, C. M. (1970). "Transmission of submillimeter waves through water clouds and fogs," *J. Atmos. Sci.*, **27**, 421–425.

Pogosyan, Kh. P. (1965). "The Air Envelope of the Earth," NASA Technical Translation, NASA TT F-287. NTIS, Springfield, Va.

Porch, W. M. et al. (1973). "Blue moon: Is this a property of background aerosol?" *Appl. Opt.*, **12**, 34–36.

Porch, W. M. et al. (1975). "Visibility of distant mountains as a measure of background aerosol pollution," *Appl. Opt.*, **14**, 400–403.

Prabhakara, C. et al. (1970). "Remote sensing of atmospheric ozone using the 9.6 μ band," *J. Atmos. Sci.*, **27**, 689–697.

Prabhakara, C. et al. (1971). "Nimbus 3 IRIS ozone measurements over Southeast Asia and Africa during June and July 1969," *J. Atmos. Sci.*, **28**, 828–831.

Priebe, J. R. (1969). "Operational form of the Mueller matrices," *J. Opt. Soc. Am.*, **59**, 176–180.

Pritchard, B. S. and Elliott, W. G. (1960). "Two instruments for atmospheric optics measurements," *J. Opt. Soc. Am.*, **50**, 191–202.

Prospero, J. M. (1968). "Atmospheric dust studies at Barbados," *Bull. Am. Meteorol. Soc.*, **49**, 645, 652.

Pueschel, R. F. and Noll, K. E. (1967). "Visibility and aerosol size frequency distribution," *J. Appl. Meteorol.*, **6**, 1045–1052.

Quenzel, H. (1969). "Influence of refractive index on the accuracy of size determination of aerosol particles with light-scattering aerosol counters," *Appl. Opt.*, **8**, 165–169.

Quenzel, H. (1970). "Determination of size distribution of atmospheric aerosol particles from spectral solar radiation measurements," *J. Geophys. Res.*, **75**, 2915–2921.

Querry, M. R. et al. (1969). "Refractive index of water in the infrared," *J. Opt. Soc. Am.*, **59**, 1299–1305.

Querry, M. R. et al. (1972). "Optical constants in the infrared for aqueous solutions of NaCl," *J. Opt. Soc. Am.*, **62**, 849–855.

Rangarajan, S. (1972). "Wavelength exponent for haze scattering in the tropics as determined by photoelectric photometers," *Tellus*, **24**, 56–64.

Rasmussen, R. A. and Went, F. W. (1965). "Volatile organic material of plant origin in the atmosphere," *Proc. Natl. Acad. Sci.*, **53**, 215–220.

Rayleigh, Lord (1871a). "On the light from the sky, its polarization and color," *Phil. Mag.*, **41**, 107–120, 274–279. Also in *The Scientific Papers of Lord Rayleigh*, Vol. 1, Dover, New York (1964).

Rayleigh, Lord (1871b). "On the scattering of light by small particles," *Phil. Mag.*, **41**, 447–454. Also in *The Scientific Papers of Lord Rayleigh*, Vol. 1, Dover, New York (1964).

Rayleigh, Lord (1884). "Presidential address." *Brit. Assoc. Rept. Montreal*, pp. 1–23. Also in *Scientific Papers of Lord Rayleigh*, Vol. 2, Dover, New York (1964).

Rayleigh, Lord (1899). "On the transmission of light through an atmosphere containing many small particles in suspension, and on the origin of the blue of the sky," *Phil. Mag.*, **47**, 375–384. Also in *The Scientific Papers of Lord Rayleigh*, Vol. 4, Dover, New York (1964).

Reid, G. C. (1975). "Ice clouds at the summer polar mesopause," *J. Atmos. Sci.*, **32**, 523–535.

Rensch, D. B. and Long, R. K. (1970). "Comparative studies of extinction and backscattering by aerosols, fog, and rain at 10.6 μ and 0.63 μ," *Appl. Opt.*, **9**, 1563–1573.

Rhine, P. E. et al. (1969). "Nitric acid vapor above 19 km in the earth's atmosphere," *Appl. Opt.*, **8**, 1500–1501.

Rishbeth, H. and Garriott, O. K. (1969). *Introduction to Ionospheric Physics*. Academic Press, New York.

Robertson, C. W. and Williams, D. (1971). "Lambert absorption coefficients of water in the infrared," *J. Opt. Soc. Am.*, **61**, 1316–1320.

Roll, H. U. (1965). *Physics of the Marine Atmosphere*. Academic Press, New York.

Romantzov, I. I. and Khvostikov, I. A. (1946). "Tropopause photography in polarized light," *Compt. Rend. Acad. Sci. USSR (Moscow)*, **53**, 703–705.

Roosen, R. G. et al. (1973). "Worldwide variations in atmospheric transmission: 1. Baseline results from Smithsonian observations," *Bull. Am. Meteorol. Soc.*, **54**, 307–316.

Rose, A. (1948). "The sensitivity performance of the human eye on an absolute scale," *J. Opt. Soc. Am.*, **38**, 196–208.

Rose, A. (1973). *Vision: Human and Electronic*. Plenum Press, New York.

Rosen, J. M. (1964). "The vertical distribution of dust to 30 km," *J. Geophys. Res.*, **69**, 4673.

Rosen, J. M. (1967). "Simultaneous Dust and Ozone Soundings over North and Central America," Final Rept., Contract NONR-710(22). University of Minnesota, Minneapolis.

Rosen, J. M. (1968). "Simultaneous dust and ozone soundings over North and Central America," *J. Geophys. Res.*, **73**, 479–486.

Rosen, J. M. (1971). "The boiling point of atmospheric aerosols," *J. Appl. Meteorol.*, **10**, 1044–1046.

Rosen, J. M. et al. (1975). "Stratospheric aerosol measurements. II: The worldwide distribution," *J. Atmos. Sci.*, **32**, 1457–1462.

Rozenberg, G. V. (1960). "Light scattering in the earth's atmosphere," *Usp. Fiz. Nauk*, **71**, 173–213. Eng. tran. in *Sov. Phys., Uspekhi*, **3**, 346–371.

Rozenberg, G. V. (1966). *Twilight. A Study in Atmospheric Optics.* Plenum Press, New York.

Ruff, I. et al. (1968). "Angular distribution of solar radiation reflected from clouds as determined from TIROS IV radiometer measurements," *J. Atmos. Sci.*, **25**, 323–332.

Rusk, A. N. et al. (1971). "Optical constants of water in the infrared," *J. Opt. Soc. Am.*, **61**, 895–903.

Sanderson, J. A. (1940). "The transmission of infra-red light by fog," *J. Opt. Soc. Am.*, **30**, 405–409.

Sanderson, J. A. (1955). "Emission, transmission, and detection of the infrared." In *Guidance*, A. S. Locke, ed. Van Nostrand Reinhold, New York.

Sanderson, J. A. (1967). "Optics at the Naval Research Laboratory," *Appl. Opt.*, **6**, 2029–2043.

Sastri, V. D. and Das, S. R. (1968). "Typical spectral distributions and color for tropical daylight," *J. Opt. Soc. Am.*, **58**, 391–398.

Schaaf, J. W. and Williams, D. (1973). "Optical constants of ice in the infrared," *J. Opt. Soc. Am.*, **63**, 726–732.

Schuster, B. G. (1970). "Detection of tropospheric and stratospheric aerosol layers by optical radal (lidar)," *J. Geophys. Res.*, **75**, 3123–3132.

Scott, W. D. et al. (1969). "The stratospheric aerosol layer and anhydrous reactions between ammonia and sulfur dioxide," *J. Atmos. Sci.*, **26**, 727–733.

Sears, F. W. (1949). *Optics*, 3d ed. Addison-Wesley, Reading, Mass.

Sekera, Z. (1951). "Polarization of skylight." In *Compendium of Meteorology*, T. F. Malone, ed. AMS, Boston.

Sekera, Z. (1956). "Recent developments in the study of the polarization of skylight." In *Advances in Geophysics*, Vol. 3. Academic Press, New York.

Sekera, Z. (1957a). "Light scattering in the atmosphere and the polarization of sky light," *J. Opt. Soc. Am.*, **47**, 484–490.

Sekera, Z. (1957b). "Polarization of skylight." In *Handbuch der Physik*, S. Flugge, ed., Springer-Verlag, Berlin.

Sekera, Z. (1963). "Radiative Transfer in a Planetary Atmosphere with Imperfect Scattering," Rept. R-413-PR on Contract AF 49 (638)-700. Rand Corporation, Santa Monica, Calif.

Sekera, Z. and Ashburn, E. V. (1953). "Tables Relating to Rayleigh Scattering of Light in the Atmosphere," NAVORD Rept. 2061. U.S. Naval Ordnance Test Station, China Lake, Calif.

Sekera, Z. and Blanch, G. (1952). "Tables Relating to Rayleigh Scattering of Light in the Atmosphere," Scientific Rept. No. 3 on Contract AF19 (122)-239. University of California, Los Angeles.

Sekhon, R. S. and Srivastava, R. C. (1970). "Snow size spectra and radar reflectivity," *J. Atmos. Sci.*, **27**, 299–307.

Sekhon, R. S. and Srivastava, R. C. (1971). "Doppler radar observations of drop-size distributions in a thunderstorm," *J. Atmos. Sci.*, **28**, 983–994.

Shallenberger, G. D. and Little, E. M. (1940). "Visibility through haze and smoke, and a visibility meter," *J. Opt. Soc. Am.*, **30**, 168–176.

Sharp, W. E. et al. (1966). "Zenith skylight intensity and color during the total solar eclipse of 20 July 1963," *Appl. Opt.*, **5**, 787–792.

Sharp, W. E. et al. (1971). "Summary of sky brightness measurements during eclipses of the sun," *Appl. Opt.* **10**, 1207–1210.

Shifrin, K. S. (1968). "Scattering of Light in a Turbid Medium," NASA Technical Translation, Publ. No. NASA TT F-477. NTIS, Springfield, Va.

Shifrin, K. S. and Zelmanovich, I. L. (1966). *Tables of Light Scattering. Part I: Tables of Angular Functions.* Hydrometeorological Publishing House, Leningrad.

Shifrin, K. S. and Zelmanovich, I. L. (1967). *Tables of Light Scattering. Part II: Coefficients of Angular Scattering.* Hydrometeorological Publishing House, Leningrad.

Shurcliff, W. A. (1962). *Polarized Light: Production and Use.* Harvard University Press, Cambridge, Mass.

Shurcliff, W. A. and Ballard, S. S. (1964). *Polarized Light.* Van Nostrand Reinhold, New York.

Silverman, B. A. et al. (1964). "A laser fog disdrometer," *J. Appl. Meteorol.*, **3**, 792–801.

Simpson, G. C. (1941a). "On the formation of cloud and rain," *Quart. J. Roy. Meteorol. Soc.*, **67**, 99–134.

Simpson, G. C. (1941b). "Sea salt and condensation nuclei," *Quart. J. Roy. Meteorol. Soc.*, **67**, 163–169.

Sinclair, D. (1947). "Light scattering by spherical particles," *J. Opt. Soc. Am.*, **37**, 475–480.

Sissenwine, N. et al. (1968a). "Mid-latitude humidity up to 32 km," *J. Atmos. Sci.*, **25**, 1129–1140.

Sissenwine, N. et al. (1968b). "Humidity up to the Mesopause," Rept. AFCRL-68-0550, Air Force Surveys in Geophysics No. 206. AFCRL, Bedford, Mass. AD 679 996. NTIS, Springfield, Va.

Smirnov, I. P. (1946). "Searchlight investigation as a method of studying atmosphere stratification," *Compt. Rend. Acad. Sci. USSR (Moscow)*, **53**, 707–710.

Smith, A. G. (1955). "Daylight visibility of stars from a long shaft," *J. Opt. Soc. Am.*, **45**, 482–483.

Smith, P. N. and Hayes, H. V. (1940). "Transmission of infra-red radiation through fog," *J. Opt. Soc. Am.*, **30**, 332–337.

Soberman, R. K. (1963). "Noctilucent clouds," *Sci., Am.*, June, pp. 50–59.

Sommerfeld, A. (1964a). *Optics: Lectures on Theoretical Physics*, Vol. IV. Academic Press, New York.

Sommerfeld, A. (1964b). *Thermodynamics and Statistical Mechanics: Lectures on Theoretical Physics*, Vol. V. Academic Press, New York.

Southworth, R. B. (1964). "The size distribution of the zodiacal particles," *Ann. N.Y. Acad. Sci.*, **119**, 54–67.

Spencer, D. E. (1960). "Scattering function for fogs," *J. Opt. Soc. Am.*, **50**, 584–585.

Speyers-Duran, P. A. and Braham, R. R., Jr. (1967). "An airborne continuous cloud relicator," *J. Appl. Meteorol.*, **6**, 1108–1113.

Spiro, I. J. et al. (1965). "Atmospheric transmission: Concepts, symbols, units, and nomenclature," *Infrared Phys.*, **5**, 11–36.

SRI (1961). *Chemical Reactions in the Lower and Upper Atmosphere*, Symposium at Stanford Research Institute, 1961. Wiley, New York.

Stanford, J. L. and Davis, J. S. (1974). "A century of stratospheric cloud reports: 1870–1972." *Bull. Am. Meteorol. Soc.*, **55**, 213–219.

Stanley, R. C. (1968). *Light and Sound for Engineers.* Hart Publishing, New York.

Stewart, H. S. et al. (1949). "The Measurement of Slant Visibility." NRL Rept. 3484. NRL, Washington, D.C.

Stewart, H. S. and Curcio, J. A. (1952). "The influence of field of view on measurements of atmospheric transmission," *J. Opt. Soc. Am.*, **42**, 801–805.

Stewart, R. W. and Hoffert, M. I. (1975). "A chemical model of the troposphere and stratosphere," *J. Atmos. Sci.*, **32**, 195–210.

Stokes, G. G. (1852). "On the composition and resolution of streams of polarized light," *Trans. Cambridge Phil. Soc.*, **9**, 399. Also in *Mathematical and Physical Papers*, Cambridge University Press, Cambridge, England (1904).

Storebø, P. B. and Dingle, A. N. (1974). "Removal of pollution by rain in a shallow air flow," *J. Atmos. Sci.*, **31**, 533–542.

Stratton, J. A. (1941). *Electromagnetic Theory*, McGraw-Hill, New York.

Stratton, J. A. and Houghton, H. G. (1931). "A theoretical investigation of the transmission of light through fog," *Phys. Rev.*, **38**, 159–165.

Strong, J. (1941). "On a new method of measuring the mean height of the ozone in the atmosphere," *J. Franklin Inst.*, **231**, 121–155.

Strong, J. (1958). *Concepts of Classical Optics.* Freeman, San Francisco.

Sutton, O. G. (1961). *The Challenge of the Atmosphere.* Harper, New York.

Sweer, J. (1938). "The path of a ray of light tangent to the surface of the earth," *J. Opt. Soc. Am.*, **28**, 327–329.

Swider, W., Jr. (1964). "The determination of the optical depth at large solar zenith distances," *Planet. Space Sci.*, **12**, 761–782.

Thielke, J. F. et al. (1972). "Multiwavelength nephelometer measurements in Los Angeles smog aerosol. II. Correlation with size distribution and volume concentration." In *Aerosols and Atmospheric Chemistry*, G. M. Hidy, ed. Academic Press, New York.

Thompson, B. J. et al. (1967). "Application of hologram techniques for particle size analysis," *Appl. Opt.*, **6**, 519–526.

Thompson, W. I., III (1971). "Atmospheric Transmission Handbook: A Survey of Electromagnetic Wave Transmission in the Earth's Atmosphere over the Frequency (Wavelength) Range 3 kHz (100 km) to 3000 THz (0.1 μm)," Rept. DOT-TSC-NASA-71-6. U.S. Department of Transportation, Cambridge, Mass.

Thompson, P. D. and O'Brien, R. (1965). *Weather.* Time-Life Books, New York.

Tonna, G. (1975). "Microphysical analysis of an artificial fog through laser scattering," *J. Appl. Meteorol.*, **14**, 1547–1557.

Tousey, R. and Hulburt, E. O. (1947). "Brightness and polarization of the daylight sky at various altitudes above sea level," *J. Opt. Soc. Am.*, **37**, 78–92.

Tousey, R. and Hulburt, E. O. (1948). "The visibility of stars in the daylight sky," *J. Opt. Soc. Am.*, **38**, 886–896.

Trabert, W. (1901). "Die Extinction des Lichtes in einem Truben Medium," *Meteorol. Z.*, **18**, 518–525.

Tricker, R. A. R. (1970). *Introduction to Meteorological Optics*, American Elsevier, New York.

Tverskoi, P. N. (1965). "Physics of the Atmosphere: A Course in Meteorology," NASA Technical Translation, NASA TT F-288. NTIS, Springfield, Va.

U.S. (1961). "Distribution Curves of Atmospheric Transmissivity for United States Coasts," Civil Engineering Rept. 4A. United States Coast Guard, Washington, D.C. AD 254 897.

USSA (1962). "U.S. Standard Atmosphere, 1962," Cat. No. NAS 1.2:At6/962. GPO, Washington, D.C.

USSAS (1966). "U.S. Standard Atmosphere Supplements, 1966," Cat. No. NAS 1.2:At6/966/supp. GPO, Washington, D.C.

Uthe, E. E. (1972). "Lidar observations of the urban aerosol structure," *Bull. Am. Meteorol. Soc.*, **53**, 358–360.

Valley, S. L., ed. (1965). *Handbook of Geophysics and Space Environment.* McGraw-Hill, New York.

van de Hulst, H. C. (1948). "Scattering in a planetary atmosphere," *Astrophys. J.*, **107**, 220–246.

van de Hulst, H. C. (1952). "Scattering in atmospheres." In *The Atmospheres of the Earth and Planets*, G. P. Kuiper, ed. University of Chicago Press, Chicago.

van de Hulst, H. C. (1957). *Light Scattering by Small Particles.* Wiley, New York.

van de Hulst, H. C. (1963). "A New Look at Multiple Scattering," Tech. Rept., Goddard Institute for Space Studies, NASA, Washington, D.C.

van de Hulst, H. C. and Irvine, W. M. (1963). "General report on radiation transfer in planets: Scattering in model planetary atmospheres," *Mem. Soc. Roy. Sci. Liege*, **7**, 78–98.

Velasquez, D. A. (1971). "Zenith sky brightness and color change during the total solar eclipse of 12 November 1966 at Santa Ines, Peru," *Appl. Opt.*, **10**, 1211–1214.

Vincenti, W. G. and Kruger, C. H., Jr. (1965). *Introduction to Physical Gas Dynamics.* Wiley, New York.

Vogt, H. (1968). "Visibility measurements using backscattered light," *J. Atmos. Sci.*, **25**, 912–918.

Volz, F. E. (1956). "Optik der Tropfen." In *Handbuch der Physik*, Vol. VIII. Springer-Verlag, Berlin.

Volz, F. E. (1969a). "Stratospheric dust striations," *Bull. Am. Meteorol. Soc.*, **50**, 16.

Volz, F. E. (1969b). "Twilights and stratospheric dust before and after the Agung eruption," *Appl. Opt.*, **8**, 2505–2517.

Volz, F. E. (1970a). "On dust in the tropical and midlatitude stratosphere from recent twilight measurements," *J. Geophys. Res.*, **75**, 1641–1646.

Volz, F. E. (1970b). "Atmospheric turbidity after the Agung eruption of 1962 and size distribution of the volcanic aerosol," *J. Geophys. Res.*, **75**, 5185–5193.

Volz, F. E. (1970c). "Spectral skylight and solar radiance measurements in the Caribbean: Maritime aerosols and Sahara dust," *J. Atmos. Sci.*, **27**, 1041–1047.

Volz, F. E. (1971). "Stratospheric aerosol layers from balloon-borne horizon photographs," *Bull. Am. Meteorol. Soc.*, **52**, 996–998.

Volz, F. E. (1972). "Infrared refractive index of atmospheric aerosol substances," *Appl. Opt.*, **11**, 755–759.

Volz, F. E. (1973). "Infrared optical constants of ammonium sulfate, Sahara dust, volcanic pumice, and flyash," *Appl. Opt.*, **12**, 564–568.

Volz, F. E. and Bullrich, K. (1961). "Scattering function and polarization of skylight in the ultraviolet to the near infrared region," *J. Meteorol.*, **18**, 306.

Volz, F. E. and Goody, R. M. (1962). "The intensity of the twilight and upper atmospheric dust," *J. Atmos. Sci.*, **19**, 385–406.

Volz, F. E. and Sheehan, L. (1971). "Skylight and aerosol in Thailand during the dry winter season," *Appl. Opt.*, **10**, 363–366.

von Hippel, A. R. (1954). *Dielectrics and Waves.* Wiley, New York.

Waggoner, A. P. et al. (1972). "Measurement of the aerosol total scatter-backscatter ratio," *Appl. Opt.*, **11**, 2886–2889.

Waldram, J. M. (1945a). "Measurement of the photometric properties of the upper atmosphere," *Quart. J. Roy. Meteorol. Soc.*, **71**, 319–336.

Waldram, J. M. (1945b). "Measurement of the photometric properties of the upper atmosphere," *Trans. Illum. Eng. Soc.* (*London*), **10**, 147–188.

Walker, M. J. (1954). "Matrix calculus and the Stokes parameters of polarized radiation," *Am. J. Physics*, April, pp. 170–174.

Walsh, J. W. (1953). *Photometry.* Constable, London. Also Dover, New York (1964).

Ward, G. et al. (1973). "Atmospheric aerosol index of refraction and size-altitude distribution from bi-static laser scattering and solar aureole measurements," *Appl. Opt.*, **12**, 2585–2592.

Warner, J. (1969). "The microstructure of cumulus cloud. Part I. General features of the droplet spectrum," *J. Atmos. Sci.*, **26**, 1049–1059.

Webb, W. L. (1966). *Structure of the Stratosphere and Mesosphere.* Academic Press, New York.

Weickman, H. K. (1957). "Physics of precipitation," *Meteorol. Monogr.*, **3**, No. 19, pp. 226–255. AMS, Boston.

Weickman, H. K., ed. (1960). *Physics of Precipitation*, Geophysical Monograph No. 5. American Geophysical Union, Washington, D.C. Available from University Microfilms, Ann Arbor.

Weickman, H. K. and aufm Kampe, H. J. (1953). "Physical properties of cumulus clouds," *J. Meteorol.*, **10**, 204–211.

Went, F. W. (1955). "Air pollution," *Sci. Am.*, May, pp. 63–72.

Whipple, F. L. (1951). "Meteors as probes of the upper atmosphere." In *Compendium of Meteorology*, T. F. Malone, ed. AMS, Boston.

Whipple, F. L. (1952). "Exploration of the upper atmosphere by meteoritic techniques." In *Advances in Geophysics*, Vol. 1, Academic Press, New York.

Whipple, F. L. (1953). "Density, Pressure, and Temperature Data above 30 km." In *The Earth as a Planet*, G. Kuiper, ed. University of Chicago Press, Chicago.

Whitby, K. T. (1971). "Aerosol measurements: New data on urban aerosols and formation mechanisms." In *Particulate Models: Their Validity and Application*, I. H. Blifford, Jr., ed., NCAR-TN/PROC-68. NCAR, Boulder, Colo.

Whitby, K. T. et al. (1972a). "The Minnesota aerosol analyzing system used in the Los Angeles smog project," *J. Colloid Interface Sci.*, **39**, 136–164. Also in *Aerosols and Atmospheric Chemistry*, G. M. Hidy, ed. Academic Press, New York.

Whitby, K. T. et al. (1972b). "The aerosol size distribution of Los Angeles smog," *J. Colloid Interface Sci.*, **39**, 177–204. Also in *Aerosols and Atmospheric Chemistry*, G. M. Hidy, ed. Academic Press, New York.

Wilkes, M. V. (1954). "A table of Chapman's grazing incidence integral," *Proc. Phys. Soc. London*, **B67**, 304–308.

Wilkniss, P. E. et al. (1975). "CO, CCl, CO, CCl_4, Freon 11, CH_4, and Rn-222 concentrations at low altitudes over the Arctic Ocean in January 1974," *J. Atmos. Sci.*, **32**, 158–162.

Williams, W. J. et al. (1972). "Distribution of nitric acid vapor in the stratosphere as determined from infrared atmospheric emission data," *J. Atmos. Sci.*, **29**, 1375–1379.

Williams, C. S. and Becklund, O. E. (1972). *Optics: A Short Course for Engineers and Scientists*. Wiley, New York.

Wilson, R. (1951). "The blue sun of 1950 September," *Mon. Not. Roy. Astron. Soc.*, **3**, 478–489.

Wilson, R. W. and Penzias, A. A. (1966). "Effect of precipitation on transmission through the atmosphere," *Nature*, **211**, 1081.

Winch, G. T. et al. (1966). "Spectroradiometric and colorimetric characteristics of daylight in the southern hemisphere," *J. Opt. Soc. Am.*, **56**, 456–464.

Winstanley, J. V. and Adams, M. J. (1975). "Point visibility meter: A forward scatter instrument for the measurement of aerosol extinction coefficient," *Appl. Opt.*, **14**, 2151–2157.

Witte, H. J. (1968). "Airborne Observations of Cloud Particles and Infrared Flux Density (8–14 μm) in the Arctic," Scientific Rept. on Contract NOnr 477(24) (NR 307-252). University of Washington, Seattle. AD 679 241. NTIS, Springfield, Va.

WMO (1973a). *Compendium of Meteorology, Vol. 1, Part 1: Dynamic Meteorology*. Publication WMO 364. World Meteorological Organization. Unipub, New York.

WMO (1973b). *Compendium of Meteorology, Vol. 1, Part 2: Physical Meteorology*. Publication WMO 364. World Meteorological Organization. Unipub, New York.

Wolfe, W. L. and Nicodemus, F. E. (1965). "Radiation theory." In *Handbook of Military Technology*, W. L. Wolfe, ed. GPO, Washington, D.C.

Wolff, M. (1964a). "The Profile of an Exponential Atmosphere Viewed from Outer Space and Consequences for Space Navigation," Rept. E-1634. MIT Instrumentation Laboratory, Massachusetts Institute of Technology, Cambridge, Mass.

Wolff, M. (1964b). "An Optical Earth Horizon Profile Based upon Tabulated Solutions of Chandrasekhar's Equations," Rept. E-1687. MIT Instrumentation Laboratory, Massachusetts Institute of Technology, Cambridge, Mass.

Woodcock, A. H. (1953). "Salt nuclei in marine air as a function of altitude and wind force," *J. Meteorol.*, **10**, 362–371.

Wright, H. L. (1936). "The size of atmospheric nuclei," *Proc. Phys. Soc. London*, **48**, 675–689.

Wright, H. L. (1939). "Atmospheric opacity: A study of visibility observations," *Quart. J. Roy. Meteorol. Soc.*, **65**, 411–442.

Wright, H. L. (1940). "Atmospheric opacity at Valentia," *Quart. J. Roy. Meteorol. Soc.*, **66**, 66–77.

Yamamoto, G. and Tanaka, M. (1969). "Determination of aerosol size distribution from spectral attenuation measurements," *Appl. Opt.*, **8**, 447–453.

Yamamoto, G. et al. (1970). "Radiative transfer in water clouds in the infrared region," *J. Atmos. Sci.*, **27**, 283–292.

Yamamoto, G. et al. (1971). "Table of Scattering Function of Infrared Radiation for Water Clouds," NOAA Technical Rept. NESS 57. Cat. No. COM-71-50312. GPO, Washington, D.C.

Yates, H. W. and Taylor, J. H. (1960). "Infrared Transmission of the Atmosphere," NRL Rept. 5453. NRL, Washington, D.C.

Zabrodskii, G. M. and Morachevskii, V. G. (1959). "Study of the transparency of clouds and fogs," *Arkticheskii Antarkticheskii Nauchn.-Issled. Inst., Tr., Leningrad,* **228**, No. 1, 68–86. Translation for AFCRL by AMS (1961), AD 287 216. NTIS, Springfield, Va.

Zelmanovich, I. L. and Shifrin, K. S. (1968). *Tables of Light Scattering. Part III: Coefficients of Extinction, Scattering, and Light Pressure.* Hydrometeorological Publishing House, Leningrad.

Zelmanovich, I. L. and Shifrin, K. S. (1971). *Tables of Light Scattering. Part IV. Scattering of Polydisperse Systems.* Hydrometeorological Publishing House, Leningrad.

Zuev, V. E. (1974). "Propagation of Visible and Infrared Radiation in the Atmosphere." Halsted Press, New York.

Author Index

Numbers in *italics* refer to the pages on which the complete titles of the references are listed.

Subject Index